Handbook of Digital System Design for Scientists and Engineers:
Design with Analog, Digital, and LSI

Author

Wen C. Lin, Ph.D.
Professor
Department of Electrical and Computer Engineering
University of California
Davis, California

CRC Press, Inc.
Boca Raton, Florida

Library of Congress Cataloging in Publication Data

Lin, Wen C
    Handbook of digital system design for scientists and engineers: design with analog, digital, and LSI

    Includes bibliographies and index.
    1. Digital electronics. 2. Electronic digital computers. I. Title.
TK7868.D5L537   621.381   80-39807
ISBN 0-8493-0670-1

This book represents information obtained from authentic and highly regarded sources. Reprinted material is quoted with permission, and sources are indicated. A wide variety of references are listed. Every reasonable effort has been made to give reliable data and information, but the author and the publisher cannot assume responsibility for the validity of all materials or for the consequences of their use.

All rights reserved. This book, or any parts thereof, may not be reproduced in any form without written consent from the publisher.

Direct all inquiries to CRC Press, Inc., 2000 N.W. 24th Street, Boca Raton, Florida 33431.

© 1981 by CRC Press, Inc.

International Standard Book Number 0-8493-0670-1
Library of Congress Card Number 80-39807
Printed in the United States

PREFACE

The objective of this handbook is to provide sufficient self-contained design information for scientists and engineers who wish to do practical design of digital systems for their applications. It should also be suitable as a text or major reference book for seniors or graduate students in Electrical Engineering, Computer Engineering, or Biomedical/Clinical Engineering, who are engaged in design of digital systems, microcomputers, or just instrumentation for their research and development projects. Without providing exercises/problems for drills or academic interest as in the format of the conventional textbooks, this book contains principally three essential types of information: (1) fundamentals — serving as a review to the readers who are somewhat familiar with the basic theories, but occasionally wish to refresh themselves with the formula, design equations, terminologies, symbols, and notations; (2) characteristics, properties, and principles of operation of the devices, modules, and building blocks that are frequently being used as components in digital system design; and (3) design procedures by means of examples that would provide the readers with some guidelines for system design.

The unique features of this handbook are that the necessary linear device characteristics and principles are included, and the balance of circuit/system design is presented. This nontraditional composition of analog and digital materials for digital system design in this book is a response to the rapid advancement in electronic technologies. Historically, a digital engineer/logic designer would know nothing about linear circuitry, and very little about digital electronic circuit design. If necessary, they can easily get help from a strong group specialized in circuit design within their department. The rapid development in solid state technology, such as Large-Scale-Integrated (LSI) circuit, has shrunken the demand for full-time circuit designers. As a result, the system designers are forced to acquire broader knowledge in linear and digital circuits as well as logic and system design. Fortunately, this is not only possible, but also can easily be achieved with the available technologies; if they are willing to acquire some fundamentals in linear and digital circuit design. It is based on this concept that this handbook is structured. Assuming that the readers may be familiar with a few but not all of the topics that have been covered, this book was written in somewhat tutorial style. In each subarea, it always starts with fundamentals and then an attempt is made to clarify the materials which were traditionally confusing or being overlooked. Finally, some design examples, if applicable, are presented.

In the first two chapters, this book basically deals with the linear circuits, devices, and modules that are frequently encountered in digital system design. Operational amplifiers, linear active filters, phase-lock loop, and basic analog computing circuits are described and analyzed from the practical point of view. Chapters III, IV, and V are devoted to pulse, time, and logic-operation circuits and devices. Here, the Binary-State-Analysis (BSA) and graphical analysis techniques are introduced and illustrated for TTL, ECL, CMOS, I²L, and Schottky logic elements. A thorough treatment on that R-S flip-flop, J-K flip-flop, D flip-flop, master-slave flip-flop, as well as on that other multivibrators, Schmitt trigger, medium-integrated-circuit devices are presented. In Chapters VI, VII, and VIII, the essential subsystems for a digital computer-based system such as, memory, Arithmetic-Logic-Unit, and Analog-Digital-Analog converters are anatomized in detail for examination and design. System noise problems are then explored in Chapter IX. Finally, design of random (control) logic using programmable LSI is presented in Chapter X.

The majority of the materials covered in this handbook were developed and tested successfully in the laboratories and classes for students in engineering and sciences during the author's tenure at Case Western Reserve University, Cleveland, Ohio. It is his pleasure that the author would like to thank his former graduate students, especially Drs. Chi-Foon Chan and Mark Michael, and Mr. Wing-Kay Leung, who helped to develop the laboratory problems. I am especially grateful to Professors Yaohan Chu, Harry W. Mergler, Wen H. Ko, E. L. Glaser, and Dr. William H. Ninke for their encouragement. My thanks goes to Miss Grace K. Lin and Mrs. Carol She′ for their skillful drawing and typing of the manuscript. Finally, I would like to thank Professor V. R. Algazi for his encouragement during the last stage of preparing the materials for this handbook.

THE AUTHOR

Wen C. Lin received his B.S.E.E. degree with highest honors (top 5% of the class) from the National Taiwan University, Taiwan, China, in 1950, and the M.S. and Ph.D. degrees from Purdue University, West Lafayette, Indiana, in 1956 and 1965, respectively.

From 1950 to 1954 he was an Engineer with the Instrumentation Laboratory, Taiwan Power Company, and from 1956 to 1961 he was an Engineer with General Electric Company and a Senior Engineer Electronic Data Processing Division, Honeywell Corporation. He joined Case Institute of Technology as an Assistant Professor of Engineering in 1965 and became an Associate Professor in 1967. He was a Professor of Engineering until 1976 in the Department of Computer Engineering and the Department of Computing and Information Sciences, and a Professor of Electrical and Computer Engineering in the Department of Electrical Engineering and Applied Physics, Case Western Reserve University, Cleveland, Ohio, until 1978. He is presently a Professor in the Department of Electrical and Computer Engineering, University of California, at Davis. Interested in pattern recognition, signal processing (speech, pictorial, medical, etc.), medical electronics and microcomputer and digital system design, Dr. Lin is a member of Sigma Xi, Eta Kappa Nu, and a Senior Member of the IEEE. He is the editor of the IEEE Press Book entitled *Microprocess: Fundamentals and Applications,* and has published over fifty technical papers.

To My
Wife, Children
and
Eldest Brother Wen-Chan

ADVISORY BOARD

Yaohan Chu, Ph.D.
Professor
Department of Computer Science
University of Maryland
College Park, Maryland

Edward L. Glaser
Vice President and Chief Technical
 Officer
Products Group
System Development Corporation
Santa Monica, California

Wen H. Ko, Ph.D.
Director
Engineering Design Center
Professor of Biomedical Engineering
 and Electrical Engineering
Case Western Reserve University
Cleveland, Ohio

H. W. Mergler, Ph.D.
Leonard Case Professor of Electrical
 Engineering
Case Western Reserve University
Cleveland, Ohio

William H. Ninke, Ph.D.
Head
Image Processing and Display Research
 Department
Bell Laboratories
Holmdel, New Jersey

TABLE OF CONTENTS

Chapter 1
A Review of Linear Circuit Fundamentals
I. Basic Laws and Theories for Linear Circuits 1
    A. Ohm's Law 1
    B. Kirchoff's Laws 1
    C. Thevenin's Theory 1
II. Examples for Analysis of a Circuit that Contains Independent and Dependent Sources 1
    A. Example 1 1
    B. Example 2 2
    C. Example 3 4
References 5

Chapter 2
Basic Modules (Building Blocks) for Linear Systems
I. Single Stage Differential Amplifier 7
    A. AC Small Signal Analysis 7
    B. DC Analysis 12
II. Operational Amplifier 13
    A. Equivalent Circuit of an Operational Amplifier 13
    B. The Four Useful and Essential Feedback Circuit Configurations 14
    C. Consideration of Gain-Bandwidth Property 24
    D. Summary — Comparison of the Four Feedback Configurations 26
III. Experimental Examples for Typical Applications 27
    A. Analog Adder/Subtractor 27
    B. Square Wave and Triangular Wave Generator 29
    C. Pulse Generator 31
    D. Second Order Active Filters 33
IV. A Collection of Circuits for General Applications 40
    A. Integrator 40
    B. Differentiator 41
    C. Differentiator-Integrator 41
    D. Comparator 43
V. Phase-Locked Loop 43
    A. Introduction 43
    B. Major Elements of the System 44
    C. Linearized Negative Feedback System Model 46
    D. Capture Range and Lock-In Range 47
    E. Comments 48
References 48

Chapter 3
A Review and Clarification of Pulse and Switching Circuit Fundamentals
I. Clarification of Basic Concepts 51
II. Linear Elements 51
    A. Resistor 51
    B. Capacitor 52
    C. Inductor 54
III. Nonlinear Elements 55
    A. Diode 55

	B.	Transistor (Bipolar) Current Control Device 57
IV.	Pulse Circuit Fundamentals 63	
	A.	Complete Time Response of a Basic Network 63
	B.	Example 1 66
	C.	Example 2 66
	D.	Example 3 66
	E.	Important Remark 70
V.	Logic Fundamentals in Brief 70	
	A.	Definitions. 70
	B.	Theorems. 70
	C.	Minimization of Switching Function by Karnaugh Maps 71
References 72		

Chapter 4
Basic Electronic Logic Elements

I.	Background. 73	
II.	Logic Circuits 73	
	A.	Introduction 73
	B.	Resistor-Transistor Logic (RTL) 74
	C.	Diode-Transistor Logic (DTL) 77
	D.	Transistor-Transistor Logic (T^2L or TTL). 78
	E.	Emitter-Coupled Logic (ECL) Nonsaturation-Type Logic 83
	F.	Schottky-TTL-Nonsaturation-Type Logic 84
	G.	Complementary Symmetric MOSFET (CMOS) 85
	H.	Integrated Injection Logic (I^2L) 89
	I.	Comparison of the Different Logic Families. 90
	J.	Interfacing Logic Elements of Different Families. 91
	K.	Some Special and Useful Logic Elements 91
	L.	Other Important Properties of the Logic Circuit 95
	M.	Conclusion 96
III.	Switching Inductive Load by a Transistor 97	
References. 98		

Chapter 5
Basic Functional Modules

I.	Flip-Flop 99	
	A.	Basic Flip-Flop Circuit. 99
	B.	Basic Integrated Circuit Flip-Flop 103
	C.	Integrated Circuit Flip-Flop 104
	D.	Clocked R-S Flip-Flop. 105
	E.	J-K Master-Slave Flip-Flop. 105
	F.	D Flip-Flop. 110
	G.	Conclusion 112
II.	Monostable Multivibrator (One-Shot) 113	
	A.	Discrete Circuit One-Shot 113
	B.	I.C. One-Shot. 115
	C.	One-Shot Using I.C. 555 Chip 116
III.	Astable (Free-Running) Multivibrator 119	
	A.	Basic Astable Multivibrator 119
	B.	Astable Multivibrator Using 555 Chip. 120
IV.	Schmitt Trigger Circuit 122	

V.	Registers and Counters 125
	A. Registers 125
	B. Counters 126
VI.	Encoders, Decoders, Multiplexers, and Demultiplexers 128
	A. Encoder 128
	B. Decoder 128
	C. Multiplexer 129
	D. Demultiplexer 129
	E. Decoder/Demultiplexer 129
VII.	Line Receivers and Drivers 130
VIII.	Debouncer 131
IV.	Consideration of Input/Output of a Module 131
X.	A Summary of Symbols of Logic Elements and Modules 131
References 133

Chapter 6
Memory Systems

I.	Introduction 135
	A. RAM 135
	B. CAM 135
	C. ROM 136
	D. RMM 136
II.	RAM — Magnetic Core Memory 136
	A. Memory Cell 136
	B. Configuration of a 4 × 1 Memory Plane 137
	C. System Configuration 138
III.	RAM — Semiconductor Memory 141
	A. Memory Cell 141
	B. System Configuration 145
IV.	Charge-Couple-Device (CCD) 146
V.	Associative or Content Addressable Memory (CAM) 150
VI.	Read-Only-Memory (ROM) 150
	A. Nonprogrammable ROM 150
	B. Field-Programmable ROM or FPROM 150
	C. Ultraviolet Rays-Erasable-Programmable ROM (EPROM) 151
	D. Electrically Alterable Read-Only-Memory (EAROM) 152
	E. Read-Only-Memory System 152
VII.	Magnetic-Bubble Memories 153
References 154

Chapter 7
Arithmetic Logic Unit (ALU)

I.	Binary Addition 157
	A. Basic Element 157
	B. Multiple-Bit Addition 158
II.	Subtraction by Complementary Arithmetic 160
	A. Background 160
	B. Subtraction by Addition of Complements 165
III.	A Typical ALU Chip 170
IV.	Binary Multiplication 170
	A. Multiplication by Iterative Addition 170
	B. Multiplication by Add and Shift 170

V. Binary Rate Multiplier...175
VI. Binary Division 177
References..178

Chapter 8
Analog-Digital-Analog Conversion
I. Introduction..179
    A. General Considerations of Analog and Digital System Design.......179
    B. A Typical System Organization of a Digital Data Acquisition System...179
II. Digital-to-Analog Conversion 180
    A. Parallel Conversion 180
    B. Serial Conversion..180
III. Analog-Digital Conversion...182
    A. Parallel Conversion..182
    B. Counter Ramp Converter 182
    C. Successive Approximation Converter 183
    D. Dual-Slope Integrator Converter 184
IV. Key Elements Commonly Used in Analog-Digital-Analog Conversion......186
    A. Reference Voltage 186
    B. Analog Comparator 186
    C. Analog Switch 187
    D. D/A Decoder Resistor Networks 188
    E. Sample and Hold Element 191
V. Special Purpose D/A Conversion Circuits 191
    A. Inverse D/A Converter 191
    B. Multiply D/A Converter......................................192
VI. Codes Employed by A/D Converters 192
    A. Binary Codes 193
    B. Complement Code...193
    C. Gray Code..193
VII. Major Errors in Conversion 194
    A. Quantization Error...194
    B. Analog Component Error.....................................195
    C. Aperture Error 195
VIII. Key Factors for System Design Consideration Using A/D/C..............196
References..197

Chapter 9
Noise in Digital Systems
I. Introduction..199
II. External or Radiation Noise 199
III. Internal Noise..200
    A. Decoupling Technique..200
    B. Grounding Technique 201
IV. Transmission-Line Reflection 202
    A. Introduction...202
    B. Transmission Lines Terminated by Linear Devices 203
    C. Transmission Lines Terminated by Nonlinear Devices 206
    D. Impedance Matching for Transmission Lines.....................211
V. Differential Line Driver and Receiver 211
References..212

Chapter 10
Random Logic Design Using Programmable Logic Elements

I. General Background 213
    A. Design with Multiplexers 213
    B. Design with ROM 219
    C. Design with PLA 221
    D. Microprogramming 224
II. Sequential Logic Design Example 230
    A. Design Procedure 230
    B. Solution 1: Hardwired-Logic 231
    C. Solution 2: ROM Implementation 236
    D. Solution 3: PLA Implementation 237

References 242

Index 243

Chapter 1

A REVIEW OF LINEAR CIRCUIT FUNDAMENTALS

The objective of this chapter is to provide the circuit fundamentals that are essential to the designers. Some nontrivial examples will be used to illustrate how the fundamental theories or laws are applied in circuit/system analysis and design. It serves as a reminder to the readers and makes this book virtually self-contained.

I. BASIC LAWS AND THEORIES FOR LINEAR CIRCUITS

A. Ohm's Law
It states e = iR, where R is resistance of a resistor, in ohms, e is the voltage across the resistor, in volts, and i is the current, in amperes.

B. Kirchhoff's Laws
Current Law
The algebraic sum of all currents at a node must be zero at any time. If the sign of a current flows into the node, which is assigned to be positive (negative), then the current flowing out of the node must be negative (positive).

Voltage Law
The sum of all voltage across each circuit element around a loop must be zero at all times. In reference with clockwise direction, if voltage-rise (from low to high) is assigned as positive, then the voltage-drop (from high to low) must be negative.

C. Thévenin's Theory
Across a pair of nodes, a linear network can be replaced by a voltage source with open-circuit voltage, E_g, in series with a resistor with a value of R_g, where R_g is the equivalent resistance between the nodes with all independent voltage and current sources, short-circuited and open-circuited, respectively. R_g can also be determined by the ratio of E_g and I_{sc}, the short circuit current of shortening the specified pair of nodes.

II. EXAMPLES FOR ANALYSIS OF A CIRCUIT THAT CONTAINS INDEPENDENT AND DEPENDENT SOURCES

A. Example 1
Problem
For the circuit shown in Figure 1, determine I_i, I_2, I_3, I_0 and V_0, if $V_i = 10$ V, $R_1 = 100\ \Omega$, $R_2 = 100\ \Omega$, $R_3 = 10\ \Omega$ and $R_4 = 500\ \Omega$, $\beta = 10$. Note that V_i is an independent voltage source, while βI_i is a dependent current source.

Solution
Label the current and voltage polarity for each branch as shown in the circuit. By Kirchhoff's current law,

$$\text{Node A:}\quad I_i + I_0 - I_2 = 0 \tag{1}$$

$$\text{Node B:}\quad \beta I_i + I_3 - I_0 = 0 \tag{2}$$

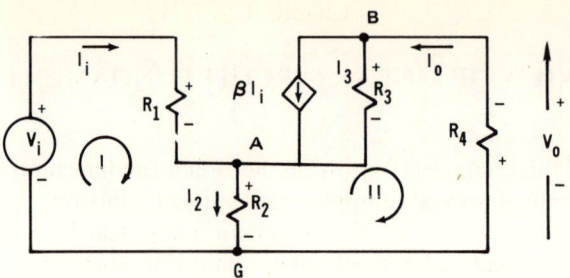

FIGURE 1. Circuit contains both dependent and independent sources.

By Kirchhoff's voltage law,

$$\text{Loop I}: V_i - I_i R_1 - I_2 R_2 = 0 \quad (3)$$

$$\text{Loop II}: I_2 R_2 + I_3 R_3 + I_o R_4 = 0 \quad (4)$$

Substituting the numerical values into Equations 1→4, we have,

$$I_i + I_o - I_2 = 0 \quad (5)$$

$$10 I_i + I_3 - I_o = 0 \quad (6)$$

$$10 - 100 I_i - 100 I_2 = 0 \quad (7)$$

$$100 I_2 + 10 K I_3 + 500 I_o = 0 \quad (8)$$

Solving Equations 5→8,

$$I_i = 9 \text{ mA}, I_2 = 91 \text{ mA}$$

$$I_3 = -8 \text{ mA}, I_o = 82 \text{ mA}$$

$$V_o = -I_o R_4 = -41 \text{ V}$$

Comment

Note that the voltage across R_2 for Loop I is a voltage-drop, thus the sign for $I_2 R$ is negative; it is, however, in Loop II a voltage-rise, thus its sign is positive. In this example, a review on the applications of the basic circuit law was carried out through a numerical example. It is important to point out that determination of the signs of the currents and voltages are usually the major problems in deriving equations.

B. Example 2
Problem

For the network shown in Figure 2(a), derive the Thévenin equivalent circuit at the terminal aa'.

Solution

By current law,

$$\text{Node A}: I_1 + \beta I_1 = I_2 \quad (9)$$

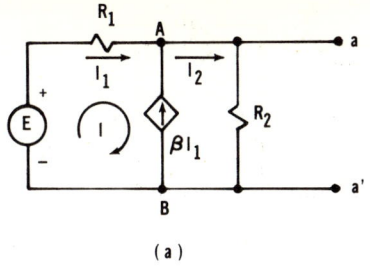

(a)

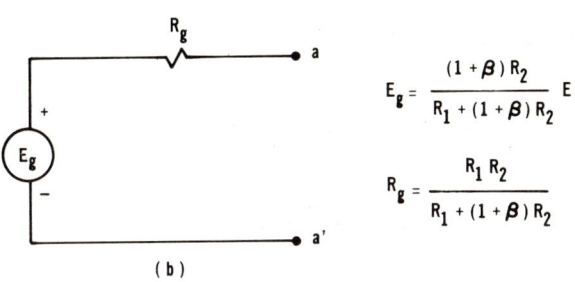

(b)

FIGURE 2. Thévenin equivalent circuit.

By voltage law,

$$\text{Loop I:} \quad E = I_1 R_1 + I_2 R_2 \tag{10}$$

From Equation 9,

$$I_1 = I_2 / (1 + \beta) \tag{11}$$

Substitute I_1, in Equation 10 by Equation 11,

$$E = \frac{R_1}{1+\beta} I_2 + I_2 R_2 = \left(\frac{R_1}{1+\beta} + R_2\right) I_2$$

$$I_2 = \frac{E}{\left(\dfrac{R_1}{1+\beta} + R_2\right)} \tag{12}$$

The Thévenin voltage,

$$E_g = I_2 R_2 = R_2 \left(\frac{E}{\dfrac{R_1}{1+\beta} + R_2}\right) = \frac{R_2(1+\beta)}{R_1 + R_2(1+\beta)} E \tag{13}$$

Let I_{sc} = the short circuit current aa'. In other words I_{sc} is I_2 with $R_2 = 0$, i.e., from Equation 12,

$$I_{sc} = I_2]_{R_2 = 0} = (1+\beta) \frac{E}{R_1}$$

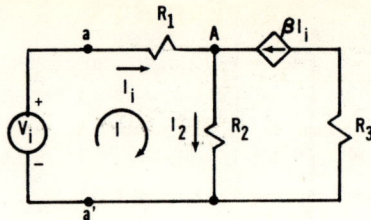

FIGURE 3. Circuit for solving input resistance.

By Thévenin theory, and Equation 13

$$R_g = \frac{E_g}{I_{sc}} = \frac{R_2(1+\beta)E}{R_1 + R_2(1+\beta)} \times \frac{R_1}{(1+\beta)E} = \frac{R_1 R_2}{R_1 + (1+\beta)R_2}$$

The Thévenin' equivalent circuit is shown in Figure 2(b).

Comment

Since the network contains a dependent current source which cannot be open-circuited, therefore derivation of R_g is more complicated and the ratio of E_g to I_{sc} is used for finding R_g. It is important to note that the Thévenin's equivalent resistance at aa' is also the resistance known as *output resistance* of the network at aa'. Oftentimes, a digital system designer needs to know, especially in interface circuit design, the input/output resistances or impedances of the given networks; therefore, one would use this technique from time to time. The next example will show how the input resistance of a network can be found.

C. Example 3
Problem

Determine the *input resistance* at terminals aa' of the network shown in Figure 3.

Solution

$$\text{NODE A:} \quad I_2 = I_i + \beta I_i = (1+\beta)I_i \tag{14}$$

$$\text{LOOP I:} \quad V_i = I_i R_1 + I_2 R_2 \tag{15}$$

Substitute I_2 in Equation 15 by Equation 14,

$$V_i = I_i R_1 + I_i(1+\beta)R_2 = I_i[R_1 + (1+\beta)R_2] \tag{16}$$

By definition the input resistance at terminal aa', i.e., $R_i \triangleq \dfrac{V_i}{I_i}$, from Equation 16,

$$R_i = \frac{V_i}{I_i} = R_1 + (1+\beta)R_2 \tag{17}$$

REFERENCES

1. **Hayt, W. H., Jr. and Kemmerly, J. E.**, *Engineering Circuit Analysis*, 3rd ed., McGraw-Hill, New York, 1978.
2. **Fitzgerald, A. E., Higginbotham, D. E., and Grabel, A.**, *Basic Electrical Engineering*, 4th ed., McGraw-Hill, New York, 1975.

Chapter 2

BASIC MODULES (BUILDING BLOCKS) FOR LINEAR SYSTEMS

The objective of this chapter is to present the fundamentals and characteristics of the most popular analog building blocks that are often used in digital system design. With emphasis on practical applications, unnecessary rigorous circuit analysis will not be presented; approximation techniques will be used whenever applicable.

I. SINGLE STAGE DIFFERENTIAL AMPLIFIER

The circuit shown in Figure 1(a) is a single-stage differential amplifier which oftentimes is the basic element of more complex circuitries both in analog — such as an operational amplifier — and digital — such as ECL (Emitter-Couple-Logic) — and transmission line driver/receiver circuits. Due to its symmetrical circuit configuration, it has the desirable properties, i.e., low DC drift, high common mode rejection and split-phase outputs. The small AC signal analysis and DC signal analysis for the circuit are described below.

A. AC Small Signal Analysis
Common Mode Analysis

Due to the integrated circuit technology, it is feasible to assume that the circuit shown in Figure 1(a) can be designed to have the following electrical properties: i.e., $R_{C1} = R_{C2} = R_C$; $Q_1 = Q_2$ or Q_1 and Q_2 have identical h-parameters. Figure 1(a) can be redrawn as Figure 1(b). Since the left and right circuits are identical, or $V_{E1} = V_{E2}$, there is no current flow in the wire connecting the emitters of Q_1 and Q_2. Therefore the wire can be cut and the circuit can be split into two halves as shown in Figure 1(c). Figure 1(d), then, shows the small signal equivalent circuit of the left-hand side. For simplicity let us assume that the h_{oe} and h_{re} of the transistors are negligible; then the circuit can be reduced further as shown in Figure 1(e). Based on Figure 1(e), we have,

$$e_1 = i_b(R_g + h_{ie}) + i_e 2R_E \qquad (1)$$

$$V_{O1} = -h_{fe} i_b R_C \qquad (2)$$

$$i_e = (1 + h_{fe}) i_b \qquad (3)$$

From Equations 1, 2, and 3,

$$\frac{V_{O1}}{e_1} = -\frac{h_{fe} R_C}{(R_g + h_{ie}) + (1 + h_{fe}) 2R_E} = -\frac{\frac{h_{fe}}{1 + h_{fe}} R_C}{2R_E + \frac{R_g + h_{ie}}{1 + h_{fe}}}$$

Since normally, $h_{fe} \gg 1$, then

$$\frac{h_{fe}}{1 + h_{fe}} \simeq 1$$

We have,

$$\frac{V_{O1}}{e_1} = -\frac{R_C}{2R_E + R_{eo}} \qquad (4)$$

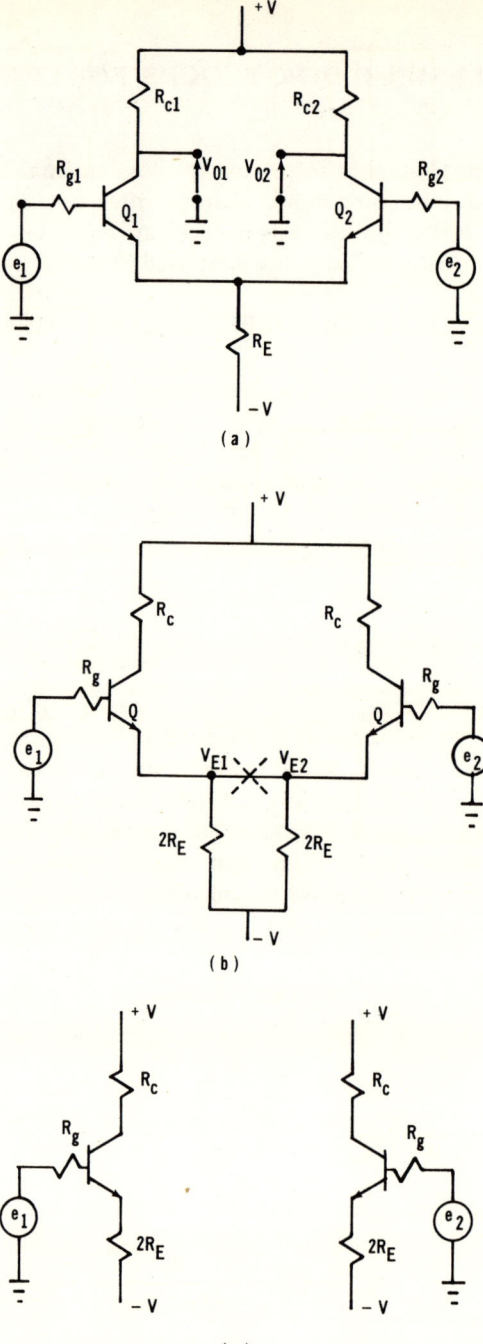

FIGURE 1. (a—i)Single state differential amplifier.

Where, R_{eo} is the equivalent output resistance at the emitter, i.e.,

$$R_{eo} \overset{\Delta}{=} \frac{R_g + h_{ie}}{1 + h_{fe}} \tag{5}$$

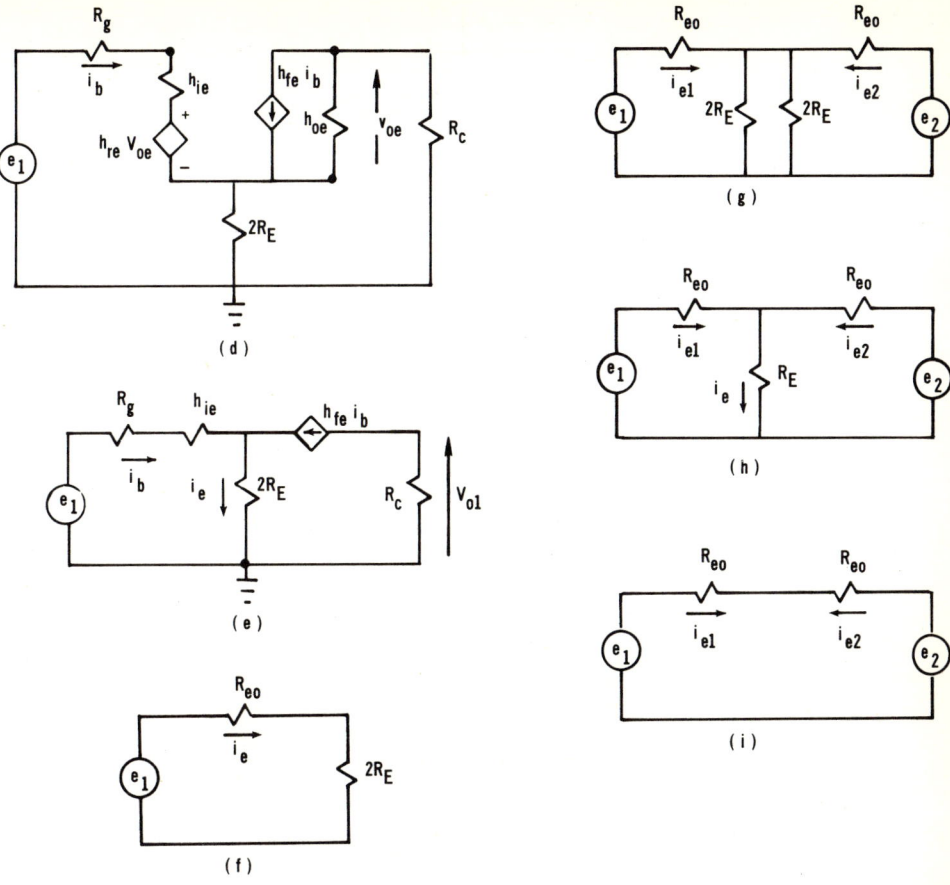

FIGURE 1 (continued)

Similarly,

$$\frac{V_{O2}}{e_2} = -\frac{R_C}{2R_E + R_{eo}} \qquad (6)$$

From Equations 4 and 6,

$$V_{O1} + V_{O2} = -\frac{R_C}{2R_E + R_{eo}}(e_1 + e_2)$$

$$\frac{V_{O1} + V_{O2}}{e_1 + e_2} = -\frac{R_C}{2R_E + R_{eo}} \qquad (7)$$

Let us define *common mode gain* (CMG),

$$CMG \triangleq \frac{\frac{1}{2}(V_{O1} + V_{O2})}{\frac{1}{2}(e_1 + e_2)}$$

Then we have,

$$\text{CMG} = -\frac{R_C}{2R_E + R_{oe}} \tag{8}$$

Differential Mode
From Equations 1 and 3,

$$e_1 = i_b(R_g + h_{ie}) + i_e 2R_E$$

$$= i_e \left[\frac{R_g + h_{ie}}{1 + h_{fe}} + 2R_E \right] \tag{9}$$

$$= i_e [R_{oe} + 2R_E]$$

According to Equation 7, an equivalent circuit can be shown in Figure 1(f). Similarly,

$$e_2 = i_e [R_{eo} + 2R_E] \tag{10}$$

Figures 1(g) and 1(h) depict the composite equivalent circuit based on Equations 7 and 8.

R_E normally is very large. In fact, most of the differential amplifiers replace R_E by a current generator supplying I_e, which has high output resistance; it is, therefore, reasonable to assume that, $R_E \gg R_{eo}$, $i_{e1} \gg i_e$, $i_{e2} > i_e$, or, $i_{e1} + i_{e2} = i_e \simeq 0$. We have, $i_{e1} = -i_{e2}$. If an equivalent circuit is shown in Figure 1(i), then

$$\begin{aligned} e_1 - i_{e1} R_{eo} &= e_2 - i_{e2} R_{eo} \\ e_1 - e_2 &= (i_{e1} - i_{e2}) R_{eo} = 2 i_{e1} R_{eo} \end{aligned} \tag{11}$$

But,

$$V_{o1} = -h_{fb} R_C i_{e1} \tag{12}$$

$$V_{o2} = -h_{fb} R_C i_{e2} = h_{fb} R_C i_{e1} \tag{13}$$

where

$$h_{fb} = \frac{i_c}{i_e} \simeq 1$$

By subtracting Equation 12 from Equation 13,

$$V_{o1} - V_{o2} = -2 h_{fb} i_{e1} R_C \simeq -2 i_{e1} R_C \tag{14}$$

Equation 14 ÷ Equation 11,

$$\frac{V_{o1} - V_{o2}}{e_1 - e_2} \simeq -\frac{i_{e1} R_C}{2 i_e R_{oe}} = -\frac{R_C}{R_{eo}} \tag{15}$$

Let us define the *differential mode gain* (DMG),

$$\text{DMG} \triangleq \frac{\tfrac{1}{2}(V_{o1} - V_{o2})}{\tfrac{1}{2}(e_1 - e_2)} \simeq -\frac{R_C}{R_{eo}} \tag{16}$$

For single-end output, we can have, Equation 12 ÷ Equation 11

$$\frac{V_{o1}}{e_1 - e_2} = \frac{h_{fb} R_C}{2R_{eo}} \simeq -\frac{R_C}{2R_{eo}} \qquad (17)$$

and Equation 13 ÷ Equation 11,

$$\frac{V_{o2}}{e_1 - e_2} = \frac{h_{fb} R_C}{2R_{eo}} \simeq \frac{R_C}{2R_{eo}} \qquad (18)$$

Equations 17 and 18 are defined as *single-end* to *differential-end voltage gain*. Note that there is phase inversion in Equation 17 while there is no inversion in Equation 18. The circuit, therefore, has the phase-splitting property. In view of Equations 17, 18, and 16, it reveals that the voltage for the single-end output is one half of the DMG.

C. Common Mode Rejection Property

Let us define *common mode rejection* (CMR),

$$\text{CMR} \triangleq \frac{\textit{differential mode gain}}{\textit{common mode gain}} = \frac{\text{DMG}}{\text{CMG}}$$

From Equations 16 and 8 we have

$$\text{CMR} \simeq \left(-\frac{R_C}{R_{eo}}\right) \div \left(-\frac{R_C}{2R_E + R_{oe}}\right)$$

$$= 1 + \frac{2R_E}{R_{eo}} \qquad (19)$$

In view of Equation 19, it reveals that as R_E increases, the CMR will increase accordingly. Normally, R_E is the output resistance of a constant current generator, and can be assumed to have large value. Therefore, the CMR of a differential amplifier is normally very high. In practice, as the amplifier is being used with the differential mode configuration, high CMR is a very desirable feature; it will reject the commonly induced noise at the input pair. As shown in Figure 2, e_s is the input signal to be amplified. The input lines 1 and 2, however, would also act as antennas picking up noise, e_1 and e_2. As a result, e_1 and e_2 will be amplified by the CMG of Equation 8, while e_s, by the DMG of Equation 16. It is evident that the higher the CMR of Equation 19, the better the signal-to-noise ratio of the amplifier.

The CMR of a given amplifier can be experimentally determined. Figure 3 shows the circuit diagram for determined CMR by, respectively, measuring the ratios

$$\frac{V_{do}}{e_s} = \text{DMG}$$

and,

$$\frac{\frac{1}{2}(V_{o1} + V_{o2})}{e_{CM}} = \text{CMG}$$

where e_s is a signal generator used as the differential input, while e_{CM} is a signal generator used to generate common-mode input signal.

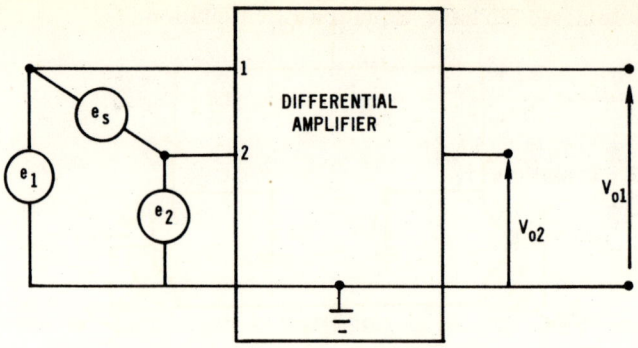

FIGURE 2. Common mode rejection.

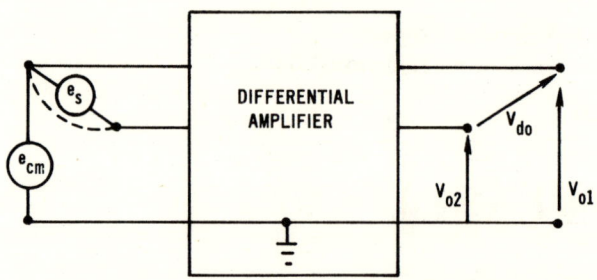

FIGURE 3. Common mode rejection measurement.

Finally, the CMR can be determined by dividing the DMG by CMG obtained from the measurement.

B. DC Analysis

In reference to Figure 4,

$$I_{b_1} R_{g_1} + V_{b_1} = I_{b_2} R_{g_2} + V_{b_2}$$

$$V_{o_1} = V_{CC} - I_{C_1} R_C \qquad (20)$$

$$V_{o_2} = V_{CC} - I_{C_2} R_C$$

$$V_{o_1} - V_{o_2} = -I_{C_1} R_C + I_{C_2} R_C \qquad (21)$$

Let

$$\beta_1 = \frac{I_{C_1}}{I_{b_1}}, \quad \beta_2 = \frac{I_{C_2}}{I_{b_2}}$$

Then Equation 21 becomes

$$V_{o_1} - V_{o_2} = -\beta_1 I_{b_1} R_C + \beta_2 I_{b_2} R_C$$

$$= R_C \beta_1 I_{b_1} \left(\frac{\beta_2 I_{b_2}}{\beta_1 I_{b_1}} - 1 \right) \qquad (22)$$

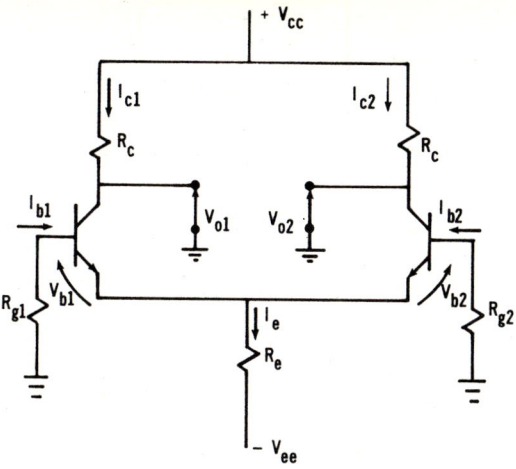

FIGURE 4. DC analysis.

From Equation 20

$$I_{b_1} R_{g_1} \left(\frac{I_{b_2} R_{g_2}}{I_{b_1} R_{g_1}} - 1 \right) = (V_{b_1} - V_{b_2}) \qquad (23)$$

From Equations 22 and 23, we have

$$V_{o_1} - V_{o_2} = R_C \beta_1 \left(\frac{\beta_2 I_{b_2}}{\beta_1 I_{b_1}} - 1 \right) \frac{(V_{b_1} - V_{b_2})}{\left(\frac{I_{b_2} R_{g_2}}{I_{b_1} R_{g_1}} - 1 \right) R_{g_1}} \qquad (24)$$

It is desirable that, $V_{o_1} = V_{o_2}$, if the amplifier is at its quiescent condition, or zero signal input. Equation 24 shows the parameters which control the equality of V_{o_1} and V_{o_2}.

II. OPERATIONAL AMPLIFIER

An operational amplifier is a multiple-stage direct-coupled amplifier. Its first stage is generally a differential amplifier and last stage is a single-end buffer amplifier. One may consider an operational amplifier as one which is a differential amplifier followed by a high-gain DC amplifier. The operational amplifier has been the essential building block of an analog or linear computing system for decades. In the event of the large-scale-integrated-circuit (LSI), technology, the unit cost of some general purpose operational amplifiers has been reduced to almost a negligible level with respect to that of the system where operational amplifiers are being used as components. A digital system designer can now consider an operational amplifier as a system component equivalent to a NAND gate. One may use it wherever necessary without being cost conscious. What follows are the basic principles of operation and practical applications of the operational amplifier. Rigorous circuit analysis will not be covered; approximation techniques will be employed whenever feasible.

A. Equivalent Circuit of an Operational Amplifier

Figure 5 depicts a simplified equivalent circuit of an operational amplifier which has

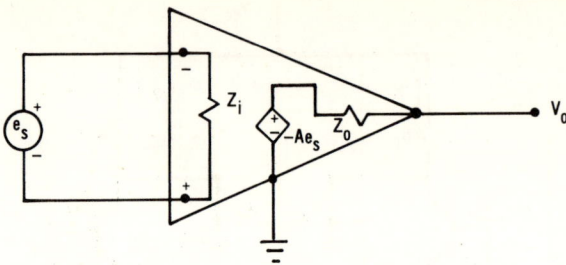

FIGURE 5. Equivalent circuit of an operational amplifier.

three essential terminals for processing signals, i.e., two differential inputs labeled, respectively, with minus and plus signs, and one output labeled with V_o. R_i is the input and R_o the output resistances. The amplifier has an open-loop-voltage-gain, A. As the polarity assigned for the input signal voltage e_s, the output voltage, $V_o = -Ae_s$. The ideal model of an operational amplifier, however, is normally defined as follows, $R_i = \infty$, $R_o = 0$, $A = \infty$, but $V_0 = 0$, when $e_s = 0$; bandwidth $= \infty$, offset current $= 0$, offset voltage $= 0$, and common mode rejection $= \infty$. In practice, the ideal model oftentimes is evidently over-simplified for some circuit analysis and design. Depending on the nature of the application, for instance, if the input, e_s, is a low-level direct current signal, the offset current and offset voltage effect on the output voltage cannot be neglected. In view of Equation 24, it reveals that mismatch of the base-emitter junction voltages V_{b1} and V_{b2} as well as that of $\beta_1 I_{b1}$ and $\beta_2 I_{b2}$ would cause $V_{01} - V_{02} \neq 0$. The former is known as *off-set voltage* and the latter is known as *off-set current*. This concept can be applied to the practical operational amplifier which will be elaborated upon next. Fortunately, in most applications, such as AC coupled circuit configuration, or a large DC signal, these problems are not significant. For applications in low-level DC analog signal or analog integration operation, however, these problems can be severe and special purpose operational amplifiers, or low-drift or chopper stabilized instrumentation amplifiers should be employed.

B. The Four Useful and Essential Feedback Circuit Configurations

In most applications, the operational amplifier is used for processing analog or continuous signals. An operational amplifier is simply a "raw" component which needs to be externally modified or tailored to meet certain requirements. Given below are the four most popular and useful feedback circuit configurations and their unique features.

Shunt-to-Shunt Feedback — Inverted Voltage Amplifier

The circuit configuration is shown in Figure 6(a). It is important to point out that the operational amplifier needs external resistor networks to serve as a bias network to supply bias currents regardless of the circuit configuration. Here, R_1, R_2, and R_b are used for biasing purposes, but R_1 and R_2, in addition, determine the voltage gain for the network. For maintaining the circuit symmetry of the differential inputs to minimize drifting effect, i.e., maintaining the output voltage equals zero with zero input,

$$R_b = \frac{R_1 R_2}{R_1 + R_2}$$

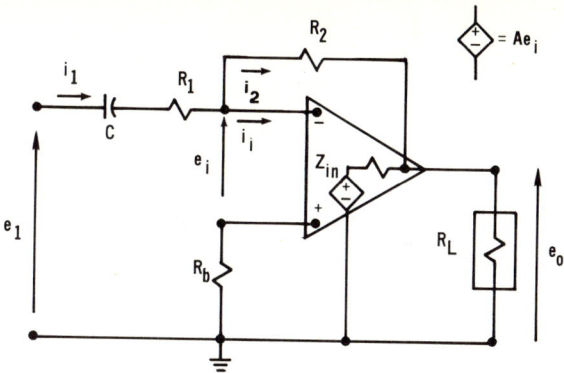

FIGURE 6(a). AC coupled shunt-shunt feedback.

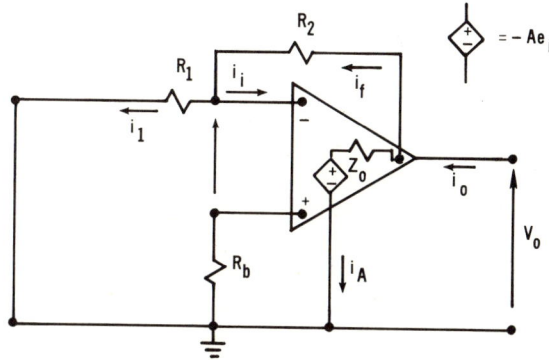

FIGURE 6(b). Output impedance of shunt-shunt feedback.

if the coupling capacitor C is not used and there is a direct current path from R_1 to ground through input signal source; otherwise, $R_b = R_2$.

For convenience without sacrificing the generality, we shall consider the capacitor coupling as a special case and therefore the analysis hereafter will be based on the circuitry without the coupling capacitor.

Voltage Gain of the Feedback Network: A_{fb}

$$i_1 = i_i + i_2$$

Since $Z_{in} \simeq \infty$, e_i, e_o are finite, and

$$e_i = \frac{e_o}{A} \simeq \frac{e_o}{\infty} = 0$$

where A is the voltage gain of the operational amplifier. Since,

$$i_i = \frac{e_i}{z_{in}} \simeq \frac{e_i}{\infty} = 0$$

then $i_1 = i_2$.

But, $e_o = e_1 - i_1 R_1 - i_2 R_2$, then,

$$e_o = e_1 - i_1(R_1 + R_2) \tag{25}$$

From the input loop,

$$e_1 = i_1 R_1 + e_i + i_i R_b = i_1 R_1 + e_i \tag{26}$$

$$i_1 = \frac{1}{R_1}(e_1 - e_i) = \frac{1}{R_1}\left(e_1 - \frac{e_o}{-A}\right) = \frac{1}{R_1}\left(e_1 + \frac{e_o}{A}\right)$$

Substituting Equation 25 into Equation 26

$$e_o = e_1 - \frac{1}{R_1}\left(e_1 + \frac{e_o}{A}\right)(R_1 + R_2)$$

$$e_o\left(1 + \frac{R_1 R_2}{R_1 A}\right) = e_1\left(1 - \frac{R_1 + R_2}{R_1}\right) = e_1\left(-\frac{R_2}{R_1}\right)$$

$$\frac{e_o}{e_1} = \frac{-R_2/R_1}{1 + \frac{1 + R_2/R_1}{A}} = -\frac{R_2}{R_1} \text{ if } A \gg \left(1 + \frac{R_2}{R_1}\right)$$

Finally,

$$A_{fb} = \frac{e_o}{e_1} = -\frac{R_2}{R_1}$$

where fb stands for feedback configuration.

Input Impedance of the Feedback Network: Z_{ifb}

Since

$$Z_{ifb} \triangleq \frac{e_1}{i_1} = \frac{i_1 R_1 + e_i}{i_1}$$

But $e_i \simeq 0$, or $e_i \ll i_1 R_1$, we have $Z_{ifb} = R_1$.

Output Impedance of the Feedback Network: Z_{ofb}

Referring to Figure 6(b), and based on the definition of output impedance, the input terminals are short circuited and a voltage source v_o is applied at the output terminals. Z_{ofb} is then defined as V_o/i_o.

$$i_o = i_f + i_A$$

$$i_f = i_1 + i_i = i_1 \tag{27}$$

$$e_i = i_f R_1 \tag{28}$$

$$V_o = i_f R_2 + e_i = i_f R_2 + i_f R_1 = i_f(R_2 + R_1) \tag{29}$$

$$V_o = i_A Z_o + (-Ae_i) \tag{30}$$

Substituting Equations 27 and 28 into Equation 30,

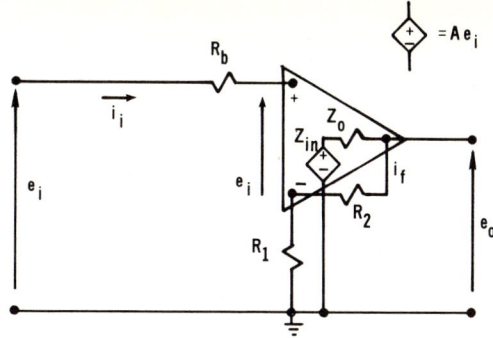

FIGURE 7(a). Shunt-series feedback.

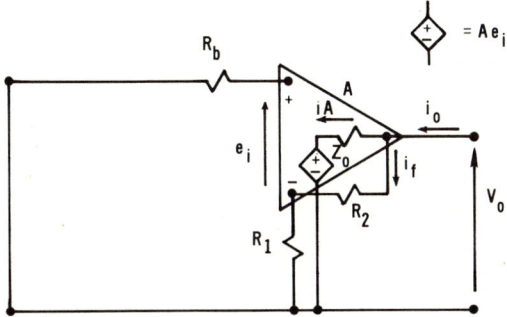

FIGURE 7(b). Output impedance of shunt-series feedback.

$$V_o = (i_o - i_f)Z_o - Ai_f R_1$$
$$= i_o Z_o - i_f(1 + AR_1) \qquad (31)$$

Substituting Equation 29 into Equation 31,

$$V_o = i_o Z_o - \left(\frac{V_o}{R_2 + R_1}\right)(1 + AR_1)$$

$$V_o \left[1 + \frac{1 + AR_1}{R_2 + R_1}\right] = i_o Z_o$$

$$V_o \left[1 + \frac{1}{R_2 + R_1} + A\frac{R_1}{R_1 + R_2}\right] = i_o Z_o$$

Since $R_2 R_1 > -1$

$$V_o \left[1 + A\frac{R_1}{R_1 + R_2}\right] = i_o Z_o$$

Thus,

$$Z_{ofb} = \frac{V_o}{i_o} = \frac{Z_o}{1 + A\left(\dfrac{R_1}{R_1 + R_2}\right)}$$

Shunt-Series Feedback — Noninverted Voltage Amplifier

Figure 7(a) shows the circuit configuration.

Voltage Gain of the Feedback Network: A_{fb}

$$e_1 = i_i R_b + e_i + (i_i + i_f) R_1$$

For the same reason as described above, $i_i \simeq 0$, $e_i \simeq 0$, or $i_f \gg i_i$, we have

$$e_1 = i_f R_1 \tag{32}$$

$$e_o = i_f R_2 + (i_i + i_f) R_1 = i_f (R_1 + R_2) \tag{33}$$

Equation 33 ÷ Equation 32

$$\frac{e_o}{e_1} = \frac{R_1 + R_2}{R_1} = 1 + \frac{R_2}{R_1}$$

$$A_{fb} = \frac{e_o}{e_1} = 1 + \frac{R_2}{R_1}$$

Input Impedance of the Feedback Network: Z_{ifb}

$$\begin{aligned} e_1 &= i_i (R_b + Z_{in}) + (i_i + i_f) R_1 \\ &= i_i (Z_{in} + R_b + R_1) + i_f R_1 \end{aligned} \tag{34}$$

$$e_o = i_f (R_1 + R_2)$$

$$A e_i = e_o + i_f Z_o = i_f (R_1 + R_2 + Z_o)$$

$$e_i = i_i Z_{in} \tag{35}$$

$$A i_i Z_{in} = i_f (R_1 + R_2 + Z_o)$$

Substituting Equation 35 into Equation 34,

$$e_1 = i_i (Z_{in} + R_b + R_1) + R_1 \frac{A i_i Z_{in}}{R_1 + R_w + Z_o}$$

$$= i_i \left[Z_{in} + R_b + R_1 + R_1 \frac{A Z_{in}}{R_1 + R_2 + Z_o} \right]$$

$$= i_i \left[Z_{in} \left(1 + A \frac{R_1}{R_1 + R_2 + Z_o}\right) + R_b + R_1 \right]$$

Since, $Z_o \ll R_1 + R_2$,

$$Z_{in} \left(1 + A \frac{R_1}{R_1 + R_2}\right) \gg (R_b + R_1)$$

We have

$$e_1 \simeq i_i Z_{in}\left(1 + A\frac{R_1}{R_1 + R_2}\right)$$

$$\frac{e_1}{i_1} \simeq Z_{in}\left(1 + A\frac{R_1}{R_1 + R_2}\right)$$

$$Z_{ifb} = Z_{in}\left(1 + A\frac{R_1}{R_1 + R_2}\right)$$

Output Impedance of the Feedback Network: Z_{ofb}
Referring to Figure 7(b),

$$i_o = i_f + i_A \tag{36}$$

$$V_o = Ae_i + i_A Z_o \tag{37}$$

$$V_o = i_f(R_2 + R_1) \tag{38}$$

$$e_i = -i_f R_1 \tag{39}$$

Substituting Equations 36 and 39 into Equation 37

$$\begin{aligned}V_o &= -Ai_f R_1 + (i_o - i_f)Z_o \\ &= i_o Z_o - i_f(Z_o + AR_1)\end{aligned} \tag{40}$$

Substituting Equation 38 into Equation 40,

$$V_o = i_o Z_o = \frac{V_o}{R_1 + R_2}(Z_o + AR_1)$$

$$\simeq I_o Z_o = \frac{V_o}{R_1 + R_2}(AR_1) \text{ since } Z_o \ll AR_1$$

$$V_o\left[1 + A\frac{R_1}{R_1 + R_2}\right] = i_o Z_o$$

$$\frac{V_o}{i_o} = Z_o \Big/ \left[1 + A\frac{R_1}{R_1 + R_2}\right]$$

$$Z_{ofb} \triangleq \frac{V_o}{i_o} = \frac{Z_o}{1 + A\frac{R_1}{R_1 + R_2}}$$

Series-Series Feedback — Voltage to Current Converter

Figure 8(a) shows the series-series feedback circuitry. As the name implies, the feedback element R_1 is a series element in the input as well as output loop. Here, R_L represents the load resistance, and R_b, the bias resistor, which value should be equal to the parallel resistance of $(R_L + Z_o)$ and R_1.

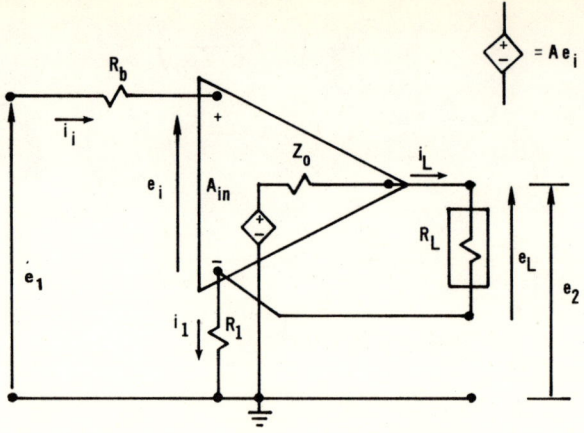

FIGURE 8(a). Series-series feedback.

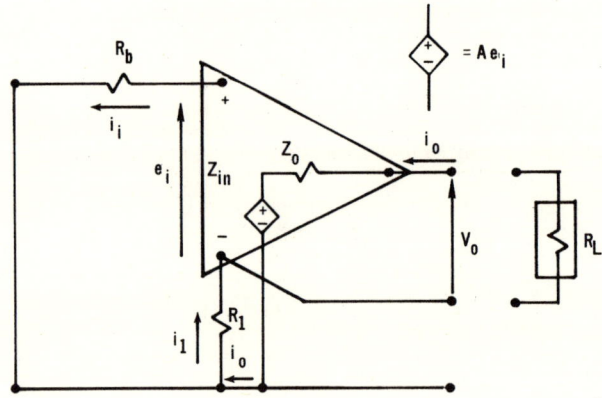

FIGURE 8(b). Output impedance of series-series feedback.

Voltage Gain of the Feedback Network: $A_{fb} \triangleq e_L/e_1$

In Figure 8(a), we have $e_1 = i_i R_b + e_i + i_1 R_1$, $i_1 = i_L + i_i \simeq i_L$ for $i_L \gg i_i$. Thus, $e_1 \simeq i_L R_1$. Since, $e_i \simeq 0$. And,

$$\frac{e_2}{e_1} \simeq \frac{i_L (R_L + R_1)}{i_L R_1} = 1 + \frac{R_L}{R_1} \tag{41}$$

But,

$$e_L = i_L R_L \tag{42}$$

$$e_L = e_2 - i_L R_1$$

Thus,

$$e_L = e_2 - \frac{e_L}{R_L} R_1$$

and

$$e_2 = e_L \left(1 + \frac{R_1}{R_L}\right) \tag{43}$$

Substituting Equation 43 into Equation 41,

$$A_{fb} = \frac{e_L}{e_1} = \frac{\left(1 + \frac{R_L}{R_1}\right)}{\left(1 + \frac{R_1}{R_L}\right)} = \frac{R_L}{R_1}$$

From Equations 42 and 43

$$\frac{i_L}{e_1} = \frac{1}{R_1}$$

The last equation constitutes the voltage (e_1) to current (i_L) conversion.

Input Impedance of the Feedback Circuit: Z_{ifb}
Since the input circuit configuration is identical with respect to the input loop of shunt-series feedback configuration with $R_2 = R_L$, we have

$$Z_{ifb} = Z_{in}\left(1 + A\frac{R_1}{R_1 + R_L}\right)$$

Output Impedance of the Feedback Circuit: Z_{ofb}
In reference to Figure 8(b), since $R_b < R_1 \ll Z_{in}$, and

$$i_1 = \frac{Z_{in} + R_b}{Z_{in} + R_b + R_1} i_o$$

we have $i_1 \simeq i_o$ or $i_i \ll i_1$.
But

$$V_o = Ae_i + i_o Z_o + i_1 R_1 \tag{44}$$

and

$$e_i = i_i R_b + i_1 R_1 \simeq i_1 R_1 = i_o R_1 \tag{45}$$

Substituting Equation 45 into Equation 44

$$V_o = Ai_o R_1 + i_o Z_o + i_o R_1$$

$$= I_o [Z_o + R_1(1 + A)]$$

$$\frac{V_o}{I_o} = Z_o + R_1(1 + A)$$

$$Z_{ofb} = Z_o + R_1(1 + A)$$

Series-Shunt Feedback — Current Amplifier
Figure 9(a) shows the series-shunt feedback circuitry. Note that the feedback element R_3 is in series within the output loop, while the voltage drop across R_3 is fed back in parallel with the input through R_2. Because the input is a current source which can be

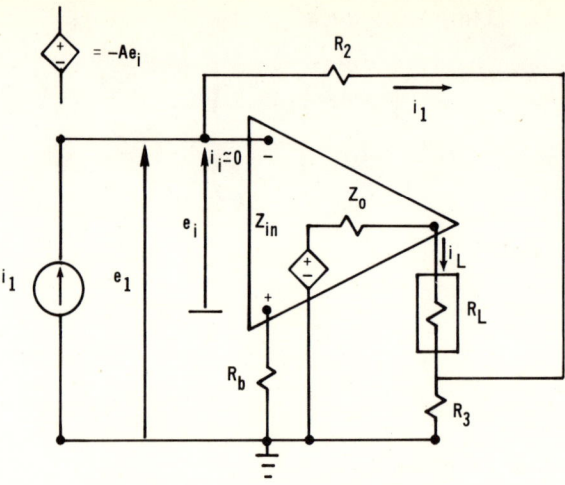

FIGURE 9(a).   Series-shunt feedback.

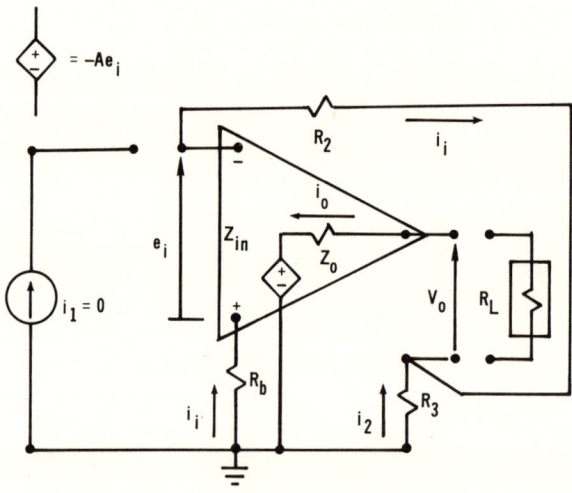

FIGURE 9(b).   Output impedance of series-shunt feedback.

assumed having very high source impedance, the bias-resistor R_b can be designed such that

$$R_b \simeq R_2 + \frac{R_3 R_L}{R_3 + R_L}$$

Current Gain of the Feedback Network: $G_{fb} \triangleq i_L/i_1$

In Figure 9(a), $e_i = i_1(R_2 + R_3) + i_L R_3$, and $-A e_i = i_L(Z_o + R_L + R_3) + i_1 R_3$. By eliminating e_i, we have

$$-A[i_1(R_2 + R_3) + i_L R_3] = i_L(Z_o + R_L + R_3) + i_1 R_3$$

$$i_L[-AR_3 - (Z_o + R_L + R_3)] = i_1 [R_3 + A(R_2 + R_3)]$$

$$\frac{i_L}{i_1} = -\frac{R_3 + \frac{A}{1+A} R_2}{R_3 + \frac{-Z_0 + R_L}{1+A}} \simeq -\left(1 + \frac{R_2}{R_3}\right)$$

Since

$$R_3 \gg \frac{R_L - Z_0}{1+A}$$

$$G_{fb} \triangleq \frac{i_L}{i_1} = -\left(1 + \frac{R_2}{R_3}\right)$$

Input Impedance of the Feedback Circuit: Z_{ifb}

Since the input loop equation of the circuit is,

$$\begin{aligned} e_1 &= i_1 R_2 + (i_L + i_1) R_3 \\ &= i_1 (R_3 + R_2) + i_L R_3 \end{aligned} \qquad (46)$$

But from the current gain equation, and assume that,

$$R_3 \gg \frac{R_L - Z_0}{1+A}$$

We have,

$$i_L = i_1 G_{fb} \simeq -\left(1 + \frac{R_2}{R_3}\right) i_1 \qquad (47)$$

Substituting Equation 47 into Equation 46,

$$e_1 \simeq i_1 (R_3 + R_2) - i_1\left(1 + \frac{R_2}{R_3}\right) R_3$$

$$\frac{e_1}{i_1} \simeq 0$$

$$Z_{ifb} \triangleq \frac{e_1}{i_1} \simeq 0$$

This is quite reasonable, since the input loop of this circuit is similar to shunt-shunt feedback circuit except that $R_1 = 0$ here. Therefore z_{ifb} for this circuit is approximately zero.

Output Impedance of the Feedback Circuit: $Z_{ofb} \triangleq V_o/i_o$

Since the input is expected to be a current source, it is disconnected for deriving the output impedance as shown in Figure 9(b).

Since,

$$\begin{aligned} i_o &= i_i + i_2 \\ v_o &= i_o Z_0 + (-Ae_i) + i_2 R_3 \\ e_i &= i_i R_2 - i_2 R_3 + i_i R_b \\ R_3 &\ll (Z_{int} + R_2) \text{ or } i_2 \gg i_i \end{aligned} \qquad (48)$$

We have,

$$i_o \simeq i_2$$
$$e_i \simeq -i_2 R_3 \simeq -i_o R_3 \tag{49}$$

Substituting Equation 49 into Equation 48,

$$v_o = i_o Z_o + A i_o R_3 + i_o R_3$$

$$= i_o [Z_o + (1+A) R_3]$$

$$\frac{v_o}{i_o} = Z_o + (1+A) R_3$$

$$Z_{ofb} = Z_o + (1+A) R_3$$

C. Consideration of Gain-Bandwidth Property

It was defined previously that an ideal operational amplifier has an infinite width of bandwidth. That is to say, an ideal operational amplifier has a constant voltage or current gain for any signal of any frequency. In practice, an operational amplifier, however, has a finite bandwidth which can usually be found in the specification sheets provided by the manufacturer. In a more detailed specification sheet, the gain-bandwidth characteristic of a typical operational amplifier is normally described by its Bode plot, or mathematically.

$$\text{Voltage gain (in dB)} = 20 \log \left| \frac{E_o}{E_i} \right| = 20 \log |A|$$

which is a function of the signal's frequency as shown in Figure 10. Note that the 3 dB point is at a frequency equal to 100 Hz, and the Bode plot shown has a gain of 100 dB with 100 Hz bandwidth. Oftentimes the gain-bandwidth of an operational amplifier is specified at 0 dB or unity gain, thus the unity-gain-bandwidth of this amplifier is 10 MHz. The gain-bandwidth information for an amplifier is very important to the system designer. To process a signal with a given bandwidth, an amplifier should be selected with a gain-bandwidth at least equal to the desirable gain-bandwidth of the signal.

It is important to point out that in practice, operational amplifiers are mostly used in conjunction with feedback circuit configurations as described in Section B of this chapter. Note that the gain with feedback, A_{fb}, is normally considerably lower than the gain without feedback which is also known as open-loop gain, A. Thanks to the discovery of the linear negative feedback theory during World War II, the gain-loss in feedback configuration is used to pay for the desirable features. That is, the feedback network is relatively free from the variation of amplifier parameters due to environmental changes or aging effects of the circuit components. As for bandwidth, the feedback configuration will result in a bandwidth virtually wider than that of open-loop configuration. As shown in Figure 10, with an open-loop gain of 100 dB, the amplifier has a bandwidth of 100 Hz. With a feedback gain A_{fb} = 10 or 20 dB, however, the amplifier has a bandwidth of 1 MHz. In other words, without feedback the amplifier can only be used to process signals with a bandwidth of 100 Hz, while with feedback, the amplifier can process signals with a bandwidth of 1 MHz. But the latter has considerable lower gain. In some cases a designer may want to have both gain and bandwidth

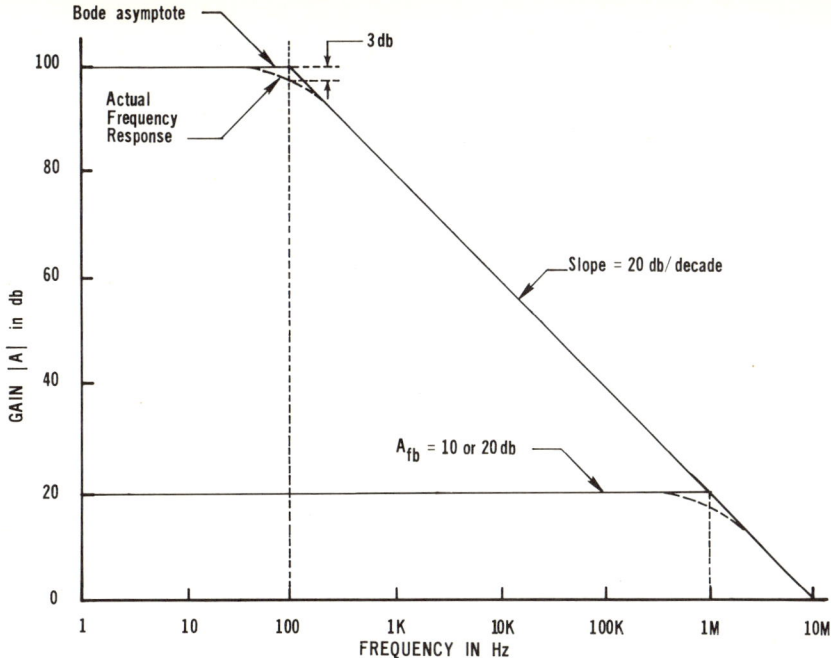

FIGURE 10. Bode diagram of an operational amplifier. (Courtesy of Texas Instruments Inc.)

at the same time. Fortunately, there are operational amplifiers having pins accessible to the designer to add external frequency compensation circuitry known as lead/lag compensation networks in linear feedback theory.

Figure 11(a) shows the test circuitry for the operational amplifier μA709 which provides the accessible pins for frequency compensation networks: R_1, C_1; R_2, C_2. The solid lines shown in Figure 11(b) are the Bode plots of the same amplifier with different values of the compensation networks. The dotted lines are the Bode plots of the amplifier with different feedback networks R_3, R_4, respectively. If the amplifier is the type with internal frequency compensation, it would have a Bode plot similar to the one shown with $R_1 = 1.5K$, $C_1 = 5000$ pF, $R_2 = 50$, and $C_2 = 200$ pF; then for $A_{fb} = 60$ dB, or $R_3/R_4 = 1000$, the amplifier would have a bandwidth of 1 kHz. Note that due to the availability of external compensation pins, with $R_1 = 0$, $C_1 = 10$ pF, $C_2 = 3$ pF, $R_2 = 50$, we can now have a bandwidth of 1 MHz in this case, i.e., the 60 dB dotted line. At this point, one might ask what would happen if one uses the same values of compensation network components but have lower A_{fb}, say $A_{fb} = 20$ dB. The result will undoubtedly be an unstable network. This is because of the fact that the Bode plot with the first set of values, i.e., $R_1 = 0$, $C_1 = 10$ pF, $C_2 = 3$ pF, $R_2 = 50$ will have a phase-shift greater than 180°, at the gain, $A_{fb} = 20$ dB. Thus it is important that the designer should use the suggested values for the compensation network for a given closed loop gain, A_{fb}; or he/she should select the ones having internal compensation networks but be satisfied with a narrower bandwidth or whatever the bandwidth result in a given A_{fb}. In general, however, the bandwidth of an amplifier is virtually increased as the amount of feedback known as Feedback Factor, i.e., (1 + Aβ), where A is the open-loop-gain; β is the feedback ratio determined by the feedback network.

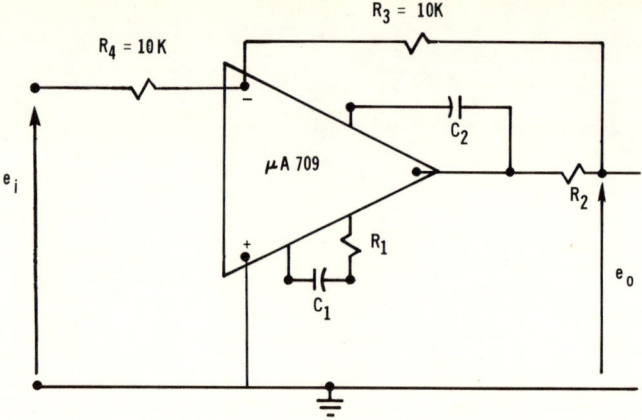

FIGURE 11(a). Operational amplifier with external frequency compensation.

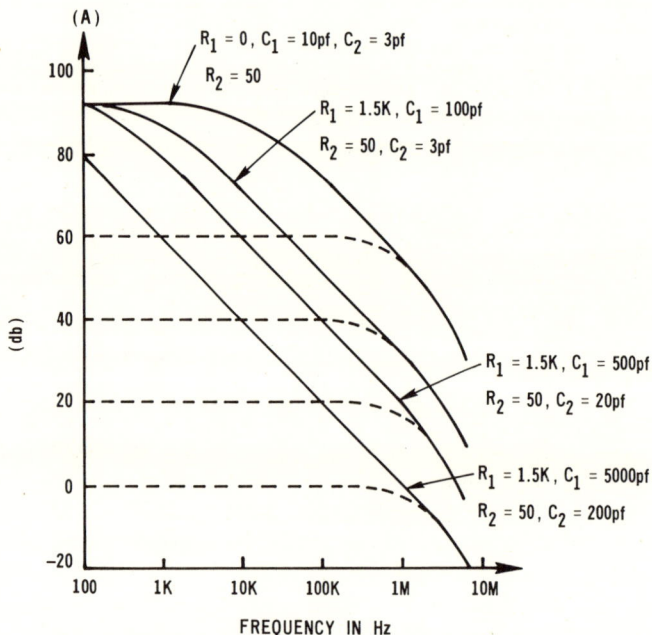

FIGURE 11(b) Bode diagram of the operational amplifier with external compensation network.

D. Summary — Comparison of the Four Feedback Configuration

It is important to point out that except for special applications, such as analog comparator, etc., an operational amplifier is normally used with one of the four kinds of feedback circuit configurations described in the preceding sections. Therefore, it is desirable to have a comparison table provided for fast references. Table 1 shows their major characteristics. The main advantage of using feedback circuitry as shown in this table is that all the key parameters such as gain and input/output impedances are not primarily functions of the characteristics of the individual amplifier; rather, they are virtually determined by the external resistive network which can be selected depending on the requirement of the applications. In other words, it can be made almost as accurate or as stable as the resistors used.

Table 1

CIRCUIT CONFIGURATION	GAIN	INPUT IMPEDANCE	OUTPUT IMPEDANCE	COMMENTS
(circuit with R_1, R_2, R_b)	Voltage Gain: $-\dfrac{R_2}{R_1}$	R_1	$\dfrac{z_o}{1 + A\left(\dfrac{R_1}{R_1 + R_2}\right)}$	Inverted Output. Controllable Input Impedance Very Low Output Impedance
(circuit with R_b, R_2, R_1)	Voltage Gain: $1 + \dfrac{R_2}{R_1}$	$Z_{in}\left[1 + A\,\dfrac{R_1}{R_1 + R_2}\right]$	$1 + A\,\dfrac{R_1}{R_1 + R_2}$	Non-Inverted Output. Very High Input Impedance Very Low Output Impedance Generally Used as Buffer Amplifier
(circuit with e_1, R_b, R_1, R_L, i_L)	$i_L = e_1/R_1$	$Z_{in}\left[1 + A\,\dfrac{R_1}{R_1 + R_2}\right]$	$z_o + R_1(1 + A)$	Voltage to Current Conversion High Input Impedance High Output Impedance R_L Must Be Floating From Ground. i_L is Independent of R_L
(circuit with i_1, R_2, R_L, R_b, R_3)	Current Gain: $\dfrac{i_L}{i_1} = -\left(1 + \dfrac{R_2}{R_3}\right)$	$\simeq 0$	$z_o + (1 + A)R_3$	Current to Current Conversion Low Input Impedance Very High Output Impedance R_L Must Be Floating From Ground.

III. EXPERIMENTAL EXAMPLES FOR TYPICAL APPLICATIONS

Based on the fundamentals developed in preceding sections, we shall now show some typical applications with examples and experimental results.

A. Analog Adder/Subtractor

Figure 12(a) shows the circuit configuration of an adder/subtractor. From Kirchhoff's current law,

$$\frac{e_1 - e_n}{R_1} + \frac{e_2 - e_n}{R_2} = \frac{e_n - e_o}{R_f} \qquad (50)$$

$$\frac{e_3 - e_p}{R_3} + \frac{e_4 - e_p}{R_4} = \frac{e_p}{R_f} \qquad (51)$$

$$\frac{R_f}{R_1}e_1 + \frac{R_f}{R_2}e_2 = e_n\left[1 + \frac{R_f}{R_1} + \frac{R_f}{R_2}\right] - e_o \qquad (52)$$

$$\frac{R_f}{R_3}e_3 + \frac{R_f}{R_4}e_4 = e_p\left[1 + \frac{R_f}{R_3} + \frac{R_f}{R_4}\right] \qquad (53)$$

Let

$$R_f\left(\frac{1}{R_1} + \frac{1}{R_2}\right) = R_f\left(\frac{1}{R_3} + \frac{1}{R_4}\right) = k \qquad (54)$$

Substituting into Equations 53 and 54, Equation 53 − Equation 54 yields,

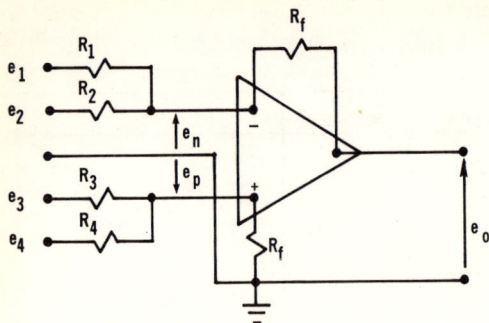

FIGURE 12(a).   Analog adder/subtractor.

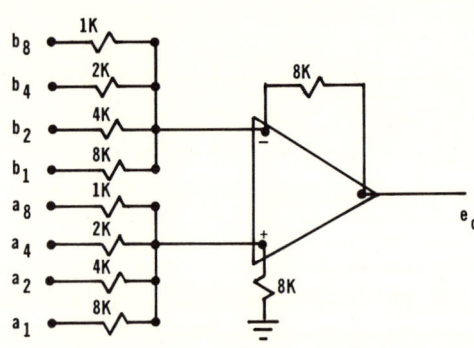

FIGURE 12(b).   Four-bit binary adder/subtractor.

FIGURE 12(c).   Experimental circuit.

e_1	e_2	e_3	e_4	e_o	
				Measured	Calculated
1	1	0	0	−1.40	−1.55
1	0	0	1	−0.45	−0.45
0	0	1	1	+1.50	+1.55
2	0	1	1	−0.50	−0.45
1	2	1	2	0	0
2	1	1	2	−0.5	−0.45
2	3	1	3	−1.0	−1.00
3	2	1	0	−3.0	−3.10

FIGURE 12(d).   Experimental results.

$$R_f \left[\frac{1}{R_1} e_1 + \frac{1}{R_2} e_2 - \frac{1}{R_3} e_3 - \frac{1}{R_4} e_4 \right]$$

$$= (e_n - e_p)[1+k] - e_o$$

$$= -\frac{e_o}{A}[1+k] - e_o$$

$$= -e_o \left[1 + \frac{1+k}{A} \right]$$

where open-loop gain,

$$-A = \frac{e_o}{e_n - e_p}$$

Let $A_o \gg (1 + k)$, we have,

$$R_f \left[\frac{1}{R_3} e_3 + \frac{1}{R_4} e_4 \right] - R_f \left[\frac{1}{R_1} e_1 + \frac{1}{R_2} e_2 \right] = e_o \quad (55)$$

Based on Equation 55, a four-bit binary adder/subtractor with analog output can easily be implemented with the circuit configuration shown in Figure 12(b).

Let $a_1, a_2, a_3, a_4; b_1, b_2, b_3, b_4$ equal to a binary value, 0 or V volts, respectively. Then we have,

$$e_o = V[2^3 a_8 + 2^2 a_4 + 2^1 a_2 + a^0 a_1] \\ -V[2^3 b_8 + 2^2 b_4 + 2^1 b_2 + b^0 b_1] \quad (56)$$

where $a_1, ..., a_4; b_1, ..., b_4$ are either 0 or 1. The output voltage e_o is the analog value of the binary adder/subtractor. This circuit is therefore a 4-bit binary adder/subtractor with a built-in digital-to-analog converter.

Experimental Example

To verify the theoretical derivation, an experimental circuit shown in Figure 12(c) was used and the experimental results against the theoretical are shown in Figure 12(d). The theoretical values are determined by,

$$e_o = \frac{1.2}{1.2} e_3 + \frac{1.2}{2.2} e_4 - \left[\frac{1.2}{1.2} e_1 + \frac{1.2}{2.2} e_2 \right]$$

$$= e_3 + 0.55 e_4 - e_1 - 0.55 e_2$$

A μA747 operational amplifier with internal frequency compensation network was used. In all cases the theoretical values agree with the experimental results within ±10%. This is within the tolerances of the resistors and measuring equipment.

B. Square Wave and Triangular Wave Generator

Figure 13(a) shows a simple circuit which will generate both triangular and square waves at the same time and frequency. As is shown, e_{tr} denotes triangular and e_{sq}, square waves, respectively. The circuit operates as follows.

Let us assume that the e_{sq}-terminal is slightly positive at the beginning. It would then produce a positive voltage,

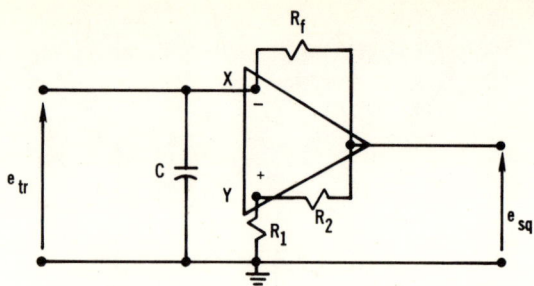

FIGURE 13(a). Square and triangular wave generator.

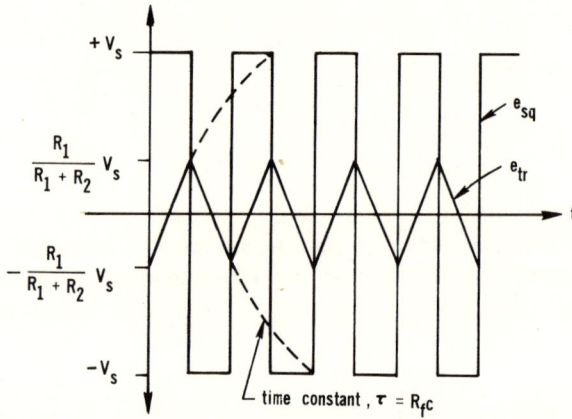

FIGURE 13(b). Square and triangular waveforms.

$$\frac{R_1}{R_1 + R_2} e_{sq}$$

at the noninverted input terminal, Y. Since this is a positive feedback configuration, it will drive almost instantly the output to become more positive until reaching the saturation voltage, V_s, of the operational amplifier, which is limited by the voltage of the power supplies. At this point, the voltage at Y will be equal to

$$\frac{R_1}{R_1 + R_2} V_s$$

As a result, capacitor C will be charged toward $+V_s$ by e_{sq} through resistor R_f. That is, the voltage at the inverted terminal X, will rise exponentially with a time constant equal to $R_f c$ toward V_s. However, as the voltage exceeds the voltage at the noninverted terminal, Y, i.e.,

$$\frac{R_1}{R_1 + R_2} V_s$$

the amplifier output flips to the other extreme or $-V_s$. This, in turn, would result in a negative voltage at Y. This process repeats itself and the circuit becomes an oscillator which yields square waves at e_{sq} and triangular waves at e_{tr}. Figure 13(b) depicts the two superimposed waveforms. The triangular waveform is the linear portion of the exponential charging (discharging) waveform, $v(t) = V(1 - e^{-t/\tau})$ where,

$$V = V_s + \frac{R_1}{R_1 + R_2}V_s = V_s\left(1 + \frac{R_1}{R_1 + R_2}\right)$$

$$\tau = R_f c$$

Based on this equation and Figure 13(b) the peak-to-peak amplitude of the triangular signal is determined by,

$$2\left(\frac{R_1}{R_1 + R_2}V_s\right)$$

and the time period of the generated waveforms is twice that of the ΔT, which is determined by

$$v(t) = \frac{2R_1}{R_1 + R_2}V_s \text{ at } t = \Delta T$$

$$\frac{2R_1}{R_1 + R_2}V_s = V(1 - e^{-\Delta T/\tau})$$

or

$$\Delta T = R_f c \ln\left(1 + 2\frac{R_1}{R_2}\right)$$

and the frequency,

$$f = \frac{1}{2\Delta T} \tag{57}$$

Experimental Example

To verify the derivation, the circuit shown in Figure 13(a) was implemented with the following circuit components: $c = 0.1\ \mu f$, $R_f = 10K$, $R_2 = 10K$, $R_1 = 1.2K$, operational amplifier = $\mu A747$, power supplies = ± 18 V, and $V_s = 16$ V.

A comparison between the calculated and experimental values is shown below:

Description	Calculated values	Measured values	Difference
Square wave amplitude	±16 volts	±16 V	None
Triangular wave amplitude	±1.7	±2.0	0.3
Frequency	2344Hz	2439Hz	95 Hz

C. Pulse Generator

The circuit shown in Figure 14(a) is a pulse generator, the pulse-width and frequency are controlled by the circuit components. This circuit is similar to that of triangular/square wave generators, even in principle of operation. The addition of a diode and a resistor R_D in parallel with R_f causes the capacitor being charged and discharged with different time constants, τ_c and τ_d; whereas the triangular/square wave generator has the same time constant, $R_f c$. For the pulse generator circuitry, the charging time constant

$$\tau_c = \frac{R_D R_f}{R_D + R_f}C$$

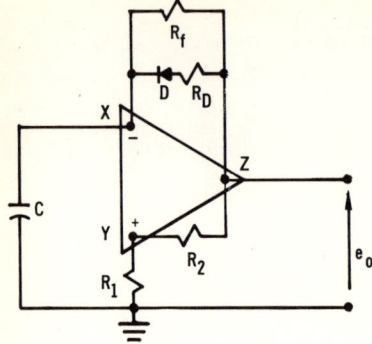
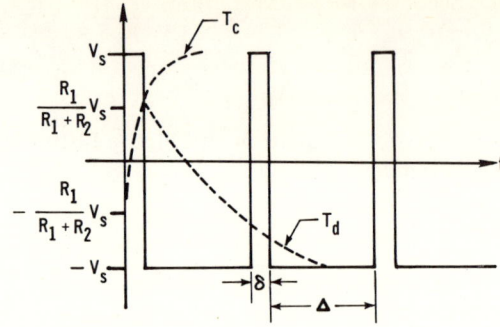

FIGURE 14(a). Pulse generator.

FIGURE 14(b). Pulse generator waveform.

while the discharging time constant $\tau_d = R_f C$, or $\tau_c < \tau_d$. Similar to the triangular/square wave generator, as the output voltage e_o at Z is positive or V_s, the diode D is forward biased, and the capacitor C is being charged through resistors R_f and R_d with a time constant of τ_c. As the voltage at X increases to slightly greater than

$$\frac{R_1}{R_1 + R_2} V_s$$

the output voltage flips from $+V_s$ to $-V_s$ which would turn off the diode D and the capacitor is then discharged through R_f with a time-constant τ_d. As shown in Figure 14(b), it can be seen that τ_c and

$$\frac{R_1}{R_1 + R_2} V_s$$

control the pulse-width δ; and τ_d and

$$-\frac{R_1}{R_1 + R_2} V_s$$

control the pulse-to-pulse duration.

Experimental Example

The pulse generator circuit was tested with the following circuit components C = 0.1 μf, R_f = 10K, R_d = 1.2K, R_1 = 1.2K, R_2 = 10K. Operational amplifier = μA747, power supplies = ±18 V, V_S = 15 V.

The experimental results are shown below:

Description	Calculated values	Measured values	Difference
Frequency (Hz)	4240	4545	305
δ(msec)	0.213	0.2	0.013
Δ(msec)	0.228	0.02	0.028

Comments

If only positive pulses are desired, a diode can be connected in series with the output terminal Z, pointing away from Z. This diode will eliminate the negative portion of the waveform. As the value of R_f is 10 or more times greater than that of R_D, the pulse-width can be controlled by varying R_D, while the frequency can be controlled by R_f.

D. Second Order Active Filters
Low-Pass Filter

Figure 15(a) shows the circuit of a second order low-pass active filter. Note that the operational amplifier itself is connected in the shunt-series feedback configuration. Figure 7(a), with the nominal feedback network $R_2 = 0$, $R_1 = \infty$. As a result, the voltage gain of this circuit is

$$\frac{e_o}{e_y} = 1 + \frac{R_2}{R_1} = 1$$

which is known as a voltage follower, i.e.,

$$e_o = e_y \tag{58}$$

where e_y is the voltage at terminal y.

Referring to Figure 15(a) again, we have, $i_2 = i_1 + i_3$ where

$$i_1 = \frac{1}{R_1}(e_i - e_x)$$

$$i_2 = \frac{1}{R_2}(e_x - e_y)$$

$$i_3 = s c_1 (e_o - e_x)$$

Thus,

$$\frac{e_x}{R_2} - \frac{e_y}{R_2} = \frac{e_i}{R_1} - \frac{e_x}{R_1} + s c_1 e_o - s c_1 e_x$$

$$e_x \left(\frac{1}{R_1} + \frac{1}{R_2} + s c_1 \right) = \frac{e_i}{R_1} + s c_1 e_o + \frac{e_y}{R_2} \tag{59}$$

But

$$e_y = \frac{\frac{1}{sc_2}}{R_2 + \frac{1}{sc_2}} e_x \tag{60}$$

From Equations 58, 59, 60

$$\frac{e_o}{e_i} = \frac{\frac{1}{c_1 c_2 R_1 R_2}}{s^2 + s \left[\frac{1}{c_1 R_1} + \frac{1}{c_1 R_2} \right] + \frac{1}{c_1 c_2 R_1 R_2}} \tag{61}$$

From s-domain or pole-zero concept, the transfer function of a second order low-pass filter

$$H(s) = \frac{w_o^2}{s^2 + 2\xi w_o s + w_o^2} \tag{62}$$

From Equations 61 and 62 we have the cut-off or roll-off frequency of the filter,

$$w_o = \frac{1}{\sqrt{R_1 R_2 c_1 c_2}}$$

damping factor,

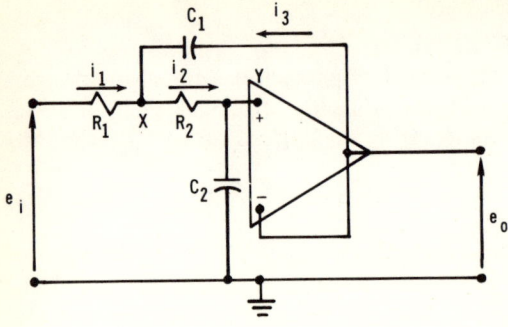

FIGURE 15(a). Low pass filter.

Frequency (Hz)	e_o* (volt)	VOLTAGE GAIN GAIN	VOLTAGE GAIN dB
10	1.0	1.0	0
20	0.97	0.97	-0.26
50	0.96	0.96	-0.35
75	0.83	0.83	-1.62
100	0.76	0.76	-2.38
150	0.60	0.60	-4.44
200	0.47	0.47	-6.56
250	0.35	0.35	-9.12
300	0.28	0.28	-11.10
500	0.14	0.14	-17.10
1000	0.05	0.05	-26.00
2000	0.02	0.02	-33.90

*e_i = 1 volt (p-p) sine wave

FIGURE 15(b). Experimental data.

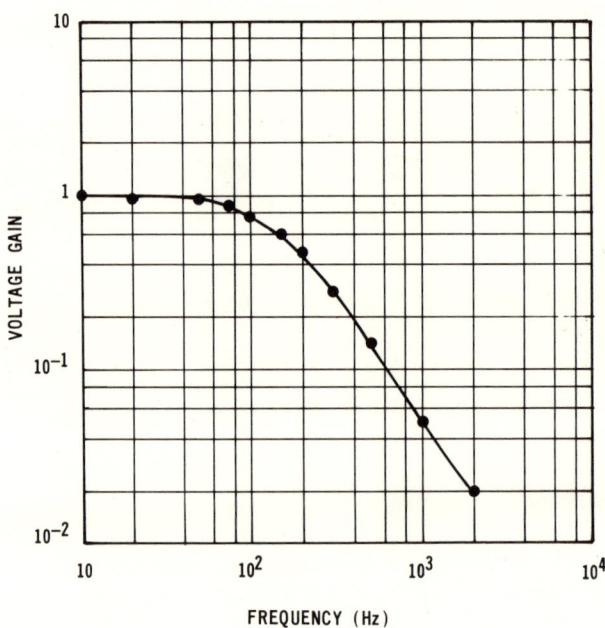

FIGURE 15(c). Frequency response (low-pass filter).

$$\xi = \sqrt{\frac{c_2}{c_1}} \; \frac{1}{2} \; \frac{R_1 + R_2}{\sqrt{R_1 R_2}}$$

For experimental verification, the following circuit components were used: $C_1 = C_2 = C = 0.0094 \, \mu F$, $R_1 = R_2 = R = 100K$, and operational amplifier = $\mu A747$.

Figure 15(b) shows the data of the frequency response for this circuit and Figure 15(c) depicts the corresponding Bode diagram. The comparison between the calculated and measured values are listed below:

Description	Calculated values	Measured values
Cut-off frequency	169	102
Gain slope	−12 dB/octave	−8.8 dB/octave
Low frequency gain	1	1

High-Pass Filter

Figure 16(a) shows the circuit of a second order high-pass filter. The derivation of the transfer function in terms of circuit components which is similar to that of the low-pass filter presented in the last section is, however, being omitted from this and the succeeding sections, except the key formulations, i.e.,

Transfer function,

$$H(s) = \frac{s^2}{s^2 + 2\xi w_o s + w_o^2} \tag{63}$$

Cut-off frequency,

$$w_o = \frac{1}{\sqrt{R_1 R_2 C_1 C_2}}$$

Damping factor,

$$\xi = \sqrt{\frac{R_1}{R_2}}$$

For experimental verification, $C_1 = C_2 = C = 0.0047\ \mu F$, $R_1 = R_2 = R = 10K$, and operational amplifier = $\mu A747$.

Figure 16(b) shows the data of the frequency response for this circuit and Figure 16(c) depicts the corresponding Bode diagram. Comparison between the calculated and measured values are listed below:

Description	Calculated values	Measured values
Cut-off frequency (Hz)	3386	3300
Gain slope	12 dB/octave	9.6 dB/octave
High frequency gain	1.0	1.0

Bandpass Filter

Figure 17(a) shows the circuit of a second order bandpass filter. Its key formulations are as follows:

Transfer function

$$H(s) = \frac{A_o (w_o/Q) s}{s^2 + (w_o/Q) s + w_o^2} \tag{64}$$

where, A_o = voltage gain at center frequency, Q = quality factor, and w_o = center frequency.

Center frequency,

$$w_o = \frac{\sqrt{2}}{CR}$$

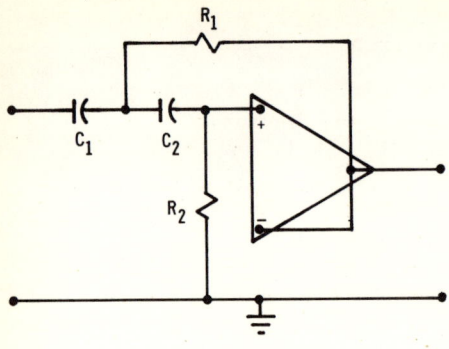

FIGURE 16(a). High-pass filter.

Frequency (Hz)	e^*_o (volt)	Gain
100	0.015	0.015
200	0.025	0.025
300	0.030	0.030
400	0.050	0.050
500	0.065	0.065
750	0.125	0.125
1K	0.200	0.200
1.5K	0.350	0.350
2.0K	0.480	0.480
2.5K	0.600	0.600
3.0K	0.650	0.650
5.0K	0.800	0.800
7.5K	0.900	0.900
10.0K	0.920	0.920
15.0K	0.950	0.950
20.0K	0.970	0.970
25.0K	0.980	0.980
50.0K	1.00	1.00

$^*e_i = 1$ volt (p–p) sine wave

FIGURE 16(b). Experimental data.

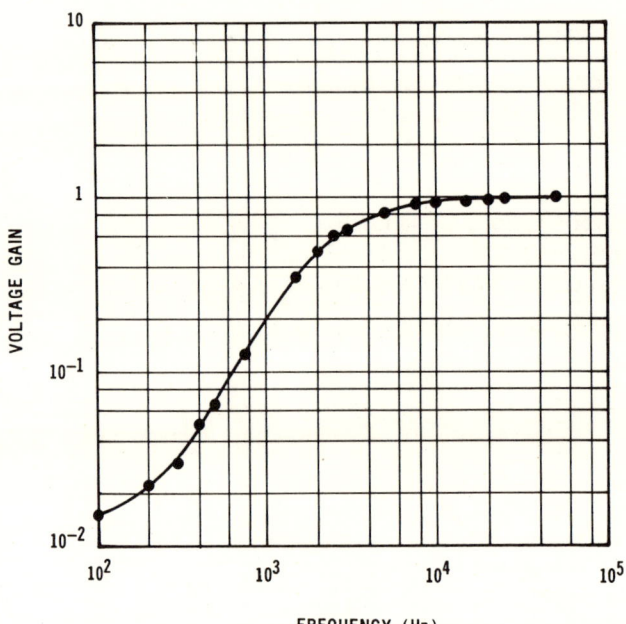

FIGURE 16(c). Frequency response (high-pass filter).

Quality factor,

$$Q = \frac{\sqrt{2}}{5 - A_{fb}}$$

Center frequency gain,

$$A_o = \frac{A_{fb}}{5 - A_{fb}}$$

A_{fb} = Operational amplifier gain with feedback = $1 + \dfrac{R_2}{R_1}$

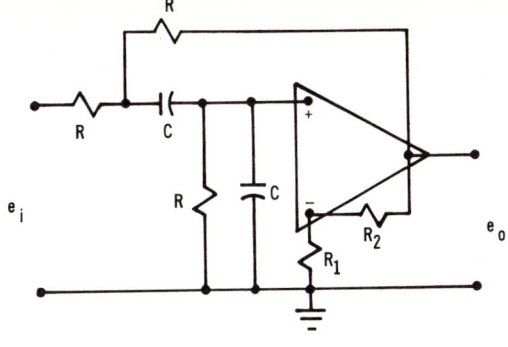

FIGURE 17(a). Bandpass filter.

Frequency(Hz)	e^*_o(volt)	Gain
10	0.025	0.025
20	0.05	0.05
50	0.10	0.10
100	0.16	0.16
200	0.31	0.31
250	0.40	0.40
400	0.60	0.60
500	0.79	0.79
600	1.00	1.00
700	1.20	1.20
800	1.40	1.40
900	1.60	1.60
1.0K	1.85	1.85
1.5K	3.70	3.70
2.0K	6.00	6.00
2.5K	4.50	4.50
3.0K	3.00	3.00
3.5K	2.40	2.40
4.0K	1.75	1.75
5.0K	1.40	1.40
10.0K	0.62	0.62
20.0K	0.30	0.30
50.0K	0.12	0.12
100.0K	0.06	0.06

$^*e_i$ = 1 volt (p-p) Sine wave

FIGURE 17(b). Experimental data.

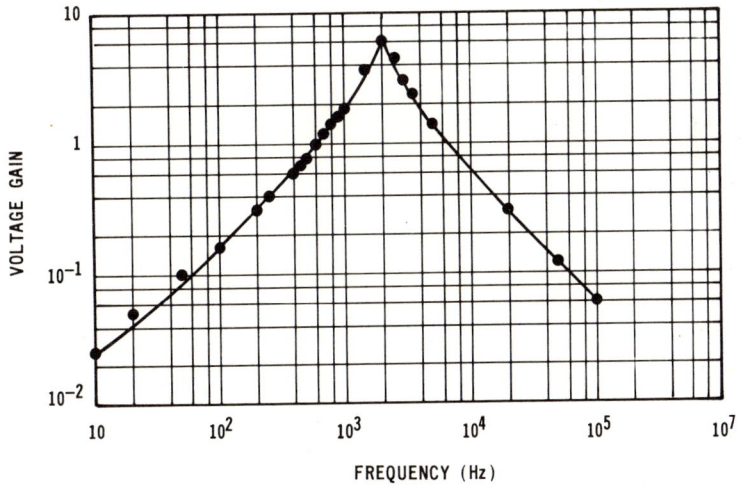

FIGURE 17(c). Frequency response (band-reject filter).

For experimental verification, R = 27K, C = 0.0047 µF, R_1 = 1.2K, R_2 = 3.9K, and operational amplifier = µA747.

Figure 17(b) shows the data of the frequency response for the circuit and Figure 17(c) depicts the corresponding Bode diagram. Comparison between the calculated and measured values are listed below:

Descriptions	Calculated values	Measured values
Center frequency (Hz)	1774	2050
Q	1.89	2.28
Bandwidth (Hz)	936	900
Center frequency gain	5.66	6.0
Low frequency slope	±12 dB/octave	+10 dB/octave
High frequency slope	−12 dB/octave	−11.2 dB/octave

Band-Reject Filter

Figure 18(a) shows the circuit of a band-reject filter. Note that the network at the input end is known as a Twin-T network which would pass through signal of all frequencies except some frequencies which the network is tuned for. Its key formulations are as follows:

Center of reject frequency,

$$w_o = \frac{1}{RC}$$

Quality factor,

$$Q = \frac{R_1}{4R}$$

and

$$C_1 = \left[\frac{1}{2Q}\right] C$$

For experimental verification the following data is used for the design:

Center of reject frequency $\quad f_o = \frac{w_o}{2} = 1$ kHz

$$Q = 5, \ C = 0.0047 \ \mu F$$

$$R = \frac{1}{w_o C} = 33.9K, \ 33K \text{ was used}$$

$$R_1 = 4RQ = 660K, \ 680K \text{ was used}$$

$$C_1 = \left[\frac{1}{2Q}\right] C = 470 \text{ pF}$$

$$\frac{C}{2} = 0.00235 \ \mu F, \ 0.0022 \text{ was used}$$

$$2R = 68K$$

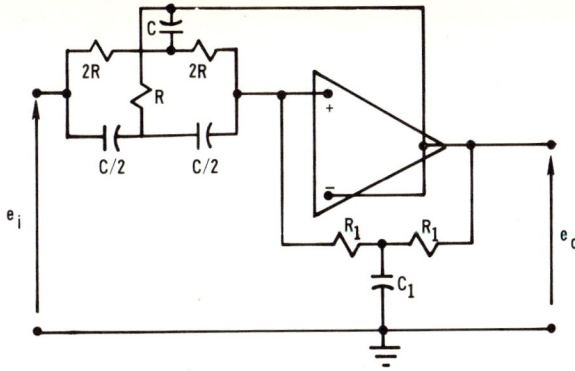

FIGURE 18(a). Band-reject filter.

Frequency(Hz)	e_o^*(volt)	Gain
10	1	1
50	1	1
100	1	1
200	1	1
500	1	1
800	0.97	0.97
1K	0.92	0.92
1.1K	0.72	0.72
1.15K	0.40	0.40
1.20K	0.14	0.14
1.25K	0.45	0.45
1.30K	0.68	0.68
1.50K	0.90	0.90
2K	0.97	0.97
5K	1	1
1.0K	1	1
100K	1	1
300K	0.9	0.9
500K	0.6	0.6
1M	0.3	0.3
2M	0.14	0.14

*e_i(= 1 volt (p-p) sine wave

FIGURE 18(b). Experimental data.

Figure 18(b) shows the experimental data of the frequency response for the circuit and Figure 18(c) depicts the Bode diagram. Comparison between the calculated and measured values are listed below:

Description	Calculated values	Measured values
Center of reject frequency (Hz)	1026	1200
Bandwidth (Hz)	200	200
Q	5.15	6
Gain (max)	1.0	1.0
Gain (min)	~0	0.14

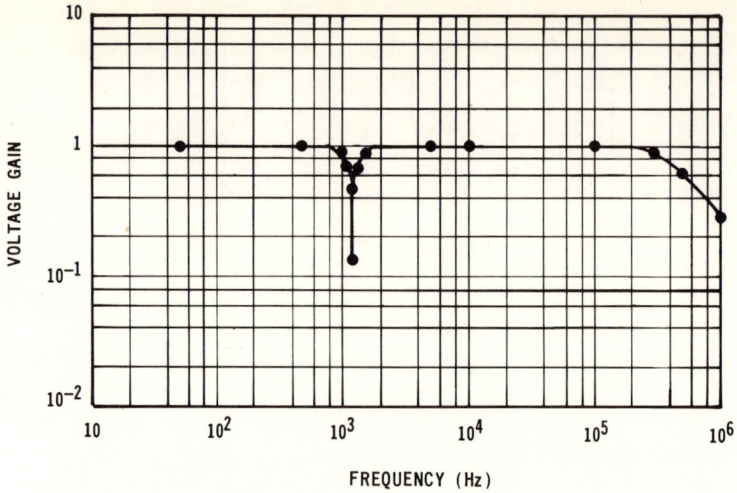

FIGURE 18(c). Frequency response (band-reject filter).

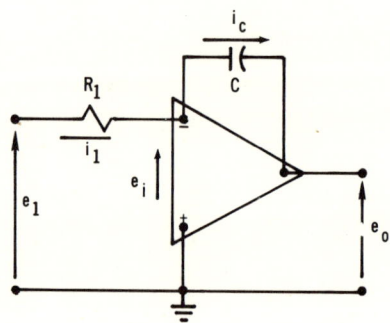

FIGURE 19. Integrator.

IV. A COLLECTION OF CIRCUITS FOR GENERAL APPLICATIONS

A. Integrator

Figure 19 shows a simple and useful analog integrator circuit. Its derivation follows

$$e_1 = i_1 R_1 + e_i$$

$$e_o = e_i - \frac{1}{C} \int i_C dt$$

$$e_o = \frac{e_o}{A} - \frac{1}{C} \int i_C dt$$

But $1 \gg 1/A$, $i_1 \simeq i_c$.
We have

$$e_o = -\frac{1}{C} \int i_1 dt \simeq -\frac{1}{C} \int \frac{e_1}{R} dt$$

$$= -\frac{1}{RC} \int e_1 dt \tag{65}$$

Note that this circuit can also be viewed as a shunted-shunted feedback circuit configuration, referring to Table 1, replacing R_1, R_2 by impedance notations as a function of s-operator

$$\frac{E_o(s)}{E_i(s)} = -\frac{Z_2(s)}{Z_1(s)} = -\frac{\frac{1}{SC}}{\frac{1}{R_1}} = -\frac{1}{R_1 CS} \quad (66)$$

$$\frac{e_o}{e_i} = -\frac{1}{jwR_1 C}$$

Equation 66 shows that this circuit has an extremely high gain at low frequency. For practice, see differentiator-integrator circuit described in Section C below.

B. Differentiator

Figure 20 shows the circuit of an analog differentiator. Following the similar procedure described in the last section, we have

$$e_1 = \frac{1}{C}\int i_1 dt + e_i \simeq \frac{1}{C}\int i_1 dt$$

$$i_1 = C\frac{de_1}{dt}$$

$$e_o = -i_2 R_2 + e_i \simeq -i_1 R_2$$

$$e_o \simeq -R_2 C \frac{de_1}{dt}$$

Alternately,

$$\frac{E_o(s)}{E_i(s)} = -\frac{R_2}{\frac{1}{SC}} = R_2 CS$$

$$\frac{e_o}{e_i} = -jR_2 Cw \quad (67)$$

Equation 67 shows that this circuit has an extremely high gain at high frequency. For practice, see differentiator-integrator circuit described in Section C below.

C. Differentiator-Integrator

Figure 21(a) shows a useful and practical circuit configuration. By properly selecting R_1, C_1 and R_2, C_2, the circuit can be used as a differentiator, or an integrator and even a first order bandpass filter. Its derivation follows.

Let

$$Z_1(s) = R_1 + \frac{1}{SC_1} = \frac{1}{SC_1}(R_1 C_1 S + 1)$$

$$Z_2(s) = \frac{R_2 \frac{1}{SC_2}}{R_2 + \frac{1}{SC_2}} = \frac{R_2}{R_2 C_2 S + 1}$$

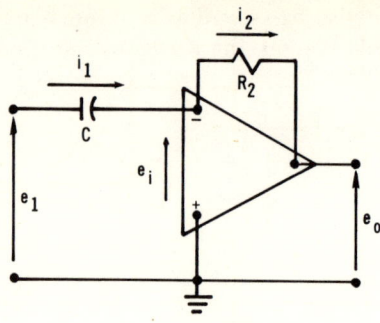

FIGURE 20. Differentiator.

then

$$\frac{E_o(s)}{E_1(s)} = -\frac{Z_2(s)}{Z_1(s)} = -\frac{R_2}{R_2C_2S+1} \cdot \frac{SC_1}{R_1C_1S+1}$$

$$= -\frac{S}{R_1C_2\left(S+\dfrac{1}{R_2C_2}\right)\left(S+\dfrac{1}{R_1C_1}\right)}$$

Let

$$w_1 = \frac{1}{R_1C_1}$$

$$w_2 = \frac{1}{R_2C_2}$$

then

$$\frac{E_o(S)}{E_1(S)} = -\frac{S}{R_1C_2\left(s+\dfrac{1}{w_1}\right)\left(s+\dfrac{1}{w_2}\right)}$$

$$\frac{e_o}{e_1} = -\frac{jw}{R_1C_2\left(jw+\dfrac{1}{w_1}\right)\left(jw+\dfrac{1}{w_2}\right)} \tag{68}$$

Figure 21(b) shows the Bode asymptotic diagram. Note that the circuit is an integrator for a signal with frequency greater than w_2. Furthermore it is a first order bandpass filter with a bandwidth equal to $w_2 - w_1$, and a differentiator for a signal with frequency less than w_1.

It was pointed out in the last two sections that the differentiator has extremely high gain for high frequency so that it is sensitive to high frequency noise, and on the other hand, the integrator is sensitive to low frequency noise. Therefore, this circuit can be used as a differentiator and in the mean while R_2C_2 is being used to attenuate the noise with the frequency higher than w_1 providing that the highest frequency component of the desirable signal is less the w_1. On the same token, the network R_1C_1 can be used to deduce noise of an integrator which is used for signals having lowest frequency component greater than w_2. For integrator or differentiator only the bandpass portion of Figure 21(b) can be eliminated by designing $w_1 = w_2$, or

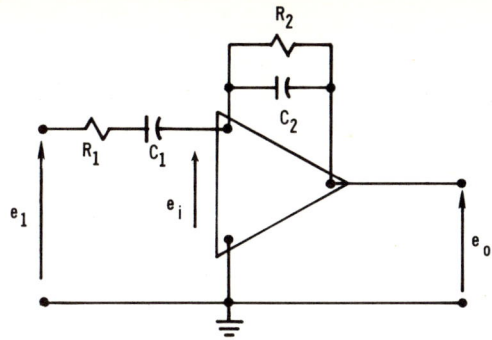

FIGURE 21(a). Differentiator-integrator.

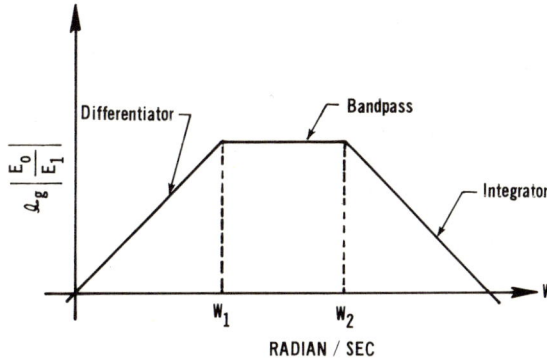

FIGURE 21(b). Bode diagram.

$$R_1 C_1 = R_2 C_2 \tag{69}$$

then, the circuit is a differentiator for $w < w_1$ and an integrator for $w > w_2$.

D. Comparator

A comparator is a special purpose operational amplifier which is normally used without any feedback circuit configuration. It is therefore designed to have wide bandwidth, high speed, or very fast response. Typically, the response time for low-to-high-level output is 20 ~ 40 nsec. It is normally used in differential input configuration. Typical applications are sense amplifiers for memory systems, voltage comparators in analog-digital-analog conversion systems, zero-crossing detector, etc. A number of comparators provide strobe input, offset balancing terminals, and can be used with single or dual power supply. The provision of strobe input is extremely useful in a noisy environment such as a magnetic core memory system.

V. PHASE-LOCKED LOOP

A. Introduction

Phase-locked loops, known as PLL, like operational amplifiers, are also an important building block in communication, control, and digital systems. It is basically a feedback system by itself. As shown in Figure 22, $V_s(t)$ denotes the incoming signal with noise; $V_o(t)$, the output signal of the voltage or current controlled oscillator; $v_e(t)$,

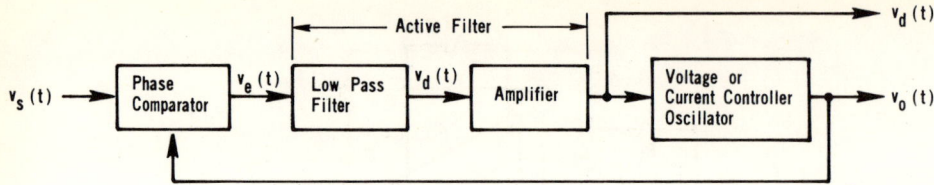

FIGURE 22. Block diagram of a phase-locked loop.

output of the phase comparator which is a function of the phase/frequency difference between $v_s(t)$ and $v_o(t)$. As the system in steady-state, or locked-in state, $v_o(t)$ will be a strong signal which has the same frequency of the $v_s(t)$ but with a phase-shift about 90°, thus the output virtually represents or replaces the incoming signal with a desirable signal-to-noise ratio. It is evident that this device is a good signal conditioning device which can recover signals being buried in noise apparently unidentifiable by the method of amplitude demodulation or discrimination.

B. Major Elements of the System
Phase Comparator and Its Mathematical Function

Figure 23(a) shows a simplified phase comparator circuitry. Its equivalent functional circuit is shown in Figure 23(b). It is basically an analog multiplier which operates as follows.

Let $v_s(t) = V_s \sin w_s t$ and $V_o(t)$ = square wave with f_o frequency and amplitude = 1, then by Fourier series analysis,

$$v_o(t) = \sum_{n=0}^{\infty} \frac{4}{\pi(2n+1)} \sin[(2n+1)w_o t]$$

and

$$v_e(t) = v_o(t) v_s(t)$$

$$= V_s \sin w_s t \left[\sum_{n=0}^{\infty} \frac{4}{\pi(2n+1)} \sin(2n+1) w_o t \right]$$

$$= V_s \sin w_s t \sin w_o t + V_s \sin w_s t \sin 3 w_o t + \ldots$$

$$= \frac{V_s}{2} [\cos(w_o - w_s)t + \cos(w_o + w_s)t]$$

$$+ \frac{V_s}{2} [\cos(3w_o - w_s)t + \cos(3w_o + w_s)t] + \ldots$$

If the cut-off frequency of the low-pass filter, w_1 has a value, $w_o - w_s < w_1 < w_o + w_s$, then the output voltage of the low-pass filter,

$$v_d(t) = \frac{V_s}{2} \cos(w_o - w_s)t \qquad (70)$$

Consider the following two possible cases:
Case I — ASSUME, $w_o \neq w_s$, then

$$v_d(t) = \frac{V_s}{2} \cos(w_o - w_s)t$$

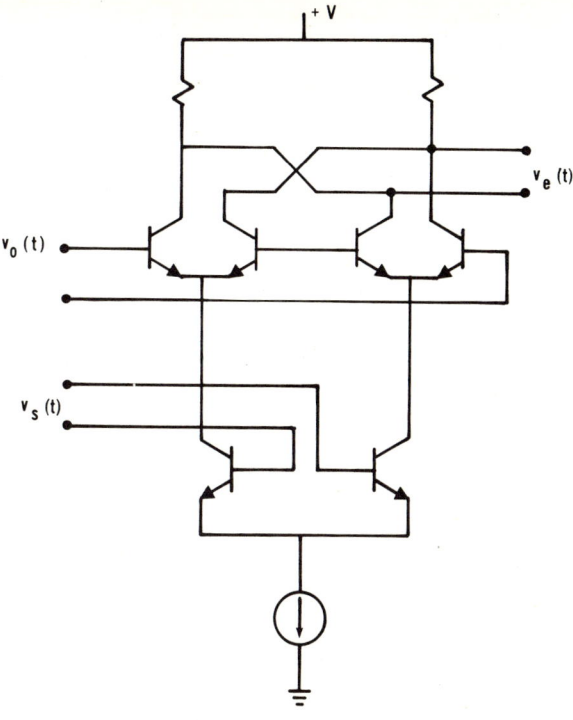

FIGURE 23(a). Phase comparator.

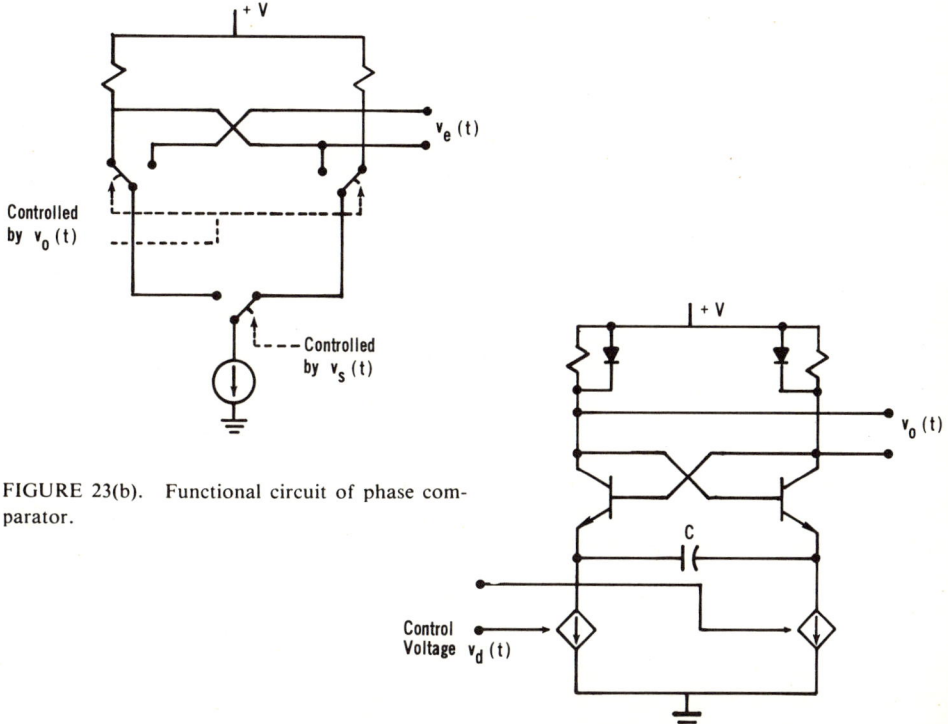

FIGURE 23(b). Functional circuit of phase comparator.

FIGURE 23(c). Voltage-controlled-oscillator.

This can be viewed as that the v_d is a time varying DC signal. Its amplitude changes from zero to $+(V_s/2)$ and back to zero, then $-(V_s/2)$, etc.; $v_d(t)$, however, is the input of the voltage controlled oscillator. That is, as $v_d = 0$ V, the oscillator will output a square wave at frequency w_o. When v_d increases, the frequency of the oscillator increases accordingly. In other words, as the v_d changes from $+(lV_s/2)$ to zero to $-(V_s/2)$, the oscillator frequency will change from $w_o + \Delta w$ to w_o to $w_o - \Delta w$. If $w_o - w_s$ is within the range of Δw, the oscillator would eventually catch-up with the signal frequency and lock-in with it.

Case II — ASSUME, $w_o = w_s$ and there is a phase difference between v_s and v_o, say, Φ. Then

$$v_e(t) = [V_s \sin w_s t][\sin(w_s t + \phi)]$$

$$= \frac{V_s}{2}[\cos\phi - \cos(2w_s t + \phi)]$$

At the output of the low-pass filter,

$$v_d = \frac{V_s}{2}\cos\phi$$

Therefore, as the oscillator locked-in with the input signal, the value of v_d represents their phase difference, $v_d = 0$ as $\Phi = 90\%$.

Low-Pass Filter-Amplifier or Low-Pass Active Filter

This section can be a passive low-pass filter followed by an amplifier or an active filter with gain. As described in the last section, the cut-off frequency, w_1, of this filter should cover the range of $w_o \pm w_s$.

Voltage Controlled Oscillator — VCO

Figure 23(c) shows a simplified VCO circuit. It is a single-capacitor free-running oscillator whose frequency,

$$f_o = \frac{V_d G_m}{2\,CV_{BE}}$$

where G_m, V_{BE} are the parameters of the controlled current generators. The diodes in the collector circuits are used to assume nonsaturation operation and limited voltage swing of the oscillator output. There are, of course, other circuitries that can serve the same purpose.

C. Linearized Negative Feedback System Model

During lock-in state, the system can be viewed and analyzed as a linear negative feedback system. Figure 24(a) shows a linearized system model. Where, ϕ_i = phase angle of the input signal, ϕ_o = phase angle of the output signal of VCO, K_d = conversion gain of phase comparator (V/rd), $F(s)$ = transfer function of the low-pass filter, A = amplifier gain, and K_o = VCO conversion gain. Since it converts voltage into frequency (f_o) and frequency into ϕ_o, while $f_o = d\phi_o/dt$, thus, ϕ_o is an integration of f_o, therefore, there is a block represented by $1/S$.

The system transfer function,

$$H(S) = \frac{AK_d F(S)}{1 + AK_d \dfrac{K_o}{S} F(S)}$$

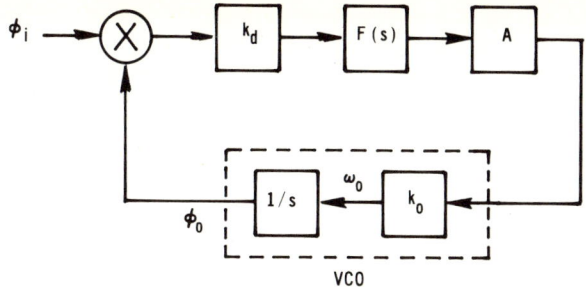

FIGURE 24(a). Linearized negative feedback system model of PLL.

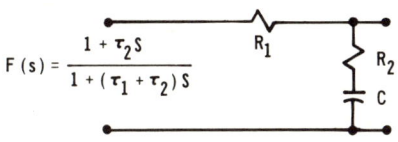

FIGURE 24(b). Lag-lead network.

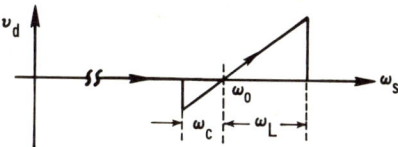

FIGURE 24(c). ω_s Increasing locking process.

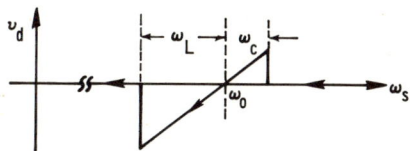

FIGURE 24(d). ω_s Decreasing locking process.

To assure system stability a lag-lead network can be used for the low-pass filter as shown in Figure 24(b), which has the transfer function,

$$F(s) = \frac{1 + \tau_2 S}{1 + (\tau_1 + \tau_2)S}$$

where, $\tau_1 = R_1C$, $\tau_2 = R_2C$.

D. Capture Range and Lock-In Range

Figures 24(c) and (d) depict a different way to view the operation of the locking process. The horizontal axis represents the signal frequency while the vertical axis, the average or DC voltage of $v_d(t)$. Figure 24(c) shows the process of w_s increases from zero toward and pass the VCO signal with a frequency equal to w_0, while Figure 24(d) shows the process of decreasing w_s.

In view of Equation 70, it reveals that $v_d(t)$ will be zero if $(w_0 - w_s) > w_1$, the cutoff frequency of the low-pass filter. If the network shown in Figure 24(b) is used,

$$w_1 = \frac{1}{\tau_1 + \tau_2}$$

As the w_s approaching w_0 and $w_0 - w_s$ entering the range between zero and w_1, the output of the low-pass filter becomes nonzero; as a result, the VCO responds to v_d adopting its frequency to w_s and the system is now locked-in. As w_s increases continuously, the VCO will lock with w_s until the v_d reaches its maximum value and VCO then cannot lock with increasing w_s and the $w_0 - w_s$ becomes greater than w_1 and v_d drops to zero. Similar processes follow with the decreasing w_s. As shown in Figures 24(c) and (d), w_c is defined as the capture frequency and w_L, the locked-in frequency. Therefore, $2w_c$ is defined as capture or acquisition range and $2w_L$, the lock or tracking range.

E. Comments

It is interesting to point out that the low-pass filter actually serves two major functions, i.e., (1) control of the capture range of the system, and (2) recapture of the signal in case the signal momentarily jumps out of the lock-range. In comparison with RC active filters, PLL has the following advantages and disadvantages.

Advantages:
1. High frequency capability
2. Independent controlling of selectivity and center frequency
3. Fewer external components
4. Ease of tuning

Disadvantages:
1. Lack of amplitude information of the input signal
2. Response to harmonics of the input signal
3. Difficult to provide automatic gain control

In view of the system diagram shown in Figure 22, PLL provides two outputs, namely $v_d(t)$ and $v_o(t)$, depending on the nature of the applications. A brief list of applications is shown as follows.

1. F.M. Demodulation:
   Broadcast FM detection
   AM/FM telemetering decoding
   FSK (Frequency Shift Keyed) demodulation
2. Frequency synchronization
3. Signal conditioning

REFERENCES

1. Eimbinder, J., Ed., *Application Considerations for Linear Integrated Circuits,* John Wiley & Sons, New York, 1970.
2. Eimbinder, J., Ed., *Design with Linear Integrated Circuits,* John Wiley & Sons, New York, 1969.
3. Eimbinder, J., Ed., *Linear Integrated Circuits: Theory and Applications,* John Wiley & Sons, New York, 1968.
4. Giles, J. N., Ed., Fairchild Semiconductor Linear Integrated Circuits Applications Handbook, Fairchild Semiconductor, Mountain View, Calif., 1967.
5. Mochytz, G. S., The operational amplifier in linear active network, *IEEE Spectrum,* Jan. 1970.
6. Naylor, J. R., Digital and analog signal applications of operational amplifiers. I, *IEEE Spectrum,* May, 1971.
7. Naylor, J. R., Digital and analog signal applications of operational amplifiers. II, *IEEE Spectrum,* June, 1971.

8. **Johnson, D. E. and Hilburn, J. L.**, *Rapid Practical Design of Active Filters,* John Wiley & Sons, New York, 1975.
9. Technical Staff, Analog Data Manual, Signetics, Sunnyvale, Calif., 1977.
10. Technical Staff, Handbook of Operational Amplifier Applications, Burr-Brown, Tucson, 1969.
11. **Grebene, A. B.**, The monolithic phase-locked loop — a versatile building block, *IEEE Spectrum,* March, 1971.
12. Signetic Linear Phase Locked Loops Applications Book, Signetics, Sunnyvale, Calif., 1972.
13. **Grebene, A. B.**, The monolithic phase-locked loop — a versatile building block, *EDN,* Oct. 1, 1972.
14. **Kesner, Don,** Take the guesswork out of phase-locked design, *EDN,* Jan. 5, 1973.
15. **Gardner, F. M.**, *Phaselock Techniques,* John Wiley & Sons, New York, 1966.
16. **Grebene, A. B. and Camenzind, H. R.**, Frequency selective I.C. using phase-lock techniques, *IEEE Solid State Circuits,* SC-4, 216, 1969.
17. **Moschytz, G. S.**, Miniaturized RC filters using phase-locked loop, *Bell Syst. Tech. J.,* 44, 823, 1965.
18. **Kalpper, J. and Frankle, J. T.**, *Phase-locked and Frequency Feed-back Systems: Principle and Techniques,* Academic Press, New York, 1972.
19. **Viterbi, A.**, *Principles of Coherent Communication,* McGraw-Hill, New York, 1971.
20. **van Trees, H. L.**, *Detection, Estimation and Modulation Theory,* Part II, John Wiley & Sons, New York, 1971.
21. **Lindsey, W. C.**, *Synchronization Systems on Communication and Control,* Prentice-Hall, Englewood Cliffs, N.J., 1972.
22. **Holmes, J. K. and Tegnelia, C. R.**, A second-order all-digital phase-locked loop, *IEEE Trans. Commun.,* January, 62, 1974.
23. **Reed, L. J. and Treadway, R. J.**, Test your PLL IQ, *EDN,* Dec. 20, 1974.
24. Technical Staff, Manual on Phase-Locked Loops, Motorola Semiconductor, Phoenix, Ariz., 1973.
25. **Gupta, S. C.**, Phase-locked loops, *Proc. IEEE,* Feb. 1975.

Chapter 3

A REVIEW AND CLARIFICATION OF PULSE AND SWITCHING CIRCUIT FUNDAMENTALS

I. CLARIFICATION OF BASIC CONCEPTS

In view of the electrical engineering curriculum, it reveals that training sequence for circuit or system designers has been direct current (DC), alternate current (AC), linear circuit, and steady state analysis of circuit and system. For this reason, having been brainwashed, designers often have difficulties in adapting themselves to deal with digital circuits and systems analysis. In digital design, most of the circuit elements used are nonlinear and techniques for transient state and nonlinear analysis of a circuit are vitally important tools to the designer. An attempt is made in this chapter to clarify some of the topics where confusion is normally experienced. Basically, for a digital designer, one should be concerned with time variable and nonlinear characteristics in addition to the four traditional parameters, i.e., current, voltage, impedance, and power. Fortunately, in digital system design, one only needs to deal with two logic states and the transient changing from one state to the other; therefore it has no reason to be so confused, had the basic concept been clarified. As for analysis for nonlinear devices, there are only three most popular techniques being used in digital design, namely, mathematical, graphical, and piece-wise linear. For linear analysis, such as the analysis of an operational amplifier described in the last chapter, the signals considered are mostly periodic, and small signals so that linear models and Fourier series analysis can be used. In digital systems or circuits, one deals mostly with large and nonperiodic signals. The traditional general expression of impedance such as,

$$Z(jw) = R + jwL - j\frac{1}{WC}$$

and linear models of voltage and current sources are seldom used. Instead, one would consider mostly the electric properties during the switching transition period and the binary steady states. Although Laplace Transform technique can be used for transient analysis, the simple time domain concept appears to be sufficient. What follows will be the details of these techniques applied to digital circuit elements or system analysis.

II. LINEAR ELEMENTS

A. Resistor

A resistor can be considered as a linear device. In other words, regardless of the magnitude of the voltage or current, it satisfies Ohm's Law: $V = IR$ for DC signal, where V: direct voltage, invariant with time; I: direct current, invariant with time; and R: resistance, a constant. Also, $v(t) = i(t) R$, where v(t): time variant voltage and i(t): time variant current. There are devices or circuit components that do not always satisfy Ohm's Law for all values of voltage and current. In such cases, graphical presentation of its voltage-current characteristics known as the V-I curve for a device would be most useful and appropriate. Although trivial, for completeness of discussion and comparison with other devices in later sections, the V-I curves of a resistor with different values are shown in Figure 1.

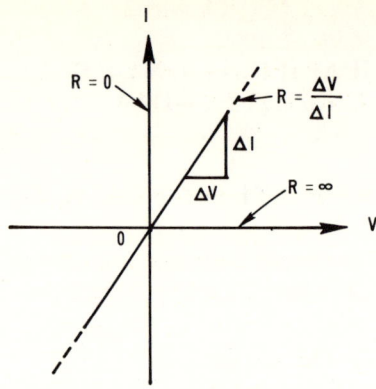

FIGURE 1. V-I curve for resistors.

B. Capacitor

For those who are familiar with problems involved only with steady state sinusoidal signals one would still use Ohm's Law and replace resistance by a capacitive reactance, X_c, i.e.,

$$v(wt) = X_c i(t) = \frac{1}{jwc} i(t)$$

where $w = 2\pi f$, f = frequency in Hertz, and C = capacitance of the capacitor in farads. However, in digital systems we mostly deal with pulse signals instead of sinusoidal. We therefore need a thorough conceptual clarification. The volt-current relationship in a capacitor is different from that in a resistor. For a direct or time invariant voltage (steady state), the current through a capacitor is always zero, while for a time variant voltage, the current is proportional to the change of the voltage with respect to time. Mathematically, Q = CV, or

$$\frac{dv(t)}{dt} = \frac{1}{C} i(t) = \frac{1}{C} \frac{d}{dt} 8(t) \tag{1}$$

where, Q = electrical charge in coulombs, C = capacitance in farads, V = constant voltage in volts, and v(t) = time varying voltage.

Also, we have

$$v(t) = \frac{1}{C} \int_{t_1}^{t} i(t)dt + v(t_1) \tag{2}$$

If i(t) is finite, then

$$\lim_{t \to t_1} v(t) = 0 + v(t_1) \tag{3}$$

Physical interpretation of Equation 3 is interesting and important. It means that *the voltage across a capacitor could not change instantaneously if the current is finite.* The following examples will clarify the concept.

Example 1

In Figure 2(a) the switch is closed at $t = t_1$. Since the maximum possible current of the circuit,

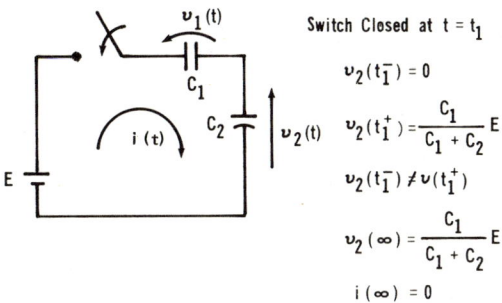

FIGURE 2(a). R-C circuit.

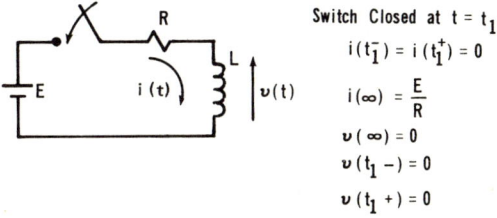

FIGURE 2(b). Pure capacitive circuit.

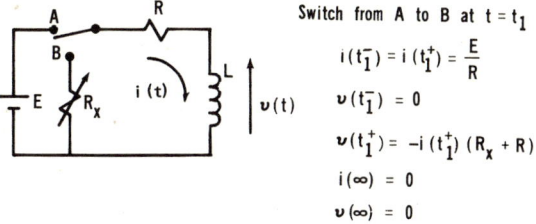

FIGURE 2(c). R-L circuit.

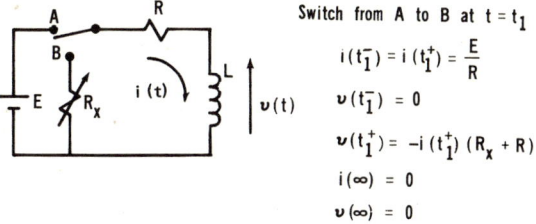

FIGURE 2(d). R-L circuit.

$$[i(t)]_{max} = \frac{E}{R}$$

The current is finite, Equation 3 is valid, and the voltage across the capacitor could not change instantaneously, then $v(t_1^-) = r(t_1^+)$, where $t_1^- \underline{\Delta}$ an infinitesimal time before t_1, and $t_1^+ \underline{\Delta}$ an infinitesimal time after t_1. The symbol "$\underline{\Delta}$" means "defined as" for *steady state* condition, $dv/dt = 0$ then from Equation 1, the current is zero. The steady state condition is usually denoted by $t = \infty$, for convenience, hence we have $i(\infty) = 0$ and the voltage across the capacitor $v(\infty) = E$ in this example.

Example 2

In Figure 2(b) the switch is closed at $t = t_1$. However, in this example, the resistance is zero, therefore the current would not be finite and the voltage $v_2(t)$ can jump from zero to

$$\frac{C_1}{C_1 + C_2} E \text{ instantaneously, and we have } v_2(\infty) = \frac{C_1}{C_1 + C_2} E, \; i(\infty) = 0$$

One may wonder how we got this result.

Based on the basic equation, $Q = CV$, one can write the loop equation:

$$E = v_1(t) + v_2(t)$$

$$= \frac{1}{C_1} q_1(t) + \frac{1}{C_2} q_2(t)$$

Since $q_1(t) = q_2(t)$, we have,

$$E = \frac{C_1 + C_2}{C_1 \cdot C_2} q_1(t)$$

$$q_1(t) = \frac{C_1 C_2}{C_1 + C_2} E = q_2(t)$$

Hence

$$v_2(t) = \frac{1}{C_2} q_2(t) = \frac{C_1}{C_1 + C_2} E$$

Most of us are familiar with the characteristics shown in Example 1, but not in Example 2. However, the latter is as important as the former. Readers are urged to bear these concepts in mind while analyzing the switching circuits.

C. Inductor

In the last section the electric properties of a capacitor were described in detail. Since the inductor is a dual element of a capacitor, we simply replace capacitance, C, by inductance, L, and current by voltage or vice versa, all the equations shown in the last section will be valid.

We have,

$$\frac{di(t)}{dt} = \frac{1}{L} v(t) \tag{4}$$

$$i(t) = \frac{1}{L} \int_{t_1}^{t} v(t)dt + i(t_1)$$

If $v(t)$ is finite,

$$\lim_{t \to t_1} i(t) = 0 + i(t_1) \tag{5}$$

which means, *the current of an inductor could not change instantaneously if the voltage across the inductor is finite.*

Example 1

In Figure 2(c) the switch is closed at $t = t_1$. We have $E = i(t)R + v(t)$, where $i(t)$ is finite, hence $v(t)$ is finite, and $i(t_1^+) = i(t_1^-) = 0$ and $v(t_1) = v(t_1^+) = E$. When $t \to \infty$, the circuit is in the steady state and $di/dt = 0$, thus,

$$v(\infty) = 0$$

$$i(\infty) = \frac{E}{R}$$

Example 2

In Figure 2(d), the switch is on A for a *long, long* time and being switched to B at $t = t_1$, we have

$$v(t_1^-) = L\frac{di(t^-)}{dt} = 0 \tag{6}$$

$$i(t_1^-) = i(t_1^+) = \frac{E}{R}$$

The loop equation with the switch on B is,

$$v(t) + i(t)(R_x + R) = 0$$

$$v(t) = -i(t)(R_x + R) \tag{7}$$

$$v(t_1^+) = -i(t_1^+)(R_x + R)$$

Since there is no generator or source in the loop with the switch on B, we know $i(\infty) = 0$, and $v(\infty) = 0$. Equations 6 and 7 show a very interesting property, i.e., the voltage across the inductor drops from zero to negative abruptly.

If $R_x = 0$, $v(t_1^+) = -i(t_1^+)R = -(t_1^+)R = \frac{E}{R}R = -E$

If $R_x = \infty$ or open circuit,

$$v(t_1^+) = -\infty!!$$

One should note that this example explains why the switching circuit sometimes yields negative spike even though only positive power supply is used. This negative spike is usually an unwelcome troublemaker. It generates noise to cause error in the digital system and sometimes causes damage to the transistors.

III. NONLINEAR ELEMENTS

A. Diode

A diode is a nonlinear device. Unlike the resistor, its V-I characteristic cannot be simply described by Ohm's Law. Instead, it can be quite accurately expressed by $I = K_1(e^{V/K_2} - 1)$, where K_1, K_2 are constants, as shown is Figure 3(a). However, for functional analysis, or for understanding how the device functions as a digital or a binary device, the V-I curve can be approximated by two pieces of straight lines as shown in

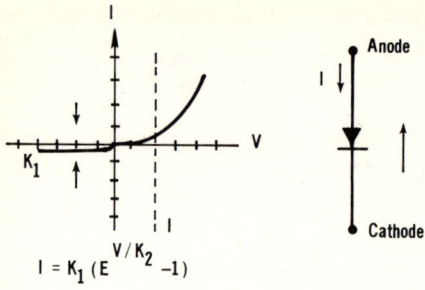

FIGURE 3(a). Diode.

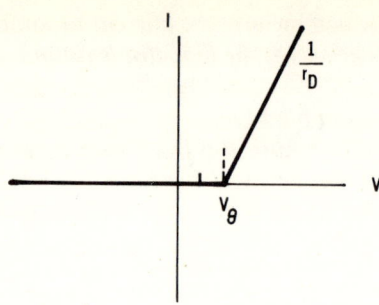

FIGURE 3(b). Piece-wise linear diode V-I curve.

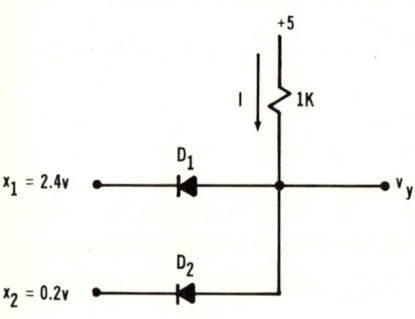

FIGURE 3(c). Resistor-diode circuit.

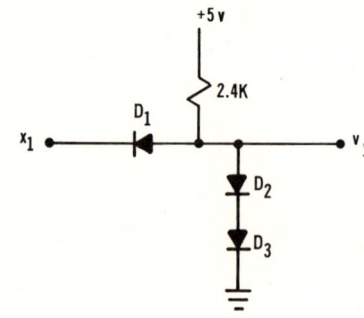

FIGURE 3(d). Resistor-diode circuit.

Figure 3(b). This technique is known as the *piece-wise linear approximation*. The curve can be interpreted as, $I = 0$ if $V \leq V_\theta$ and

$$I = \frac{V - V_\theta}{r_D} \text{ if } V > V_\theta$$

where V_θ: threshold voltage $\simeq 0.7$ V, and r_D: diode equivalent forward resistance $\simeq 50$ Ω. Physically it means that when the voltage across the diode is equal to or less than V_θ, the diode is practically open circuit, otherwise it is closed with a resistance of 50 Ω in series. In a diode-resistor logic network, one would be mostly interested in knowing in which states the diodes are. Sometimes the problems are not so trivial as is shown here. The following two examples may show some of the problems.

Example 1

For the diagram shown in Figure 3(c), let x_1 = logical 1 = 2.4 V, x_2 = logical 0 = 0.2 V, it appears that diodes D_1 and D_2 will be both conducting since their anodes are connected to 5 V while the cathode is at 2.4 V and 0.2 V, respectively; and the voltage difference between the cathode and anode for both diodes exceeds the threshold voltage $V_\theta \simeq 0.7$ V. However, actually D_2 is ON and D_1 is OFF. Although an experienced person can see this immediately, let us pretend to be a novice and set up a truth table:

Case	D_1	D_2
1	Off	Off
2	On	Off
3	On	On
4	Off	On

Case 1 is impossible because the voltage difference between terminals of both diodes exceeds V_θ.

Case 2 will result in

$$V_y = 2.4 + V_\theta + Ir_D = 2.4 + V_\theta + \frac{5 - V_y}{1k} \cdot r_D$$

since $r_D \ll 1$ K, $V_y \simeq 2.4 + 0.7 = 3.1$ V and $3.1 - x_2 > V_\theta$, thus, D_2 could not be OFF.

In Case 3, we have $V_y \simeq x_2 + V_\theta = 0.2 + 0.7 = 0.9$V and $V_y \simeq x_1 + V_\theta = 2.4 + 0.7 = 3.1$ Volts, but V_y cannot have two values at the same time; thus this is impossible.

Case 4 will result in $V_y = 0.9$ V which will keep D_1 from conducting, since $V_y - X_1 = 0.9 - 2.4 = -1.5 < V_\theta$. Hence, Case 4 is the only possible state for this circuit.

Example 2

In Figure 3(d), let x_1 = logical 1 = 2.4 V, if D_1 is ON, then $V_y \simeq 2.4 + 0.7 = 3.1$. Since D_2 and D_3 are in series, the minimum required voltage for D_2 and D_3 to be conducting is $V_y = 2 V_\theta \simeq 1.4$ V. Thus D_2 and D_3 have to be conducting. But if D_2 and D_3 are conducting, $V_y \simeq 0.7 + 0.7 = 1.4$, which will require V_y to have two values at the same time, thus this will not be the case. Now, if D_2 and D_3 are conducting, then $V_y \simeq 1.4$, which will keep D_1 from conducting if $x_1 = 2.4$. This is possible, therefore, if $x_1 = 2.4$ V, D_1 will be off, and D_2 and D_3 will be ON. Now let x_1 = logical 0 = 0.2 V, then $V_y \simeq x_1 + V_\theta = 0.2 + 0.7 = 0.9$ V, which is less than $2V_\theta$. Thus, D_2 and D_3 will be OFF, and D_1 will be ON.

Comments

These two examples show the basic concept which will be used over and over again when the logic designer has to know how the integrated circuit logic elements such as DTL, TTL, etc., function. For convenience, let us define this method of analysis as *binary-state-analysis* or BSA which is basically derived from the piece-wise linear V-I curve mode.

B. Transistor (Bipolar) Current Control Device

The transistor is also a nonlinear device that can be used as a switch, or as a basic element for an amplifier. Since we are dealing with digital circuits, transistor use as a switch or a binary-state device will be emphasized here.

In a switching circuit, a transistor can be treated analogously as a push-button switch as shown in Figure 4(a). For a push-button switch, one has to apply a force on the button to complete the electric circuit, the force required will depend on the strength of the spring inside the switch. For a transistor, the base terminal can be thought of as the button. But instead of applying a mechanical force at the base, it requires a proper amount of current flowing through the base in order to turn on the transistor switch. The following example shows how a transistor can be used as a switch.

Example 1
Binary-State-Analysis (BSA) for a Transistor Switch

A transistor can be considered as two diodes, i.e., Base-Emitter and Base-Collector, being connected back to back with current gain property. That is, in the circuit diagram, I_C/I_B = current gain = β which is normally much greater than 1. For simplicity, let us neglect the leakage current of the transistor. As shown in Figure 4(a), since

$$V_y = 5 - I_C \cdot 1K = 5 - \beta I_B 1K, \tag{8}$$

we have, $V_y = 5$ V if $I_B = 0$. That is, if the transistor is considered as a switch, then the switch is open when the current $I_B = 0$. Now let $\beta = 20$. If $I_B = 0.25$ mA, then $I_C = \beta I_B = 5$ mA. Substituting into Equation 8, we have $V_y = 0$ and transistor switch is now closed. Consider $I_B = 0.1$ mA, from Equation 8, $V_y = 5 - 20 \cdot 10^{-3} \cdot 1K = 3$ V. This means that the transistor switch is neither completely ON nor OFF. It behaves like a conventional switch having a poor contact. Of course, this is not desirable. However, $(V_y)_{min} = 0$, which cannot be negative; thus from Equation 8, $(I_c)_{max} = 5$ mA. That is, the collector current of the transistor is limited by the power supply and the collector resistance.

Now, consider $I_B = 1$ mA, with the transistor in its ON state. Although the control current I_B demands $I_C = \beta I_B = 20$ mA, the circuit however can supply no more than 5 mA. This is equivalent to one who may push the button harder and yet the switch is still only on the ON state. One can never make a lamp brighter by pushing the switch harder. A transistor being overdriven like this is defined as in "saturation" state. Most of the logic circuit is designed such that the transistor is either in the OFF or saturation state except the extremely high speed logic circuit such as ECL, emitter-coupled-logic or current mode switching logic circuit. It is to be noted that when a push-button switch is ON, $V_y = 0$; but for a transistor switch, $V_y \simeq 0.1 - 0.2$ V when it is driven to saturation. This voltage is called *saturation voltage* which is not desirable in processing analog signal but acceptable for logic network realization.

In practice, the control current I_B is usually generated with a voltage source as shown in Figure 4(b).

Let x_1 = logical 1 = 2.4 V. Since Base-Emitter is a diode junction, the BSA technique used in the last section can be applied here. We have,

$$I_B = \frac{2.4 - V_\theta}{2K} \simeq \frac{2.4 - 0.7}{2K} = 0.85 \text{ mA}$$

and $\beta I_B = 20 \cdot 0.85$ mA = 17 mA > 5mA. Thus the transistor is saturated and the switch is turned ON. Now let x_1 = logical 0 = 0.2 V, which is less than the threshold voltage V_θ, thus, $I_B \simeq 0$, and the transistor switch is turned OFF, $V_y = 5$ V.

Graphical Analysis Method

To demonstrate how the graphical analysis technique can be used for a nonlinear device, let us consider the same circuit shown in Figure 4(b). Figures 4(c) and (d),

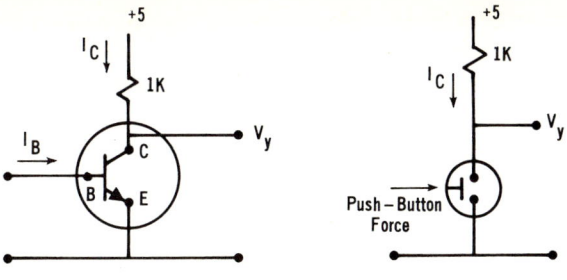

FIGURE 4(a). Resistor-transistor circuit.

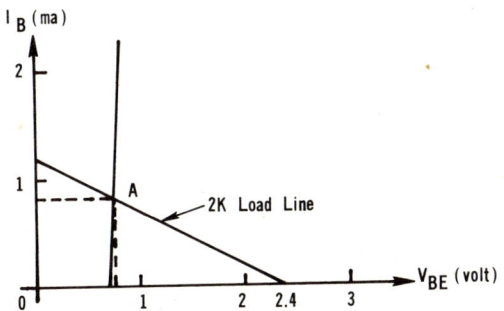

FIGURE 4(c). Transistor input V-I curve.

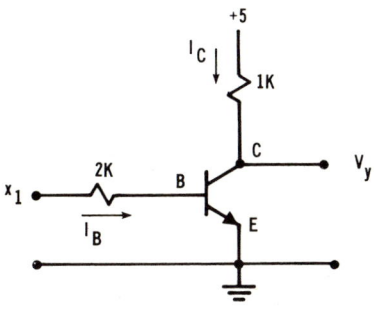

FIGURE 4(b). Resistor-transistor circuit.

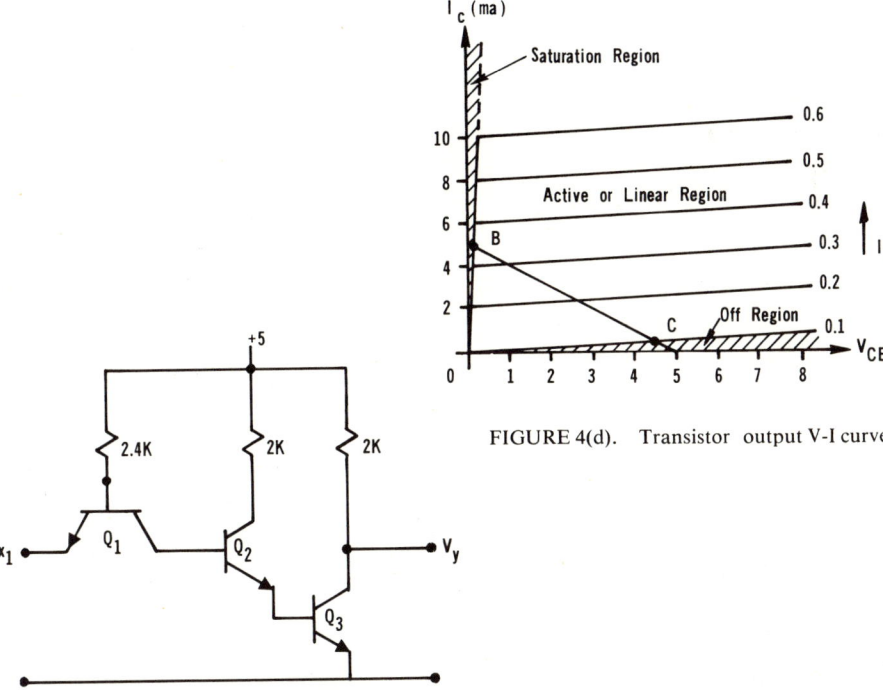

FIGURE 4(d). Transistor output V-I curve.

FIGURE 4(e). Resistor-transistor circuit.

depict the input and output piece-wise-linear V-I characteristic curves of the transistor with the corresponding load lines for the resistors. Note that in Figure 4(c), the intersection point A is a $V_{BE} \simeq 0.7$ and $I_B \simeq 0.85$ mA; and in Figure 4(d), points B and C, respectively, are the saturation and OFF operating points where at B: $I_c \simeq 5$ mA, $V_{CE} \simeq 0.2$ V, and at C: $I_c \simeq$ leakage current of the transistor $V_{CE} \simeq 5$ V. It is interesting to point out that point B will not move as I_B exceeds 0.5 mA. That is, the transistor is in saturation state when $I_B > 0.35$ mA. The advantage of using saturation-state and off-state for a transistor to implement logic one and zero is now apparent. It provides the assurance of two distinct states for binary or digital operation. Logic elements operate at saturation mode do have one disadvantage, however. They have slower switching speed in comparison with nonsaturation logic, such as ECL and Schottky TTL which will be described in the later sections.

Example 2

It is desirable to determine the states of transistors Q_1, Q_2, Q_3 shown in Figure 4(e) with X_1 = logical 1 = 2.4 V, and x_1 = logical 0 = 0.2 V. Basically, the circuit has the diode junctions of the Base-Collector of Q_1, and the Base-Emitter junctions of Q_2 and Q_3 in series, which is shunting the diode junction of the Base-Emitter of Q_1. The analysis of Example 2 in the last section can be used here. We have, for x_1 = 2.4 V, the three diode junctions, i.e., Base-Collector of Q_1 and Base-Emitter of Q_2 and Q_3 are conducting, thus, Q_2 and Q_3 are saturated, and $V_y \simeq 0.2$ V. Similarly, x_1 = 0.2 V, Base-Emitter junction of Q_1 is ON, and Q_2 and Q_3 are OFF, thus $V_y = +5$ V.

C. Field-Effect-Transistor (FET) Voltage Control Device

An FET is basically a voltage control device, it draws practically no DC current at the input. Tables 1 and 2 show a summary of the classification, symbol, and major electrical properties of the FET.

FETs when used in digital circuitry have the following major advantages: (1) high fan-out due to high input impedance, (2) low component count, and (3) extremely low power dissipation.

The following examples show how the FET can be used in a digital system.

Example 1

The diagram shown in Figure 5 is a simple switching circuit using a junction N-channel depletion type FET. From Table 2, the FET-switch will be OFF if $V_{GS} < -V_p$, the pinch-off voltage; and will be ON if $V_{GS} = 0$. Let the FET be 2N3822, then, one may use $V_{DD} = +15$, $V_{GG} = -15$, $R_S = R_G = 1$ M, $R_D = 15$K, and define x_1 = logical 1 = +15 V. Then $V_y = 0$ V if $x_1 = +15$; $V_y = 15$ V if $x_1 = 0$.

Example 2

The circuit shown in Figure 5(b) is a simple complementary IGFET, also known as CMOS. It is called "Complementary" because both P-channel (Q_1)-and N-channel (Q_2) are used at the same time. The devices are enhancement types. The electrical characteristic of an N-channel IGFET is shown in Table 2. For P-channel just change the sign of I_D and V_{GS} from positive to negative, i.e., the device will be ON if V_{GS} is negative. It will be OFF if $V_{GS} = 0$. Now, let x_1 be logical 1 = +V, then for Q_1, $V_{GS} = 0$. For Q_2, $V_{GS} = +V$. Thus, Q_1 will be OFF and Q_2 will be ON, $V_y = 0$. If x_1 = logical 0 = 0 V, then, for Q_1, $V_{GS} = -V$; for Q_2, $V_{GS} = 0$; thus Q_1 will be ON and Q_2 will be OFF, $V_y = +V$. The circuit is analogously operating as a single pole double throw switch with its arm connected to V_y. It is interesting to note that, if the load of this circuit is composed of other FET gates, the circuit has practically no resistive load (the

Table 1

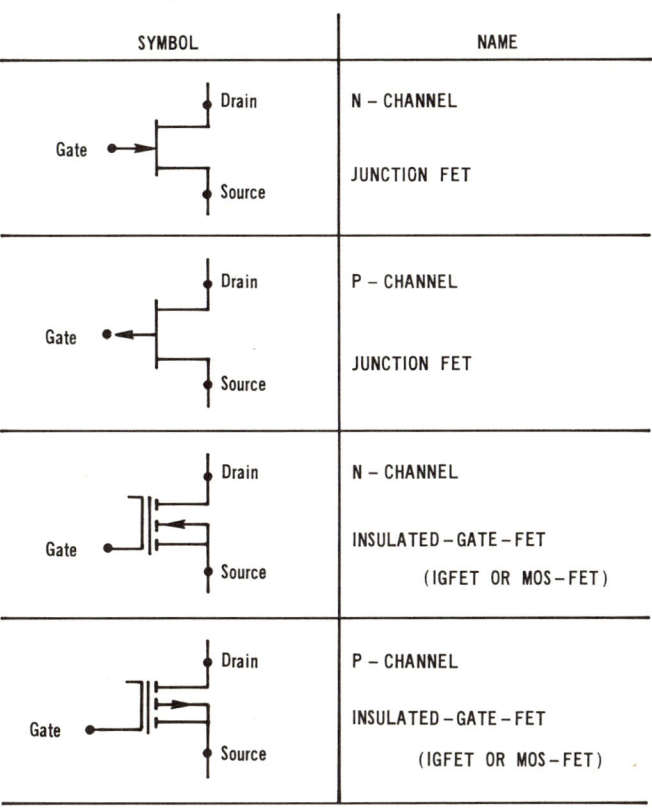

input resistance of an IGFET is about 10^{19} Ω), and dissipates zero power at the steady state. However, since the input of IGFET is capacitive, the circuit will dissipate power during the state transition time. Thus the circuit is extremely attractive when low power dissipation is primarily important.

Example 3

Although in any digital system the major circuit elements are binary in nature, the system usually is interfaced with the physical world which is mostly analog in nature. Therefore, a digital system usually contains analog-digital and digital-analog converters. This example is to show how a FET can be used as an analog switch for switching a low level analog signal.

As described in the last section, when a conventional transistor (bipolar transistor) is turned on, a saturation voltage of 0.1 ~ 0.2 V would exist between collector and emitter. The magnitude of this voltage is sensitive to temperature and individual device. This would introduce considerable error if it is used as a switch for switching a low level analog signal. However, since the path between drain and source of an FET in the neighborhood of $V_{DS} = 0$ behaves just like a linear resistor, the FET becomes a good analog switch. Figure 6 shows a typical low level output characteristic of an FET. Notice that for $V_{GS} = 0$, the V-I curve is horizontal which means the device has very high resistance. For $V_{GS} = -10$, the device behaves like a resistor of 400 Ω, as long as $-0.2 \leq V_{DS} \leq 0.2$. As the voltage V_{DS} increases, the V-I curve in Figure 6(a) will not be linear like a resistor. Thus, in Figure 6(b), as long as the signal voltage E_s is within ± 0.2V, and R_L is >> 400 Ω but much less than the off-resistance of the FET, the device is a good analog switch.

Table 2

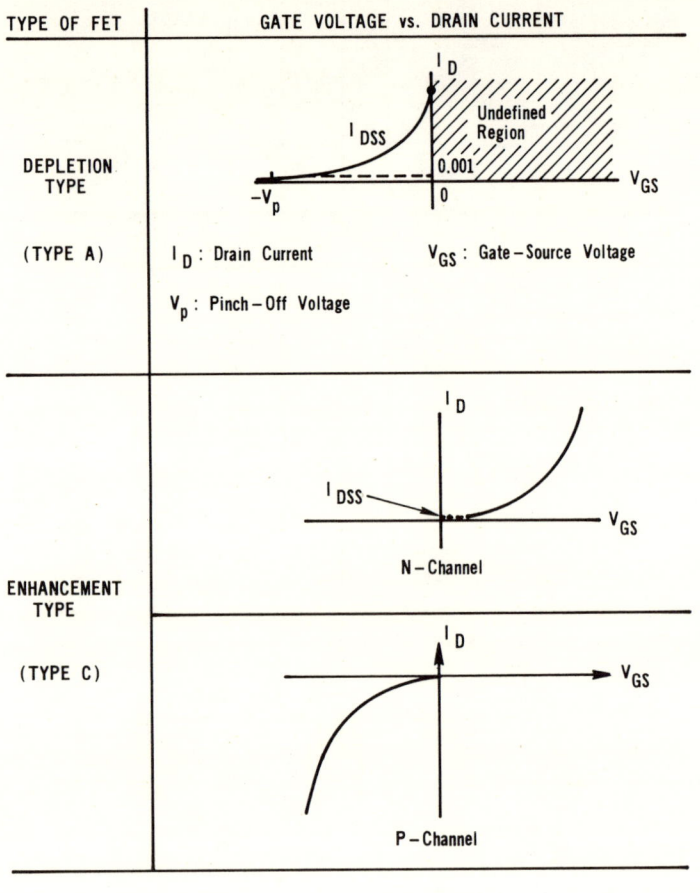

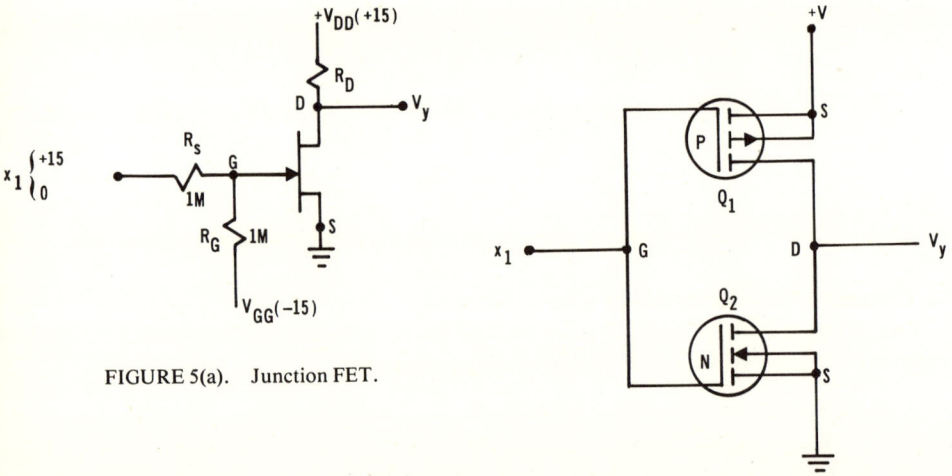

FIGURE 5(a). Junction FET.

FIGURE 5(b). CMOS inverter.

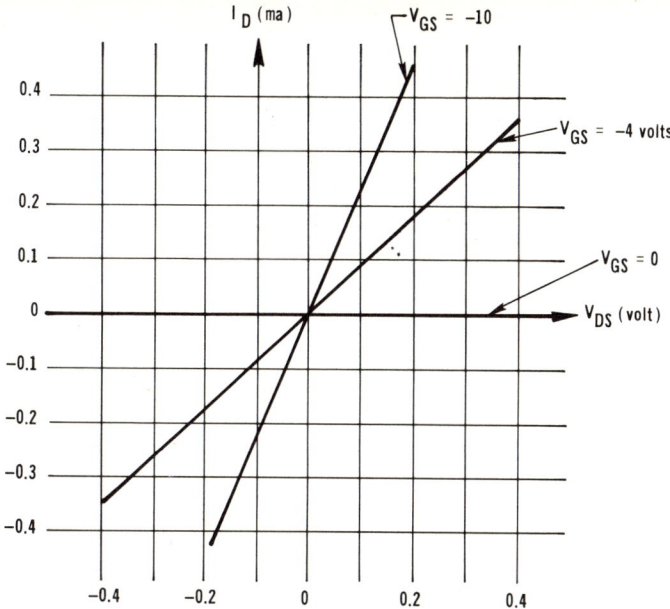

FIGURE 6(a). FET low level output characteristic.

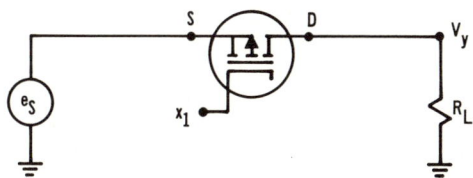

FIGURE 6(b). FET as an analog switch.

IV. PULSE CIRCUIT FUNDAMENTALS

In this section, an attempt at clarifying the time response of a network to a pulse is to be made. Specifically, how an R-C network responds to a pulse or a switching signal will be described and a review on R-C networks mixed with diode follows. A thorough understanding of their electrical properties with respect to time is essential and it would make the study of multivibrators in the later sections a painless and pleasant one.

A. Complete Time Response of a Basic Network

The circuit shown in Figure 7 is a good old friend of yours. If the switch S is switched from A to B at t = 0, we have the following conditions:

$$V_C(0^-) = E_a$$

$$V_C(0^+) = E_a$$

$$V_C(\infty) = E_b$$

$$i(0^-) = 0$$

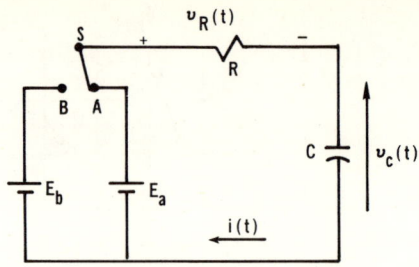

FIGURE 7(a). R-C circuit.

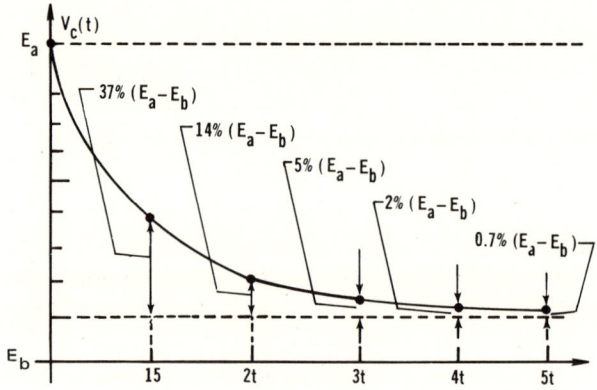

FIGURE 7(b). $V_c(t)$ for $E_b < E_a$.

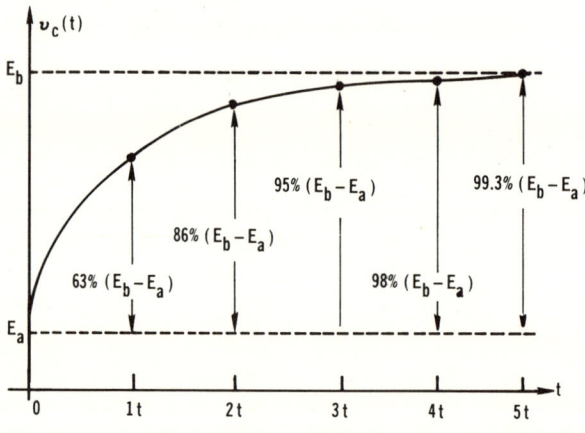

FIGURE 7(c). $v_c(t)$ for $E_b > E_a$.

$$i(0^+) = \frac{E_b - V_C(0^+)}{R}$$

$$V_R(0^-) = 0$$

$$V_R(0^+) = E_b - V_C(0^+) = E_b - V_C(0^-) = E_b - E_a$$

$$V_R(\infty) = E_B - V_C(\infty) = 0$$

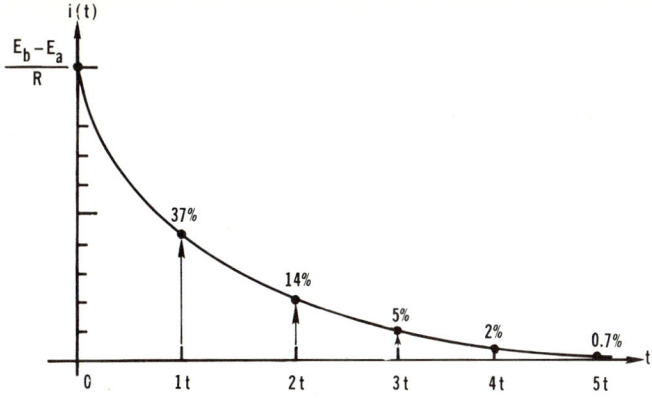

FIGURE 7(d). i(t) for $E_b > E_a$.

However, we are interested in knowing how the circuit behaves for $0^+ < t < \infty$. In this circuit for $0^+ < t < \infty$, the loop equation is,

$$E_b = V_R(t) + V_C(t)$$

$$= Ri(t) + \frac{1}{C}\int i(t)dt$$

which is simply a first order differential equation, whose general solution is,

$$i(t) = K_1 + K_2 \epsilon^{-\frac{t}{RC}} \qquad (9)$$

for $t = 0^+$

$$i(0^+) = K_1 + K_2 \qquad (10)$$

for $t = \infty$

$$i(\infty) = K_1 + 0 = K_1 \qquad (11)$$

Substitute Equation 11 into Equation 10, $K_2 = i(0^+) - K_1 = i(0^+) - i(\infty)$. Hence,

$$i(t) = i(\infty) + \{i(0^+) - i(\infty)\} e^{-t/RC} \qquad (12)$$

where RC $\underline{\Delta}$ circuit time constant, $i(0^+)$ $\underline{\Delta}$ initial value, and $i(\infty)$ $\underline{\Delta}$ final value. Equation 12 can be generalized and be applicable to any circuit which can be lumped into one resistor and one energy storage element (memory element) which can be either a capacitor or an inductor.

The generalized equation can be written as,

$$x(t) = x(\infty) + \{x(0^+) - x(\infty)\} e^{-t/\tau} \qquad (13)$$

$x(t)$ can either be voltage or current, τ = time constant = RC for resistor-capacitor network, and τ = L/R for resistor-inductor network.

B. Example 1

Applying Equation 13 to the circuit shown in Figure 7, one may easily determine the complete time response of $V_c(t)$ and $i(t)$ in the following way:

$$V_C(t) = V_C(\infty) + \{V_C(0^+) - V_C(\infty)\} e^{-t/RC} \tag{14}$$

$$= E_b + (E_a - E_b) e^{-t/RC}$$

and $i(t) = i(\infty) + i(0^+) - i(\infty) e^{-t/RC}$

$$= 0 + \left\{ \frac{E_b - E_a}{R} \right\} - 0 e^{-t/RC} \tag{15}$$

$$= \frac{E_b - E_a}{R} e^{-t/RC}$$

Since, $e^{-1} \simeq 0.37 = 37\%$, $e^{-2} = 0.14 = 14\%$, $e^{-3} \simeq 0.05 = 5\%$, $e^{-4} \simeq 0.02 = 2\%$, and $e^{-5} = 0.007 = 0.7\%$ the time waveform of Equations 14 and 15 can easily be sketched as shown in Figures 7(b), 7(c), and 7(d).

Notice that the key points for determination of the complete time response of the circuit are: (1) initial value, (2) final value, and (3) time constant.

C. Example 2

In Figure 8(a) determine and sketch the output waveform for (a) $V_{in}(t)$ is a pulse with amplitude equal to 5 V and pulse width equal to 10 msec, and (b) $V_{in}(t)$ is a pulse with amplitude changes from -5 to $+5$ V and pulse width 1 msec.

Solution

$t = RC = 1$ msec. For $0^+ < t < 10$ msec, $[V_o]_{initial} = 0$ V, $[V_o]_{final} = 5$ V, and thus, $V_o(t) = 5 + (0 - 5)e^{-t/10^{-3}}$, $V_o(t = 10$ msec$) = 5 - 5e^{-(t-10 \times 10^{-3})/10^{-3}} = 5 = 5e^{-10} = 5$. For $t > 10$ msec, $[V_o]_{initial} = 5$ and $[V_o]_{final} = 0$. Thus, $V_o(t > 10$ msec$) = 5e^{-(t-20 \text{ msec})/10^{-3}}$ $t = 1$ msec. For $0^+ < t < 1$ msec, $[V_o]_{initial} = -5$, $[V_o]_{final} = +5$, $V_o(t) = 5 + (-5 - 5)e^{-t/10^{-3}}$, and $V_o(t) = 5 - 10e^{-t/10^{-3}}$. For $t > 1$ msec, $[V_o]_{initial} = V_o(t = 1$ msec$) = 5 - 10e^{-1} = 5 - 3.7 = 1.3$ V.

The waveforms are shown in Figures 8(b) and (c). It is now clear that if the time constant of the circuit is much, much greater than the pulse-width of the pulse, the input pulse will not show at the output.

D. Example 3

For the circuit shown in Figure 9(a), determine and sketch the waveforms of $V_c(t)$, $z(t)$, and $y(t)$ if $x_1 = 5$ V and x_2 is a pulse with $\delta = 1$ msec as shown.

Solution

Let the pulse of x_2 occur at $t = 0$, then we have

$$y(0^-) = \left(\frac{5 - 0.7}{4K} \right) \times 2k = 2.15 \text{ V}$$

where $0.7 =$ diode voltage drop $V_c(0^-) = z(0^-) - x_2(0^-) = 2.15 + 0.7 - 0 = 2.85$ V, $V_c(0^+) = V_c(0^-) = 2.85$ V, $z(0^-) = 2.85$ V, and $z(0^+) = V_c(0^+) = 2.85 + 5 = 7.85$ V.

Since the circuit would not know when the pulse would be terminated, it assumes δ

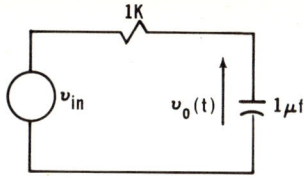

FIGURE 8(a). R-C circuit.

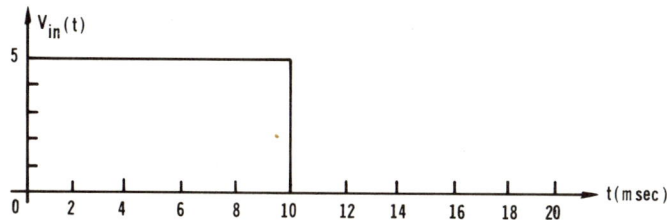

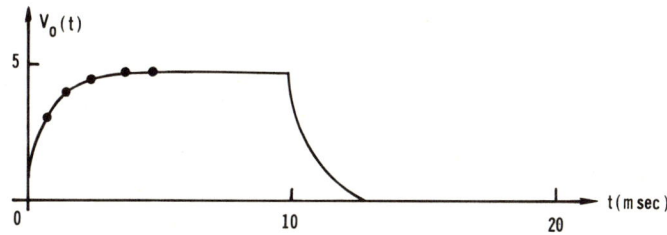

FIGURE 8(b). Signal pulse-width greater than 5τ.

$= \infty$, thus $z(\infty) = 2.85$ and $y(\infty) = z(\infty) - 0.7 = 2.15$. Now, we shall determine the time constant of the circuit. To do that, we simply short-circuit all the independent voltage sources, i.e., x_1 and x_2, then we have (1) the time constant $t = (0.1\ \mu F)\ 2K = 0.2$ msec if the diode is OFF and (2) $t = (0.1\ \mu F)\ 2K//2K = 0.1$ msec if the diode is ON.

Since during the period of $0 < t < 1$ msec the diode is conducting, the time constant $t = 0.1$ msec should be used. Then from Equation 13 $V_C(t) = V_C(\infty) + \{V_C(0^+) - V_C(\infty)\}e^{-t/\tau}$ Where, $V_C(\infty) \underline{\Delta}$ the final value of $V_C(t)$ if $\delta = \infty$, and $V_C(\infty) = z(\infty) -$, $x_2(\infty) = 2.85 - 5 = -2.15$.

Thus,

$$V_C(0^+ < t < 1) = -2.15 + \{2.85 + 2.15\}e^{-t/0.1}$$

$$= -2.15 + 5e^{-t/0.1} \qquad (16)$$

$$z(0^+ < t < 1) = z(\infty) + \{z(0^+) - z(\infty)\}e^{-t/0.1}$$

$$= 2.85 + (7.85 - 2.85)e^{-t/0.1}$$

$$= 2.85 + 5e^{-t/0.1} \qquad (17)$$

and

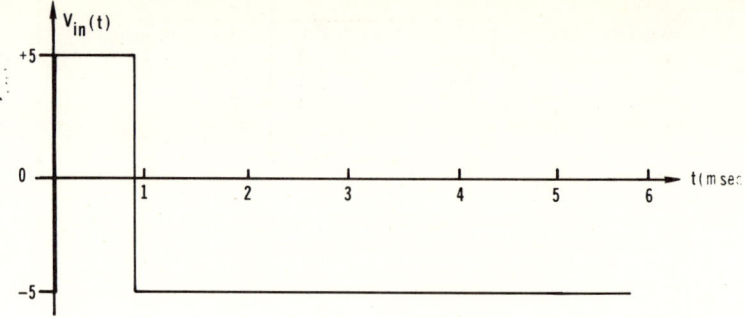

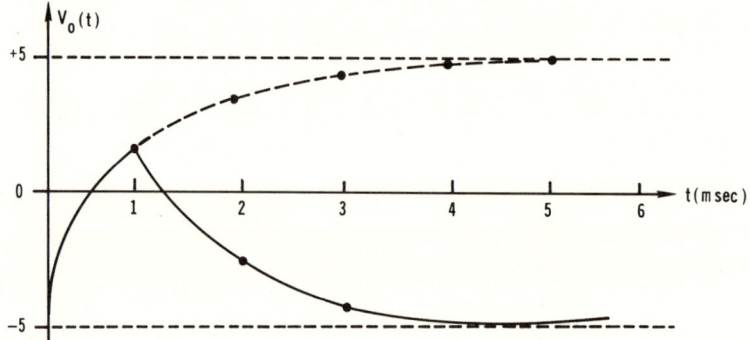

FIGURE 8(c). Signal pulse-width less than 5τ.

$y(0^+ < t < 1) = 2.15 + 5e^{t/0.1}$.

Now, let us analyze the period of $t > \delta = 1$ msec.

From Equations 16, 17, and 18, $V_C(1^-) = -2.15$, $V_C(1^+) = V_C(1^-) = -2.15$, $z(1^-) = 2.85$, $y(1^-) = 2.15$, and $z(1^+) = x_2(1^+) + V_C(1^+) = 0 - 2.15 = -2.15$, $y(1^+) = 0$ since the diode is OFF at $t = 1$. But for $1^+ < t < \infty$, $V_C(\infty) = x_1(\infty) = 5$ if the diode is not conducting; $V_C(\infty) = z(\infty) = 2.85$ if the diode is conducting. That is, V_C will change from -2.15 to either 5 V or 2.85 depending on whether the diode is conducting or not. According to the BSA, the diode will be turned on when $V_C \leq 0.7$. Therefore, the problem becomes even more interesting. The V_C is heading toward $z = x_1 = 5$ before V_C reaches 0.7 and changes its course heading to $z = 2.85$ immediately after $V_C = 0.7$. The problem is now to determine at what time $V_C = 0.7$. Let $V_C(t_1) = 0.7$, since for $1^+ < t < t_1$ the diode is OFF, then

$$V_C(1 \text{ msec} < t < t_1) = V_C(\infty) + \{V_C(0) - V_C(\infty)\}e^{-t/\tau}$$

$$= 5 + \{-2.15 - 5\}e^{-(t-1)/0.2} \qquad (18)$$

$$= 5 - 7.15e^{-(t-1)/0.2}$$

or, $0.7 = 5 - 7.15e^{-(t_1 - 1)/0.2}$, $t_1 = 1.1$ msec; and
$z(1 < t < t_1) = x_2(1 < t < t_1) + V_C(1 < t < t_1)$

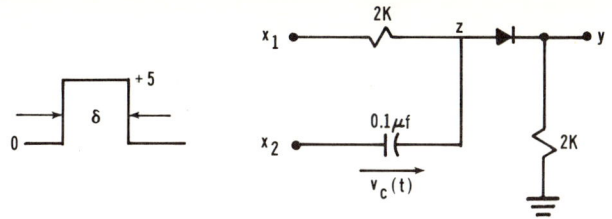

FIGURE 9(a). R-C-Diode circuit.

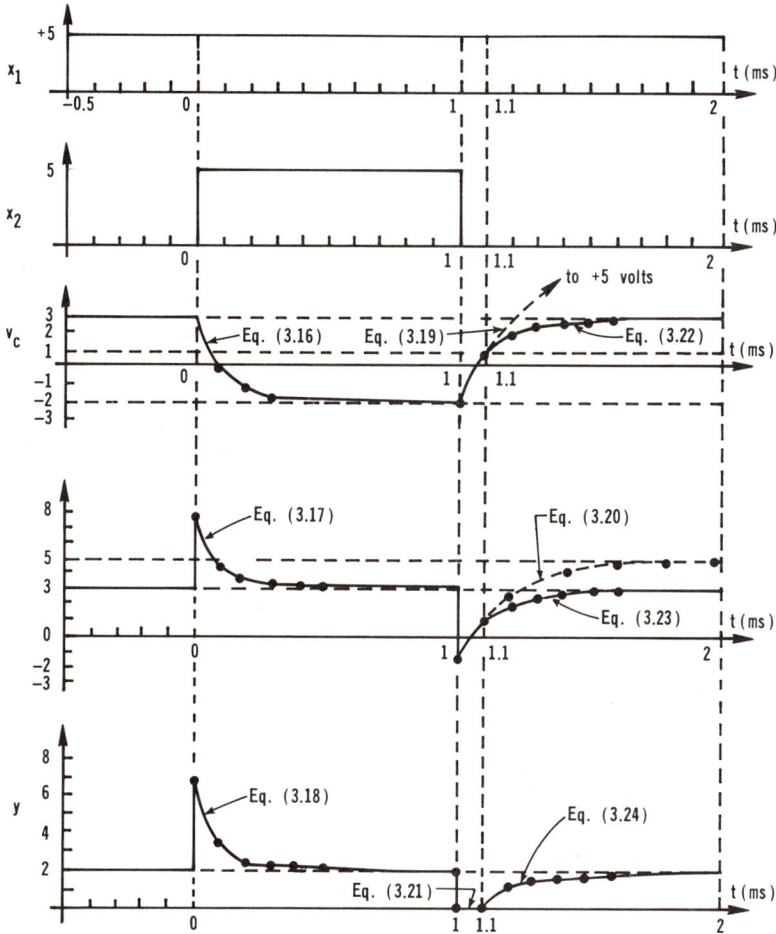

FIGURE 9(b). Waveforms of Example 3.

$$= 0 + V_C(1 < t < t_1) \quad (19)$$

$$= 5 - 7.15e^{-(t-1)/0.2}$$

$$y(1 < t < t_1) = 0 \quad (20)$$

Now, since for $t_1 < t < \infty$ the diode is conducting, we have,

$$V_C(t > t_1) = V_C(\infty) + V_C(0) - V_C(\infty) \, e^{-(t-t_1)/\tau}$$

$$= 2.85 + 0.7 - 2.85 \, e^{-(t-t_1)/0.1} \qquad (21)$$

$$= 2.85 - 2.15 e^{-(t-t_1)/0.1}$$

and $z(t > t_1) = x_2(t > t_1) + V_c(t > t_1)$

$$z(t > t_1) = x_2(t > t_1) + V_C(t > t_1)$$

$$= 0 + V_C(t > t_1) \qquad (22)$$

$$= 2.85 - 2.15 e^{-(t-t_1)/0.1}$$

$$y(t > t_1) = z(t > t_1) - 0.7 \qquad (23)$$

$$= 2.15 - 2.15 e^{-(t-t_1)/0.1}$$

The waveforms of x_1, x_2, V_C, z, and y are plotted in Figure 9(b).

E. Important Remark

The reader may feel that the first two examples are too simple and the last example is unnecessarily complex. However, the first two examples are designed to provide the reader with a good review of the fundamentals of timing circuits; and the third example is specially designed to cover the analysis techniques that one will need to understand the principle of operation of the sequential circuit elements described in the later sections. The reader is therefore urged to reread them and to understand each step described in these examples thoroughly.

V. LOGIC FUNDAMENTALS IN BRIEF

In this section, a brief review on logic fundamentals will be presented for completeness of this handbook and for the convenience of ones who may need to refresh the fundamentals for themselves from time to time. First, a summary of the definitions and theorems for logical operations is presented.

A. Definitions

1. Switching variable $\triangleq$ Any Upper Case Letter, assume only two possible values: "1" and "0".
2. Logical AND operation $\triangleq$ ".".
3. Logical OR operation $\triangleq$ " + ".
4. Logic Inversion $\triangleq$ ",", or a bar on the top of a letter, i.e., "$\overline{A}$."

B. Theorems

1. $1 \cdot 1 = 1$
2. $0 \cdot 0 = 0$
3. $1 + 1 = 1$
4. $0 + 0 = 0$
5. $1 \cdot 0 = 0 \cdot 1 = 0$
6. $1 + 0 = 0 + 1 = 1$

7. $0' = 1, 1' = 0$
8. $x \cdot 1 = x, x + 1 = 1$
9. $x \cdot 0 = 0, x + 0 = x$
10. $x + x = x, x \cdot x = x$
11. $(x)' = x', (x')' = x$
12. $x + x' = 1, x \cdot x' = 0$
13. $x + y = y + x, x \cdot y \underline{\Delta} xy = yx \underline{\Delta} y \cdot x$
14. $x + xy = x, x(x + y) = x \cdot x + x \cdot y = x$
15. $xy' + y = x + y$
16. $(w + x)(y + z) = wy + wy + wz + xz$
17. $(x + y)(y + z)(z + x') = (x + y)(z + x')$
18. $xy + yz + zx' = xy + zx'$
19. $(x + y + z + ...)' = x'y'z'...$
20. $(xyz ...)' = x' + y' + z' + ...$
21. $f(x_1, x_2, ..., x_n, +, \cdot)' = f(x_1', x_2', ..., x_n', \cdot, +)$
22. $f(x_1, x_2, ..., x_n) = x_1 f(1, x_2, ..., x_n) + x'_1 f(0, x_2, ..., x_n)$
23. $f(x_1, x_2, ..., x_n) = [x_1 + f(0, x_2, ..., x_n)][x'_2 + f(1, x_2, ..., x_n)]$

C. Minimization of Switching Function by Karnaugh Map

3-Variable Karnaugh Map

Switching Function: $F = xyz' + x'y'z + xyz + xy'z$.
Karnaugh map:

z \ xy	00	01	11	10
0	0	0	1	0
1	1	0	1	1

F

Minimized Switching Function: $F = xy + y'z$

4-Variable Karnaugh map

Switching Function: $F = w'xy'z' + w'x'yz + w'xy'z + wxy'z + w'xyz + wxyz + wx'yz$
Karnaugh map:

yz \ wx	00	01	11	10
00	0	1	0	0
01	0	1	1	0
11	1	1	1	1
10	0	0	0	0

F

Minimized Switching Function: $F = w'xy' + xz + yz$

REFERENCES

1. **Millman, J. and Taub, H.** *Pulse and Digital and Switching Waveforms,* McGraw-Hill, New York, 1965.
2. **Millman, J.** *Microelectronics: Digital and Analog, Circuits and Systems,* McGraw-Hill, New York, 1979.
3. **Krieger, M.,** *Basic Switching Circuit Theory,* Macmillan, New York, 1967.
4. **McCluskey, E. J.,** *Introduction to the Theory of Switching Circuits,* McGraw-Hill, New York, 1965.
5. **Marcus, M. P.,** *Switching Circuit for Engineers,* 2nd ed., Prentice-Hall, Englewood Cliffs, N.J., 1967.
6. **Miller, R. E.,** *Switching Circuit Theory,* Vols. I and II, John Wiley & Sons, New York, 1965.
7. **Peatman, J. B.,** *The Design of Digital Systems,* McGraw-Hill, New York, 1972.
8. **Blakeslee, T. R.,** *Digital Design with Standard MSI and LSI,* 2nd ed., John Wiley & Sons, New York, 1979.

Chapter 4

BASIC ELECTRONIC LOGIC ELEMENTS

I. BACKGROUND

In general, a digital system designer is given a verbal specification of a system to be designed and implemented with electric hardwares. The designer would normally start with a system block diagram according to the specifications. For each block, the desired inputs and outputs are defined, and its internal characteristic or transfer function is described by switching functions. Eventually, the mathematical descriptions are implemented by electronic elements. In view of the definitions and switching function described in Chapter 3, Section V, it is revealed that there are only three basic operations, i.e., "AND", "OR", and "INVERT" in any switching functions. Therefore, it requires only three basic electronic operational elements to realize thoese operations and thus the switching functions. Those basic elements are called gates. By adding some basic functional blocks such as flip-flop, clock, one shot, etc., to the switching network, blocks and then the whole system will be realized. In the following two sections, these electronic elements will be described.

II. LOGIC CIRCUITS

A. Introduction

Having studied the properties of the basic circuit elements in Chapter 3, we shall now describe how the different kinds of logic circuits or gates function. The memory logic circuits or multivibrators will be studied in the later chapters. In this section, we shall limit ourselves to studying the gates for logic network. From the Background Section above, we know that the basic logic elements for the realizing of any switching function are: Inverter, OR, and AND gates; or just simply AND-INVERT (NAND) gate only; or OR-INVERT (NOR) gate only. Before studying the function of the logic circuits, it is necessary to define a few terms which are important to describe the properties of any logic circuits.

Since the values of the variables in the switching functions that we are dealing with are limited to binary, one may use any device which has two distinct states to realize the switching variables. One may assign logical ONE to the state in which a device is ON and logical ZERO when the device is OFF, or conversely. But generally, the voltage of a logic circuit is used to represent the logic levels. For example,

$$\left.\begin{array}{l} \text{High voltage, Say } v \geq 2.4 \text{ V}, \triangleq \text{ logical 1} \\ \text{Low voltage, Say } v \leq 0.5 \text{ V}, \triangleq \text{ logical 0} \\ \quad 0.5 < v < 2.4 \text{ V}, \triangleq \text{ Undefined} \end{array}\right\} \quad (1)$$

or,

$$\left.\begin{array}{l} \text{Low voltage, Say } v \leq 0.5 \text{ V}, \triangleq \text{ logical 1} \\ \text{High voltage, Say } v \geq 2.4 \text{ V}, \triangleq \text{ logical 0} \\ \quad 0.5 < v < 2.4 \text{ V}, \triangleq \text{ Undefined} \end{array}\right\} \quad (2)$$

where, the symbol $\underline{\triangle}$ means "is defined as". The system using Equation 1 is called, *positive logic* system; The system using Equation 2 is called, *negative logic* system.

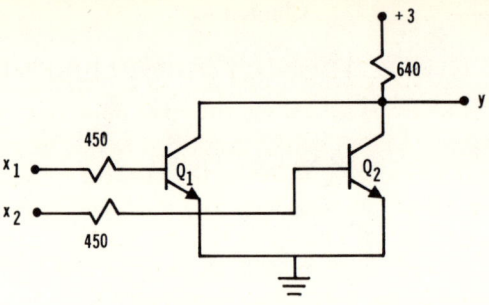

FIGURE 1. R-T logic circuit.

B. Resistor-Transistor Logic (RTL)

The circuit shown in Figure 1 is a typical RTL circuit. Let the input signals x_1 and x_2 be either 3 V or 0.2 V. Then, by applying the Binary-State-Analysis technique discussed in Chapter 3, Section III.B one may construct Table 1.

Table 1

x_1 (V)	Q_1	x_2(V)	Q_2	y(V)
0.2	OFF	0.2	OFF	3
3	ON	0.2	OFF	0.2
0.2	OFF	3	ON	0.2
3	ON	3	ON	0.2

In this table, let 3 V $\underline{\Delta}$ H $\underline{\Delta}$ logical 1 and 0.2 V $\underline{\Delta}$ L $\underline{\Delta}$ logical 0, then Table 1 can be rewritten as shown in Table 2.

Table 2

x_1	x_2	y
0(L)	0(L)	1(H)
1(H)	0(L)	0(L)
0(L)	1(H)	0(L)
1(H)	1(H)	0(L)

This can easily be translated into a switching function, i.e.,

$$y = \overline{x_1 + x_2} \qquad (3)$$

Equation 3 is a NOR switching function, which is a *positive logic* system, and the circuit is a NOR-gate. Now, consider the following: 3 V $\underline{\Delta}$ H $\underline{\Delta}$ logical 0 and 0.2 V $\underline{\Delta}$ L $\underline{\Delta}$ logical 1; Table 2 becomes Table 3.

Table 3

x_1	x_2	y
1	1	0
0	1	1
1	0	1
0	0	1

From Table 3 we have the corresponding switching function

$$y = \overline{x_1 \cdot x_2} \qquad (4)$$

Equation 4 becomes a NAND function, and this is a *negative logic* system and the same circuit becomes a NAND-gate!! It is important to point out that the logic designer should bear in mind that one cannot define the circuit as an OR-gate or an AND-gate; neither as a NOR-gate or a NAND-gate unless the positive or negative logic system is first defined. Sometimes a logic designer may find it advantageous to think in a mixed positive-and-negative logic system known as Kintner's system. In this case, one may just label the logic circuit with symbols such as H(high) and L(low) to carry out the logical design. For example, from Table 2, we have Table 4.

Table 4

x_1	x_2	y
L	L	H
H	L	L
L	H	L
H	H	L

Now, if one used *positive logic* at the input and *negative logic* for the output, then this table in logical form will be Table 5,

Table 5

x_1	x_2	y
0	0	0
1	0	1
0	1	1
1	1	1

which realizes an OR-function.

The author does not intend to confuse the reader with this concept, but rather to show the flexibility of logic assignment and to pave an easy way for the reader to convert a switching function into an integrated logic circuit. It is important to point out that this concept is not limited for RTL, but also for TTL, DTL, and ECL, which will be described in the later sections.

In this paragraph we shall introduce another term called *transfer curve* which is very useful in studying the electric properties of a logic circuit. This term can be best described with an inverter circuit.

The circuit shown in Figure 2(a) is an *RTL inverter* because if x = HIGH, then y = LOW and vice versa. Figure 2(b) is a sketch of the input vs. output transfer curve. Let the current gain of the transistor be four, threshold voltage, v_θ = 0.7; then (I) I_c = 0, y = 3 for $0 < x \leq 0.7$, (II) $0 < I_c < 4.3$ mA, y < 3 for $0.7 < x < 1.2$, and (III) I_c = 4.3 mA, y = 0.2 for x > 1.2. There are three distinct regions, i.e., in Region (I), y = H, x = L; in Region (III), y = L, x = H; in Region (II), the circuit is undefined.

For positive logic, it would normally define that logical 1 is greater than 1.2 V, and logical 0 is less than 0.5 V; otherwise the circuit is undefined. Evidently, there is no digital system that can tolerate any of the logic circuits remaining in the undefined state. With the help of the *transfer curve* concept, a few more terms can now be easily defined.

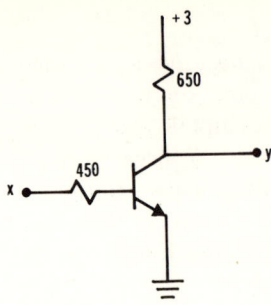

FIGURE 2(a). R-T circuit.

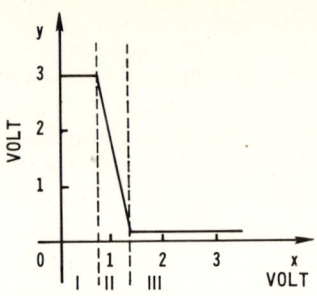

FIGURE 2(b). R-T circuit output V-I curve.

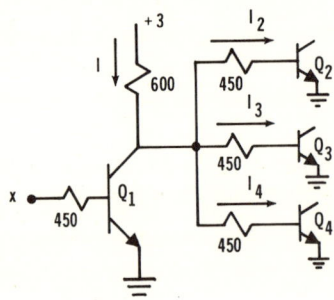

FIGURE 3(a). R-T circuit.

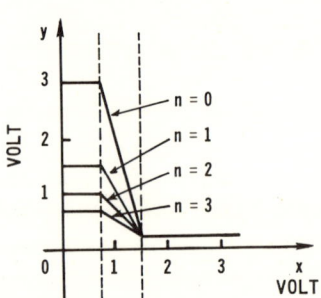

FIGURE 3(b). R-T circuit output V-I curve.

Fan Out

Figure 3(a) shows an RTL inverter driving another three identical inverters, and Figure 3(b) shows the *transfer curves* of the driver which is driving one, two, and three gates, respectively. Notice that for $x \geq 1.2$, Q_1 is saturated and $y = 0.2$ so that Q_2, Q_3, and Q_4 will be all OFF, the driver behaves the same as shown in Figure 2. However, as $x \leq 0.5$, Q_1 will be OFF and Q_2, Q_3, and Q_4 will be ON and y will no longer be 3 V, since $y = 3 - I \cdot 650 = 3 - (I_2 + I_3 + I_4) 650$ and

$$I = \frac{3 - V_\theta}{650 + \frac{450}{n}} = \frac{3 - 0.7}{650 + \frac{450}{n}}$$

where n = 1,2,3, or the number of the driven circuits. Thus, for $x \leq 0.5$, $y = 1.65$ V, for n = 1; y = 1.3 V, for n = 2; and y = 0.8 V, for n = 3. Notice that for n = 1, y > 1.2 which is in the logical 1 state, the inverter is still working properly. But for $n \geq 3$, $y \leq 1.2$ which is in the undefined state, the network will not work properly. We now define that the maximum number of gates a circuit can drive properly is called FAN-OUT of the circuit. Hence, this RTL inverter has a FAN-OUT = 1. This information is extremely important to the system designer, if the designer assigns more gates than the specified FAN-OUT of a driving circuit, the system will not function properly.

Noise Immunity

In Figure 2 let us assume that logical 1 $\triangleq$ 3 V and logical 0 $\triangleq$ 0.2 V. That is, when x = L → 0.2 V, y = H → 3 V or x = H → 3 V, y = L → 0.2 V. Consider x = 0.2, if there is a noise of +1 V which can either be a DC level drifting or an induced noise

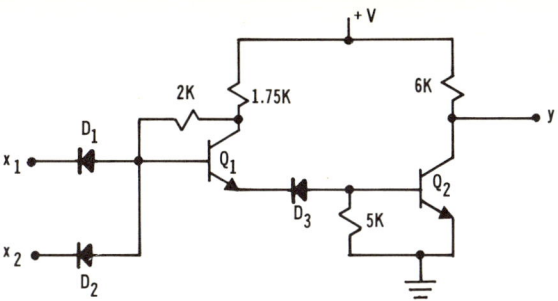

FIGURE 4. R-D-T logic circuit.

spike, at the input terminal, then x' signal + noise = 1.2 V which will result in y = 1.5 V according to the transfer curve. That is, the output is driven into the undefined region and the circuit will not function properly. However, the system will be all right if the noise is ⩽ 0.5 V. In this case, 0.5 is defined as the *noise immunity* of the circuit. A similar concept can be applied to the high state except that the negative DC drift or negative spike will drive the circuit into the undefined region. There is a family of logic circuits called Zener-Transistor Logic or High Threshold Logic (HTL) available which may have 5 to 10 V noise immunity. It is designed to be used for noisy environments.

Because of the simplicity of the RTL, the author has used it as a vehicle to define these terminologies. However, the definitions are true for all other types of logic circuits.

C. Diode-Transfer Logic (DTL)

The circuit shown in Figure 4 is a typical DTL circuit. By applying the BSA techniques described in Chapter 3, Section III one can easily verify Table 6 (positive logic is assumed),

Table 6

x_1	x_2	D_1	D_2	Q_1	D_3	Q_2	y
[0]	[0]	ON	ON	OFF	OFF	OFF	[1]
[1]	[0]	OFF	ON	OFF	OFF	OFF	[1]
[0]	[1]	ON	OFF	OFF	OFF	OFF	[1]
[1]	[1]	OFF	OFF	ON	ON	ON	[0]

where [0] ≜ logical 0, [1] ≜ logical 1. Thus, y = $x_1 \cdot x_2$, the circuit is a NAND-gate for positive logic. It is important to note that, as x_1 = H(high), D_1 is OFF, which practically draws no current. The reader may recall that in RTL, the driven circuit draws considerable current from the driving circuit causing a FAN-OUT problem, and here DTL seems to reduce the problems to none. Unfortunately, there is a problem when y = [0] and it is driving many other gates. Since when y = [0], all the gates connecting to it will be conducting and the conducting current of each diode of the driven circuits will be flowing into y terminal which becomes a current sink. This will cause the voltage at y to rise and eventually to exceed the noise immunity margin resulting in malfunction if the number of the driven circuits exceeds the FAN-OUT of the circuit.

At this point, one may ask what is the function of the 5K resistor shunting the base-emitter junction of Q_2. Actually the resistor is for the purposes of reducing the leakage current and speeding up the switching time of Q_2. For Binary-State-Analysis (BSA) this resistor can be neglected.

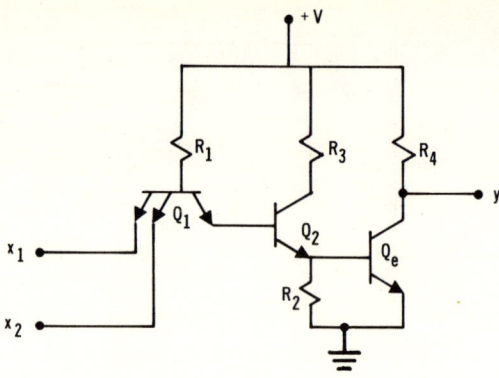

FIGURE 5. Transistor-transistor logic circuit.

D. Transistor-Transistor Logic (T²L or TTL)

The circuit shown in Figure 5 is a typical TTL circuit. The circuit is almost the same as shown in Figure 4(e) of Chapter 3, except that in this circuit, Q_1 has two emitters. Applying BSA technique, Q_1 can be viewed as three fictitious diodes D_1, D_2, and D_3. The anodes of all three diodes are connected together to resistor R_1, and their cathodes are connected to x_1, x_2 and the base of Q_2, respectively. One can then verify the states listed in Table 7

Table 7

x_1	x_2	D_1	D_2	D_3	Q_2	Q_3	y
[0]	[0]	ON	ON	OFF	OFF	OFF	[1]
[1]	[0]	OFF	ON	OFF	OFF	OFF	[1]
[0]	[1]	ON	OFF	OFF	OFF	OFF	[1]
[1]	[1]	OFF	OFF	ON	ON	ON	[0]

where, $y = \overline{x_1 \cdot x_2}$, a NAND-gate for positive logic. Since TTL are widely used, the following example is provided for the reader to get more familiar with it.

Example 1

Joe was a "Puritan" of a logic designer. He knew very little about circuits. One day, he used the circuit shown in Figure 6(a) to implement the switching function: $y = \overline{x_1 \cdot x_2} = \overline{x_1} + \overline{x_2}$ and positive logic was chosen. According to his calculation, when the photocell being exposed to light,

$$x_2 = \frac{\lambda}{100K + \lambda} \cdot 5 = \frac{7}{107} \cdot 5 = 0.33 \text{ V} \longrightarrow [0]$$

When it's dark,

$$x_2 = \frac{\lambda}{100K + \lambda} \cdot 5 = \frac{1M}{1.1M} \cdot 5 = 4.5 \text{ V} \longrightarrow [1]$$

However, he found that the circuit failed to realize the switching function, and had the responses shown in Table 8,

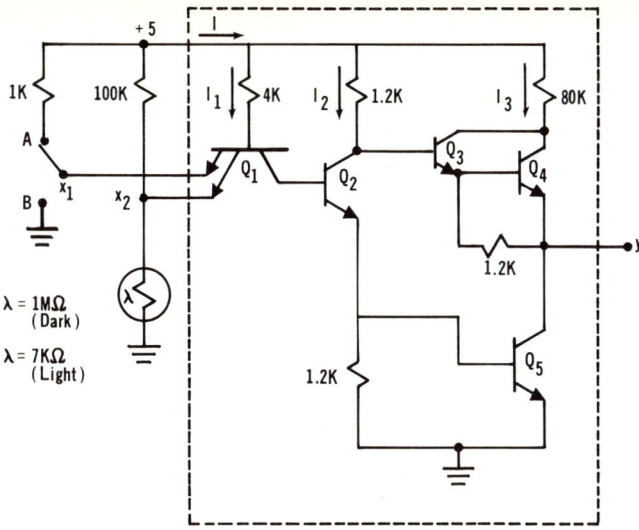

FIGURE 6(a). TTL circuit.

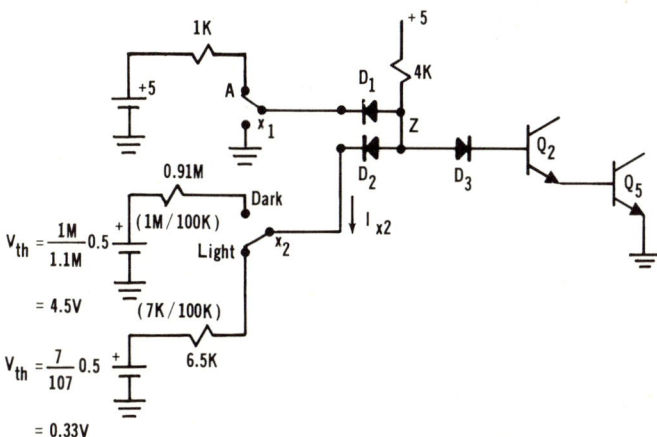

FIGURE 6(b). TTL circuit.

Table 8

x_1		x_2		y	Remark
State	Volt	State	Volt		
B	0	Light	0.33	[1]	Expected
B	0	Dark	4.5	[1]	Expected
A	5	Light	1.4	[0]	Unexpected
A	5	Dark	4.5	[0]	Expected

where, [1] $\underline{\Delta}$ logic one, [0] $\underline{\Delta}$ logic zero.

At first Joe thought that he had a defective gate, but after he had tried a few brand new ones, and even tried with different brands, he was convinced that something must be terribly wrong. Then he called his friend John for help, since John was a circuit expert although he did not know much about logic design. Having studied the circuit

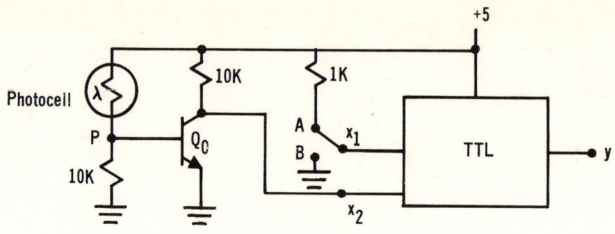

FIGURE 6(c).   TTL circuit.

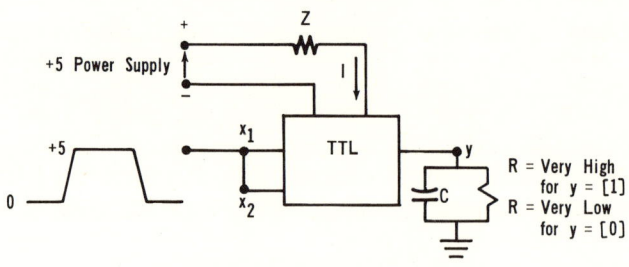

FIGURE 6(d).   TTL circuit.

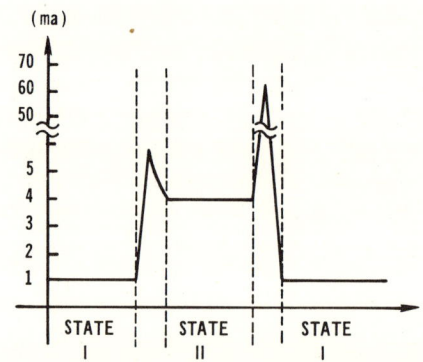

FIGURE 6(e).   TTL dynamic analysis.

for a while, John said, "Of course it does not work, the photocell has high output resistance!" John drew a circuit, shown in Figure 6(b), to explain Case 3 in the table, which was x_1 at A, and x_2 at the state of light. He used three fictitious diodes to replace Q_1 and he applied the Thévenin theory to get the equivalent circuit of the photocell as shown in the figure. Then he explained that if the circuit is functioning properly, D_2 should be ON and D_1, D_3, Q_2, and Q_5 should be OFF. If that was the case, then

$$Z = 5 - I_{x_2} \cdot 4K \tag{5}$$

with

$$I_{x_2} = \frac{5 - V_\theta - 0.33}{4K + 6.5K} = 0.38 \text{ mA}$$

Thus, $Z = 5 - 0.38 \text{ mA} \cdot 4K = 3.48$ V. However, the threshold voltage for D_3, Q_2, and Q_5 is only 3 $V_\theta \simeq 2.1$, and the potential of Z will keep D_1 OFF but turn ON D_3,

Q_2 and Q_5 which will result in $Z = 2.1$ V and still keep D_1 OFF. As $Z = 2.1$ V, D_2 will still be ON, but now $x_2 = 2.1 - V_\theta$ of $D_2 = 2.1 - 0.7 = 1.4$ V!! Thus, we have the condition as shown in the third line of the table.

Later, John suggested a solution, the circuit was shown in Figure 6(c). He expected the circuit would respond as shown in Table 9.

Table 9

| x_1 | Photo C | Q_0 | x_2 | Q_1 | | | Q_2 | Q_3 | Q_4 | Q_5 | y |
				D_1	D_2	D_3					
0	Light	ON	0.2	ON	OFF	OFF	OFF	ON	ON	OFF	[1]
0	Dark	OFF	5	ON	OFF	OFF	OFF	ON	ON	OFF	[1]
5	Light	ON	0.2	OFF	ON	OFF	OFF	ON	ON	OFF	[1]
5	Dark	OFF	5	OFF	OFF	ON	ON	OFF	OFF	ON	[0]

Joe took John's suggestion and found that John was right. Then Joe said to himself that he wished that simple circuit analysis techniques were taught in the logical design course.

Example 2

Joe designed a simple digital system with a few NAND-gates and MSI binary counters. After it was fabricated, he found that the counters did not function properly. John came to help him but they could not find the trouble. Later, Joe's high school classmate, Carl, came to visit him. Carl did not go to college but he had been an electronic technician for years. He took a careful look at the system and said, "My boss, who has been an engineer for many years, always makes me put a 0.1 μF capacitor across $V_{cc}(+5)$ pin, 14, and ground pin, 7. Why don't you try that?" Joe took Carl's advice, and the system worked. Then John became curious. He investigated the basic circuit of TTL, the inverter which is equivalent to connecting x_1 and x_2 of Figure 6(a) together as the input terminal; and he assumed that the inverter is driving a load of a capacitor in parallel with a resistor, which simulates the equivalent load of the TTL being driven. He knew that this is a good model to investigate how a circuit behaves in the system. He also knew that the voltage across a capacitor could not change instantaneously. As shown in Figure 6(d) he assumed an input signal changed from 0 V to 5 V and back to zero again. Since the circuit response for steady states, such as when the input is [0] or [1], is the same as shown in Row 1 or Row 4 of Table 9 in the last example, it is not repeated here. However, he analyzed the circuit during the transition period in the following manner. With reference to Figure 6(a), for input signal changing from low to high (Row 1 to Row 4 in Table 9), we have,

Q_1: $\begin{cases} \text{Base-Emitter junction changes from ON to OFF} \\ \text{Base-Collector junction changes from OFF to ON} \end{cases}$

Q_2, Q_5: Changes from OFF to ON
Q_3, Q_4: Changes from ON to OFF

and the loading capacitor C then discharges through Q_5. However, all transistors have to pass through their active region regardless of whether they are switching from ON to OFF or from OFF to ON. "Active Region" is defined as,

$$0 < \frac{I_C}{I_B} = \beta < \frac{(I_C)_{max}}{I_B}$$

$$\text{with } (I_C)_{max} = \frac{V_{CC}}{R_C}$$

where, I_C: collector current, I_B: base current, β: current gain, V_{CC}: voltage of the power supply, R_C: resistance of the collector resistor, and $(I_C)_{max}$: maximum possible collector current while the transistor is saturated. Therefore, there is a short period in which all transistors are in the active region. Some of them are switching from ON to OFF, others from OFF to ON, and they all draw current from the power supply. Application of BSA follows. Refer to Figure 6(a) but remember that the inputs x_1 and x_2, in this example, have been tied together. Consider steady state I, i.e., Input = [0]. Then I = $I_1 + I_2 + I_3$ with

$$I_1 = \frac{5 - 0.7}{4K} = 1.08 \text{ mA}$$

Since y = [1], R = very high, I_2 and I_3 can be neglected.

Now, consider steady state II, Input = [1],

$$I_1 = \frac{5 - 2.1}{4K} = 0.73 \text{ mA}$$

$$I_2 = \frac{5 - \text{saturation voltage of } Q_2 - V_\theta \text{ of } Q_5}{1.2K}$$

$$= \frac{5 - 0.2 - 0.7}{1.2K} = 3.4 \text{ mA}$$

$$I_3 \simeq 0$$

$$I = I_1 + I_2 + I_3 = 0.73 + 3.4 = 4.13 \text{ mA}$$

For the transition period (y = [0] to y = [1]), since loading capacitor C momentarily behaves as a short circuit.

$$I_3 \frac{5 - \text{saturation voltage of } Q_4}{80} = \frac{5 - 0.2}{80} = 60 \text{ mA}$$

Figure 6(e) shows the approximate current waveform for the power supply. Note that there is a "spike current" during each transition. During the first transition period, all transistors are conducting but the capacitor is discharging and demanding no current from the supply, and the maximum possible current will be the sum of current required for both steady states, i.e., 1.08 + 4.13 = 5.21 mA. During the second transition period, the circuit has to charge the capacitor to high voltage. As calculated, it requires 60 mA charging current from the power supply, plus the transition current for all transistors. Hence, a much larger current spike would be generated.

John suddenly realized that if the output impedance of the power supply and wires, $Z \simeq 10 \, \Omega$ (Figure 6(d)), there would be a 0.6 V voltage drop which would cause a problem. A capacitor of 0.1 μF connected directly across Pins 14 and 7 will, of course, filter out the spiking noise.

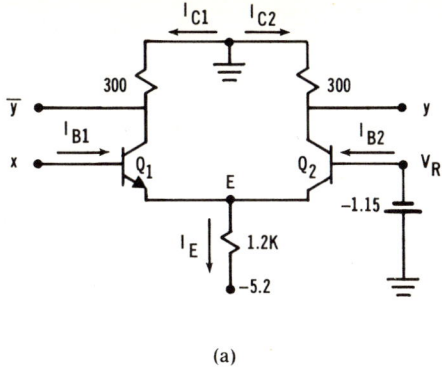

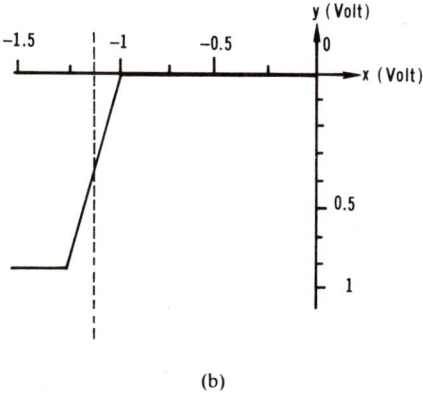

FIGURE 7. Analysis of current mode logic circuit.

E. Current-Mode Logic (Emitter-Coupled-Logic — ECL) Nonsaturation Type Logic
Introduction

All of the logic circuits discussed so far are saturation types which have definitely two distinct states and normally are not sensitive to the parameters variation of switching devices such as current gain, etc. However, because of the excess charges in the diode junctions of the devices, the saturation-type logic gates are inherently slow. For extremely high speed application, the saturation logic gates will not be working satisfactorily. In this section, the ECL circuit which is a nonsaturation-type logic will be analyzed. First, an emitter-coupled-pair circuit, the heart of the ECL, should be studied. In Figure 7(a), Q_1 and Q_2 operate like a balance. That is, while in an equilibrium state, both sides will be at the same height from the ground. If one side, say the left side, is heavier, then the left side will be low and the right side will be high in the air. Analogously, if x is more positive then V_R, then Q_1 will pull more weight than Q_2 and y will be relatively low and y will be high in voltage. In other words, the diode junctions of the base-emitter of Q_1 and Q_2 are either both conducting or just one is conducting, depending on the voltage difference between x and V_R. For example, if x = -1.5, V_R = -1.15, then Q_2 is conducting and E = $-1.15 - 0.7 = -1.85$. Since the difference between x and E is less than the threshold voltage, V_θ, Q_1 is OFF. In this case

$$I_E = \frac{E - (-5.2)}{1.2K} = \frac{3.35}{1.2K} = 2.8 \text{ mA}$$

which is approximately equal to I_{c2}. Since Q_1 is OFF, $I_{c1} = 0$. Thus $y = 0 - I_{c2}300 = -0.84$ V. Now let $x = -0.7$, Q_1 will be conducting and $E = -0.7 - 0.7 = -1.4$. The voltage difference between E and V_R is less than V_0. Hence, Q_2 is OFF. In this case, $y = 0$ V. Figure 6(b) shows the transfer curve of x vs. y. Notice that the maximum possible collector current occurs when Q_2 or Q_1 is saturated, i.e.,

$$(I_{C_2})_{max} = \frac{5.2 - \text{saturation voltage of } Q_2}{1.2K + 300} = \frac{5}{1.5K} = 3.3 \text{ mA}$$

However, as Q_1 is OFF and Q_2 is ON, the $I_{c2} = 2.8$ mA as shown in the last page. Thus $I_{c2} < (I_{c2})_{max}$, the transistor Q_2 can never be in the saturation state. Similarly, one can show that Q_1 can never be saturated if $x < -0.7$ V. Therefore, the circuit is always working either in OFF or in the active region which assures high speed switching.

Emitter-Coupled Logic (ECL)

Figure 8(a) shows a typical ECL two inputs NOR/OR gate circuit. Q_2 and Q_3 form the heart of the circuit. Diodes D_1, D_2 and the 4.89K provide the temperature compensation voltage reference, V_R, Q_4, and Q_5 are emitter-followers which are always active and serve as output buffers. It is interesting to point out that this device has two outputs, i.e., NOR and OR. Therefore, the circuit shown can be used as either an OR gate or an NOR gate, or both. Figure 8(b) depicts the input/output transfer curves. Accordingly, for positive logic, we have logic 1 = -0.9 V, logic 0 = -1.8 V. Since the device is operated in active region, it is by far the fastest logic available to the designer. Note that the voltage swing, however, for this logic is only from 0.9 to 1.8 V. As a result, this logic has poor noise immunity in comparison with TTL. The following table shows the steady states for each transistor with different inputs.

X_1	X_2	Q_0	Q_1	Q_2	Q_3	OR	NOR
-0.9	-0.9	I_r	I_H	I_H	I_L	V_H	V_L
-1.8	-0.9	I_r	OFF	I_H	I_L	V_H	V_L
-0.9	-1.8	I_r	I_H	OFF	I_L	V_H	V_L
-1.8	-1.8	I_r	OFF	OFF	I_H	V_L	V_H

Where: $I_r \triangleq$ reference current, $I_H \triangleq$ high current level, $I_L \triangleq$ low current level, and $V_H \triangleq$ high voltage level, $V_L \triangleq$ low voltage level.

F. Schottky-TTL-Nonsaturation-Type Logic

Schottky Transistor-Transistor Logic is an improved TTL circuitry. By means of Schottky-diode-clamp technique, it prevents the TTL being driven into saturation region. As shown in Figure 9(a), a Schottky diode which has short recovery time and very low forward voltage drop, 0.1 ~ 0.2 V, is connected between the base and collector terminals. Figure 9(b) shows the graphical analysis of the basic circuit. Consider the circuit without the clamping diode. Assume that the transistor is overdriven and thus operates as a saturation logic. That is, it would be in either saturated state or OFF state. In saturation state, $V_{BE} \simeq 0.7$ V, and $V_{CE} \simeq 0.2$ V, and $V_{BC} = 0.7 - 0.2 = 0.5$ V. Now, as the Schottky diode is connected as shown, it would clamp the collector voltage to the base voltage. As a result, the voltage at collector would be 0.1 to 0.2 V lower than the base voltage due to the voltage drop of the Schottky diode. Thus the voltage between collector and emitter would be 0.6 to 0.5 V as shown in the output V-I curve of the transistor, which will keep the device away from its saturation region.

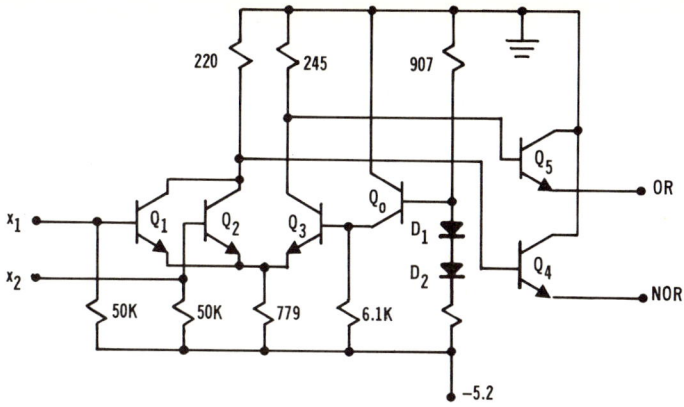

FIGURE 8(a). A typical ECL circuit.

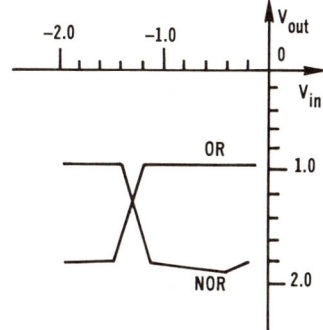

FIGURE 8(b). Input/output transfer curves.

In contrast with TTL, the "S" shape "base-bar" is used to denote the Schottky TTL. There are two classes of Schottky TTL families, i.e., regular and low power. Figure 9(c) and (d), respectively, show the typical circuit for regular and low power Schottky TTL. The former typically consumes 20 mW power with a propagation delay time about 3 ∼ 5 nsec and the latter, 2 mW with operation speed about 5 nsec.

G. CMOS (Complementary Symmetric MOSFET)

In Chapter 3, Section III.C, the basic characteristics of FET have been described. We shall now present the analysis of CMOS logic elements. There are two basic elements in the CMOS logic family, namely, the inverter and transmission gate. Their analyses follow.

Inverter

Figure 10(a) shows a CMOS inverter circuit which comprises two MOSFETS, i.e., P-channel (Q_1) and N-channel (Q_2), and both of them are enhancement types. For clarification, the electric characteristics for each type are shown, respectively, in Figures 10(b) and (c). In view of Figure 10(a), it reveals that unlike the conventional circuit, the inverter contains no resistor. In other words, the load resistor of the N-channel MOSFET (the lower one) is now being replaced by the P-channel MOSFET. Since they are all nonlinear devices, the load line is no longer a linear line. We shall

86 Handbook of Digital System Design for Scientists and Engineers

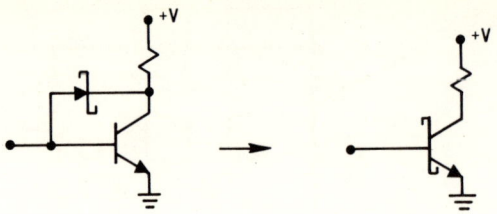

FIGURE 9(a). Schottky transistor circuit and symbol.

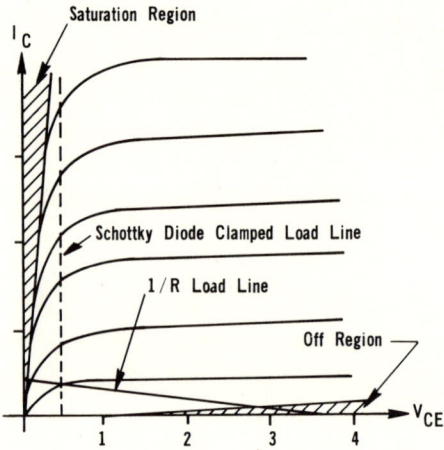

FIGURE 9(b). Schottky diode clamped operation.

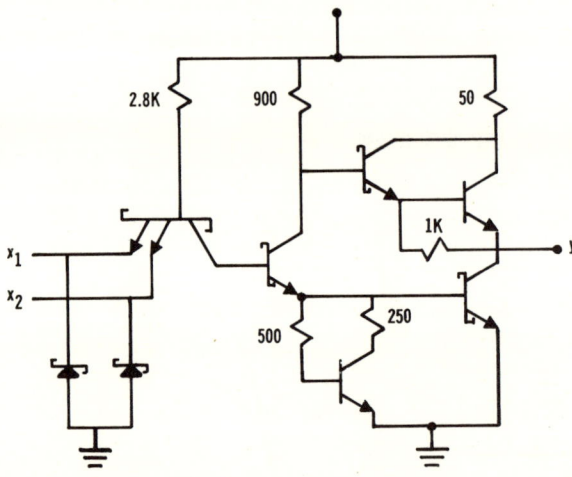

FIGURE 9(c). Regular Schottky TTL.

now apply first, the BSA (Binary-State-Analysis) technique and next the graphical analysis method. Consider the binary input levels of the inverter to be 0 and +5 V, and it is driving another CMOS logic circuit. Based on the electrical characteristics shown in Figures 10(b) and (c), and recall that the nonconducting static resistance of CMOS is extremely high, one can easily verify the response of the circuit shown below:

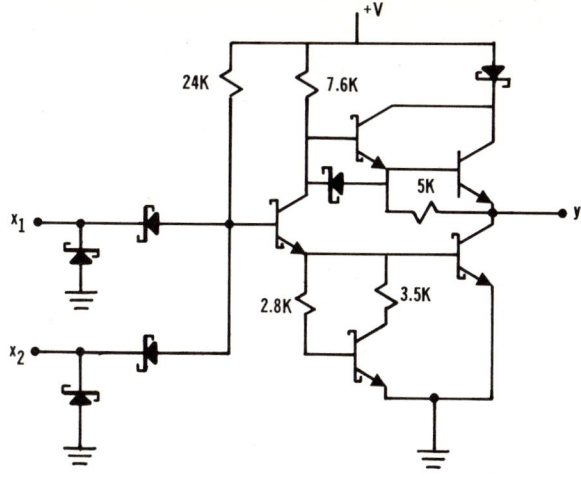

FIGURE 9(d).   Low power Schottky TTL.

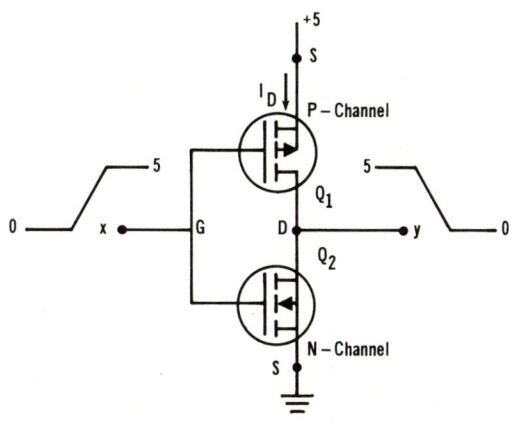

FIGURE 10(a).   CMOS inverter.

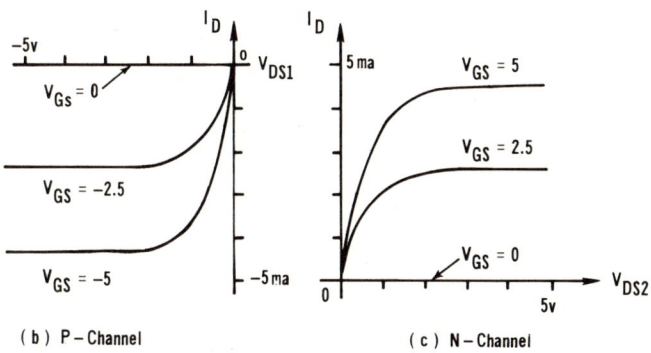

FIGURE 10(b,c).   CMOS V-I curves.

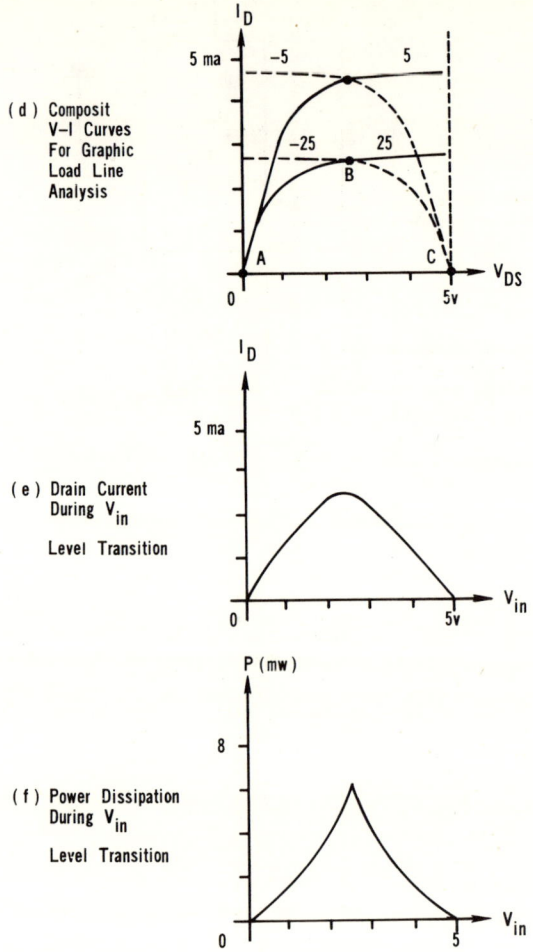

FIGURE 10(d,e,f). CMOS inverter graphical analysis.

x	V_{GS1}	Q_1	V_{GS2}	Q_2	y	I_D	Power dissipation
0	−5	ON	0	OFF	5	0	0
5	0	OFF	5	ON	0	0	0

It is interesting to point out that the circuit appears to be dissipating zero power. For static analysis, this conclusion is very true; for dynamic operation, however, it is an entirely different story. By means of graphical analysis, the dynamic operation of the circuit can clearly be shown. Figure 10(d) shows the graphical technique. Here, the V-I curve of $I_D - V_{DS}$ of P-channel (Q_1) in dot-lines is superimposed on the $I_D - V_{DS}$ curves of N-channel (Q_2), so that the load line or the operating point locus for variable input voltage, can be graphically determined. Typically, Points A, B, and C are operating points corresponding to V_{in} equals 5, 2.5, and 0, V, respectively. That is, at point A, we have $V_{in} = 5$ V, Q_2 ON and Q_1 OFF. Point A is the intersection of the V-I curve of Q_2 with $V_{GS2} = 5$, and the V-I curve of Q_1 with $V_{GS1} = 0$. Accordingly, the I_D flows and the power dissipation of the circuit occurs during the input voltage transitions between logical zero and logical one, as shown in Figures 10(e) and (f). It is now apparent that the CMOS logic circuit consumes visually no power during steady

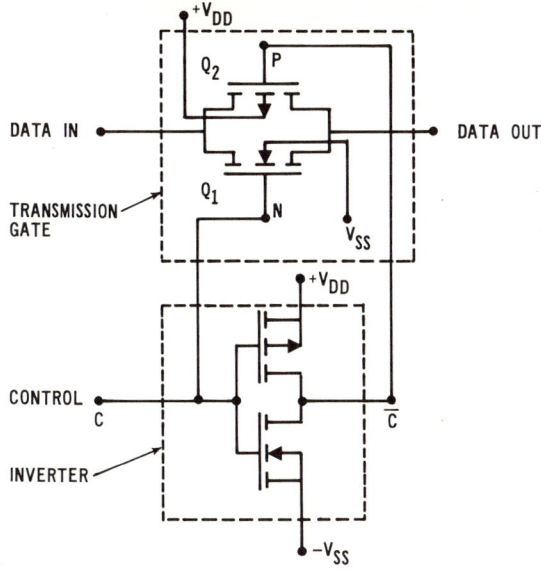

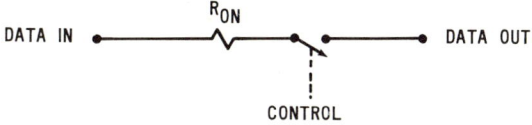

IN	C	C̄	Q_1	Q_2	OUTPUT
1	0	1	OFF	OFF	DISCONNECTED FROM INPUT
0	0	1	OFF	OFF	DISCONNECTED FROM INPUT
1	1	0	OFF	ON	1
0	1	0	ON	OFF	0

FIGURE 10(g). CMOS transmission gate.

state; it does, however, consume power during the transition state. Therefore, we may conclude that CMOS logic element consumes extremely low stand-by power and its power dissipation is proportional to the switching frequency of the signals as well as other factors, such as power supply voltage, etc.

Transmission Gate

Transmission gate is a unique basic element in CMOS logic family. It is a bilateral switch which can be used for both digital and analog switch when an inverter is used as a control device in conjunction with it. Figure 10(g) shows the circuit diagram, equivalent circuit and a binary-state-analysis table. An important limitation that a designer should bear in mind is, the voltage level of the data or signal to be transmitted should never exceed the range of V_{SS} and V_{DD}, i.e., V_{SS} < signal voltage swing < V_{DD}.

H. Integrated Injection Logic (I²L)

Integrated injection logic (I²L) is another kind of bipolar logic. In comparison with TTL, it consumes less electric power and has higher packing density. Figures 11(a) and (b) show a single I²L element and its equivalent. Figure 11(c) shows a typical NOR gate and a table which shows its BSA result. It is important to point out that the logical

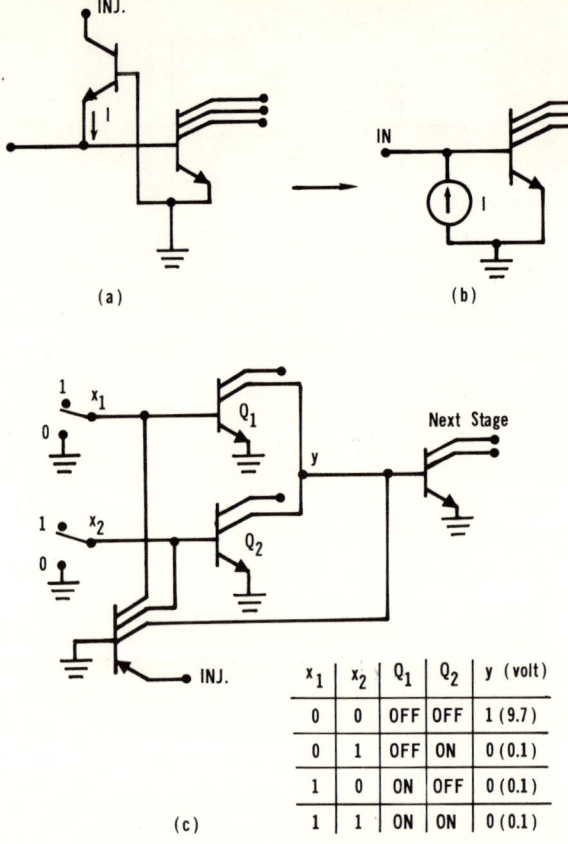

FIGURE 11. Integrated injection logic.

level or voltage swing of I²L is within the range of 0.1 to 0.7, which is considerably lower than that of TTL. Its operation speed is, however, slower than TTL but faster than CMOS.

I. Comparison of the Different Logic Families

The following table shows the electric properties of the logic families described in the preceding sections. It would offer the designer a guideline in selecting logic elements for his (her) applications.

	Standard TTL	74LS low power Schottky	CMOS	ECL	I²L
Propagation delay	10 nsec	9 nsec	25 nsec	1 nsec	5 nsec
Quiescent power per gate	10 mW	2 mW	10 nW	30 mW	1 mW
Noise immunity	1 v	0.8 v	2 v	0.4 v	N/A[a]
Fan out	10	20	50	>50	N/A[a]

[a] N/A = Data not available

J. Interfacing Logic Elements of Different Families

From preceding sections we have introduced logic elements of different families. We have found that each family has its own uniqueness; therefore, in some applications, the designer will have to deal with the problems of interfacing logic elements of different families. Although there are now devices available specifically designed for these applications, an understanding of the problems and their solutions would be essential to the designers.

Interfacing Between CMOS and TTL

There are two problems for interfacing between CMOS and TTL logic elements: First, for TTL, the binary logic levels are: logic $1 \geq 2.4$ V, logic $0 \leq 0.5$ V, while for CMOS, logic $1 \geq 70\%$ of ΔV, logic $0 \leq 30\%$ of ΔV. Where ΔV is the voltage difference between V_{DD} and V_{SS}. For instance, if $V_{DD} = 5$, $V_{SS} = 0$, then logic $1 \geq 3.5$ V and logic $0 \leq 1.5$ V. Therefore, the incompatibility of logic levels exists between the two logic families. Secondly, the CMOS is required to sink sufficient current from TTL in logic-0 state and still maintains at a voltage level less than or equal to 0.5 V to assure the logic-0 level for TTL. Figures 12(a) and (b) show the level conversion circuits for interfacing TTL to CMOS and CMOS to TTL, respectively. In Figure 12(a), the open collector TTL will provide V_{DD} volt in HIGH state and 0.2 V in LOW state, while Figure 12(b) shows that the sink current should be sufficiently low to assure LOW state input to TTL. As a result, it is recommended that one should either use high current CMOS buffer or low power TTL which has sufficiently high input impedance at LOW state input. As shown in Figure 12(b), from the graphical analysis, it is important that the resistor R shown for TTL is sufficiently large enough to assure proper value of I_{sink} so that V_{LOW} of the CMOS output is less than 0.5 V.

Interfacing Between CMOS and ECL

The problem for CMOS and ECL interface again is the incompatibility of logic levels. Figures 13(a) and (b) show, respectively, the level conversion circuit for ECL to CMOS and CMOS to ECL. Referring to Figure 13(a), when the voltage at the emitter Q_1 is higher than that of Q_2; Q_3 and Q_4 conduct, the output of Q_4 will be LOW. However, as the voltage at the emitter of Q_2 is higher than that of Q_1, Q_3 and Q_4 will be in OFF state and the output of Q_4 will be HIGH. By properly selecting the values of R_1, R_2, R_3, and R_4, the output of Q_4 will swing between zero and V_{cc}. Now let us consider Figure 13(b), since the output of CMOS swings between ground and V_{DD}, R_1 and R_2 can be selected so that the base voltage of Q_1 will swing between logic high, say -0.9 V and logic low, say some voltage lower than -1.8 V; recall that the ECL normally swings between -0.9 V and -1.8 V. Nevertheless, it is important that the designer is aware of the interfacing problems. One could either design the necessary interface networks or use some devices such as RCA 10125 and 10124 or others available in the market for logic level conversion.

K. Some Special and Useful Logic Elements

Open-Collector Gate

Open-collector logic element is one of the most popular and useful devices in logic or digital system design. Figure 14(a) shows the circuit diagram, and its BSA is shown below:

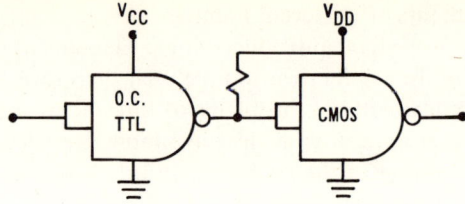

FIGURE 12(a). TTL to CMOS interface.

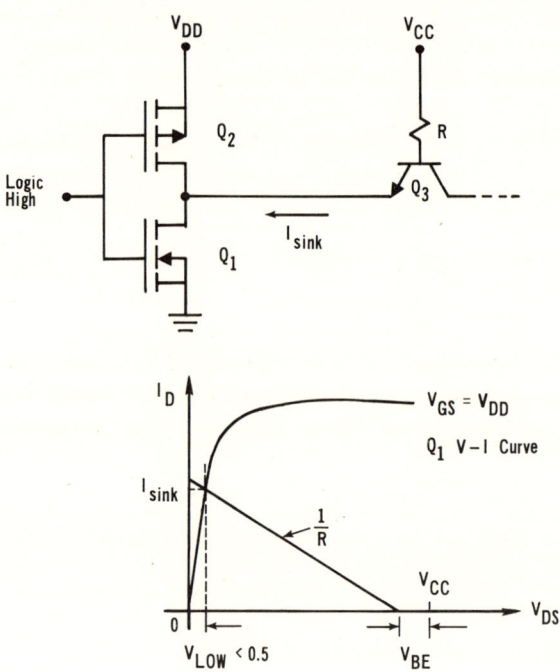

FIGURE 12(b). CMOS to TTL.

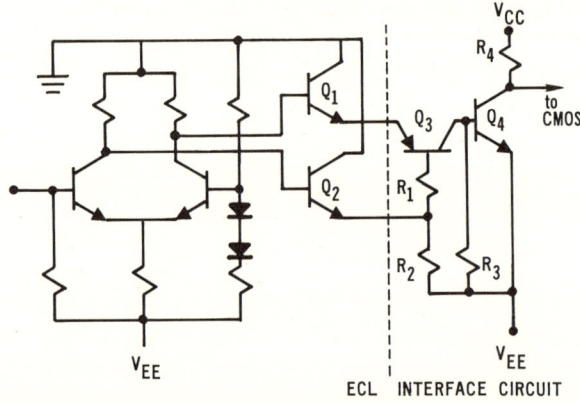

FIGURE 13(a). ECL to CMOS.

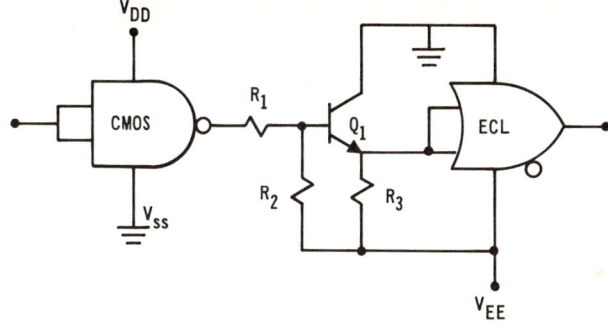

FIGURE 13(b). CMOS to ECL.

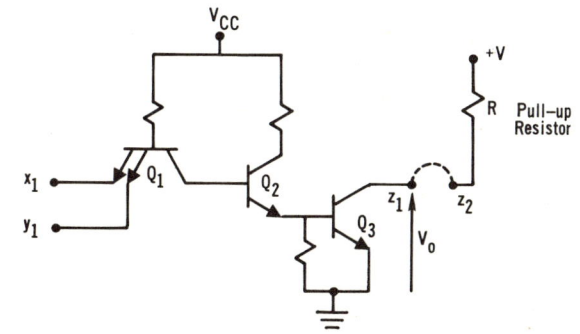

FIGURE 14(a). Open collector logic element.

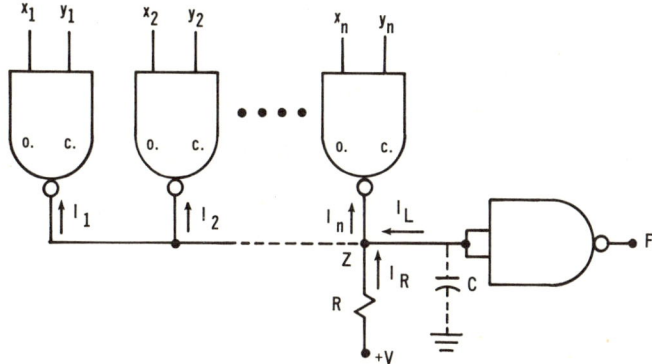

FIGURE 14(b). Switching function realization with open collectors.

				V_o	
x_1	y_1	Q_2	Q_3	z_1, z_2 opened	z_1, z_2 closed
0	0	OFF	OFF	0V	+V
0	1	OFF	OFF	0V	+V
1	0	OFF	OFF	0V	+V
1	1	ON	ON	0V	0.2 V

Note that the circuit will not function properly without the external resistor R, which is usually called "pull-up" resistor. Based on this feature, the outputs of a number of open-collector gates can be directly tied together as shown in Figure 14(b) to realize the switching function, $F = x_1y_1 + x_2y_2 + \ldots + x_ny_n$. Otherwise, an n-Fan-In OR-gate would be required. The value of the pull-up resistor, R, however, cannot be an arbitrary one. In order to assure that the voltage at z satisfies the logic levels of 0.5 V and 2.4 V, the value of the resistor cannot be out of the range of R_{max} and R_{min}. Calculation follows.

Let us define,
R_{max} = Maximum allowable value of R
R_{min} = Minimum allowable value of R
I_{OFF} = Reverse of leakage current of each gate
I_{SAT} = Saturated circuit current of each gate
I_L = Sink current from the output gate

Then, to assure $V_z \geq 2.4$, we have,

$$R_{max} = \frac{V - 2.4}{I_R} = \frac{V - 2.4}{(n + 1)I_{OFF}}$$

Similarly, to assure $V_z \leq 0.5$, $V - I_R R_{min} = 0.5$ V. Here, the worst case is that when the output of one and only one of the open-collector gate is LOW. That is, $I_R + I_L = I_{SAT} + (n - 1)I_{OFF}$, or $I_R = I_{SAT} + (n - 1)I_{OFF} - I_L$. Thus,

$$V - [I_{SAT} + (n - 1)I_{OFF} - I_L]R_{min} = 0.5 \text{ V}$$

$$R_{min} = \frac{V - 0.5}{I_{SAT} - (n - 1)I_{OFF} - I_L}$$

It is important to point out that, by increasing the supply voltage V, the voltage swing of V_z can be increased accordingly. A few words on the rise time of V_z. If C in Figure 14(b) is the input capacitance of the output gate, then the time constant, RC will determine the rise-time of V_z. For logic circuit where the speed is critical, the value of R should then be as low as possible. For power dissipation-conscious circuit, however, the value of R should be as high as possible.

Three-State or Tri-State Gate

Three-state or tri-state gate is another useful and popular gate. Figure 15(a) shows a typical circuit diagram of a tri-state gate. Its BSA is shown below:

$\overline{E}$	x	Q_3	Q_4	Q_5	D	Q_6	Q_7	Q_8	Out
1	0	ON	ON	OFF	ON	OFF	OFF	OFF	High impedance
1	1	ON	ON	OFF	ON	OFF	OFF	OFF	High impedance
0	0	OFF	OFF	OFF	OFF	ON	ON	OFF	1
0	1	OFF	OFF	ON	OFF	OFF	OFF	ON	0

Note that the output has three states, namely, High Impedance, logic 0, and logic 1. As the NOT ENABLE is low, the gate functions as a convertional inverter. When the NOT ENABLE is at logic 1 state, the output enters high impedance state, i.e., the gate is basically disconnected from its output. A typical application is shown in Figure 15(b). By using the tri-state gate, m devices of n-bit data can be connected to the same data bus of n lines. By setting the $\overline{E}_i$ of the ith device low and all others high, the ith

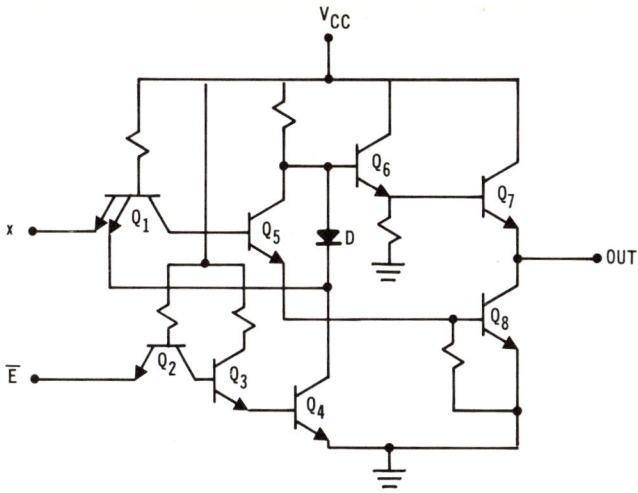

FIGURE 15(a). Circuit diagram of a tri-state gate.

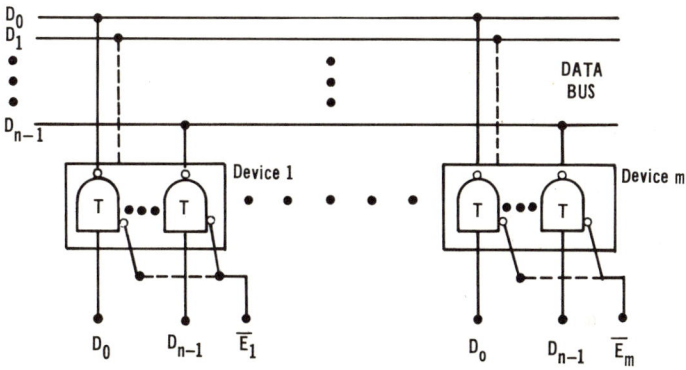

FIGURE 15(b). A typical application of tri-state gate.

device will be connected to the bus, and the n lines of data from i^{th} device will enter into the bus; other devices will be floating from the bus. This is the most desirable feature for multiplexing many devices to one bus line.

L. Other Important Properties of the Logic Circuit
Time Response of the Logic Circuit

All descriptions of logic circuits so far are limited to a steady state response. The time characteristic has not yet been explored. We shall now get acquainted with some important terminologies and leave the more challenging timing problems to the later chapter where the sequential logic design is being described.

Figure 16(a) shows a simple RTL inverter. Figure 16(b) shows the typical time response of the inverter. However, the terminologies introduced here are valid for other logic circuits. In Figure 16(b), we define:

t_d (time delay) $\underline{\Delta}$ the time between the input ON-SET time and time when the output drops to 10%

t_f (fall time) $\underline{\Delta}$ time required for output drops from 90% to 10%

t_{on} (turn-on time) $\underline{\Delta}$ $t_d + t_f$

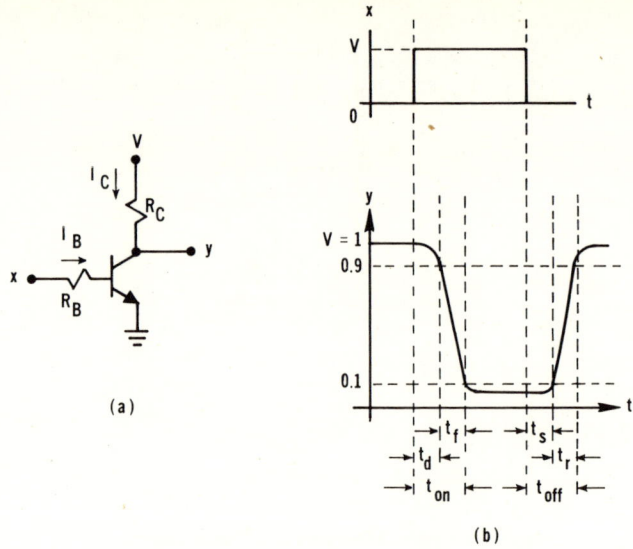

FIGURE 16. Time response characteristics.

t_s (storage time) $\underline{\Delta}$ time between when the input turns off and the time when the output rises to 10%

t_r (rise time) $\underline{\Delta}$ time required for output rises from 10% to 90%

t_{off} (off-time) $\underline{\Delta}$ $t_s + t_r$

Propagation time $\underline{\Delta}$ the time required for a unit transmitting binary information from input to output, ½ ($t_{on} + t_{off}$)

Power Consumption

Power consumption of a logic circuit is the electrical power being converted into heat within the unit. It is important for a system designer to predict the total heat that may be generated within the system. To understand the problem, consider the circuit shown in Figure 16(a). When x = [1], the transistor is turned ON, and the total power consumption,

$$(P_t)_{on} = \frac{V^2}{R_B} + \frac{V^2}{R_C} + \left(\frac{V}{R_C}\right) V_{SAT}$$

where, V_{sat} $\underline{\Delta}$ the saturation voltage. When x = [0], the transistor is turned OFF, and the power consumption in this case, $(P_t)_{off} \simeq 0$. However, in a circuit such as Figure 15(a), $(P_t)_{off}$ will not be zero. This is because in either state, there are some transistors in the ON state and some in the OFF state.

M. Conclusion

The BSA techniques have been introduced and their applications for understanding the principle of operation in different logic circuits were described in the preceding sections. The main theme of this chapter is to contrast the modeling of linear and nonlinear circuit elements; and we only need to know about the steady state and transition state responses of logic circuits. The BSA techniques introduced are general, which can be applied to other types of logic families. Finally, the table shown below provides a qualitative comparison of the major characteristics of the logic circuits. For detailed information, one can refer to the manufacturers' bulletins.

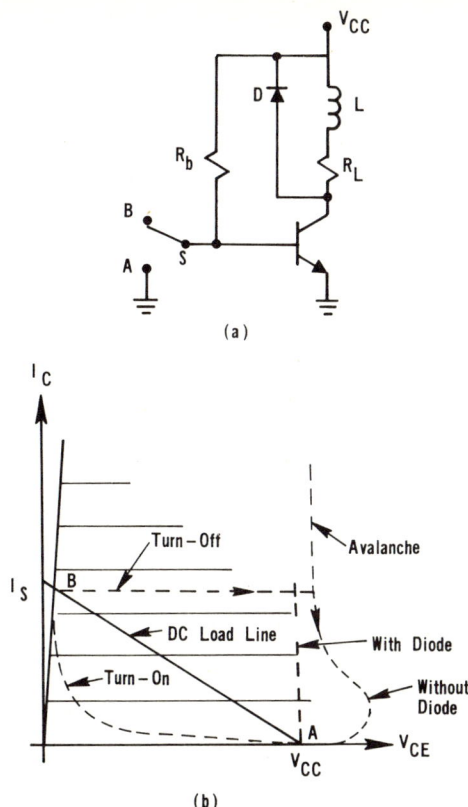

FIGURE 17. Switching an inductive load.

Properties	RTL	DTL	TTL	ECL	CMOS
Speed	Low	Low	High	Ultra-high	Medium
Noise	Poor	Good	Good	Fair	Good
Fan-out	Poor	Fair	Fair	Good	Excellent
Consumption	Medium	Low	Low	High	Very low

III. SWITCHING INDUCTIVE LOAD BY A TRANSISTOR

In digital system design, a designer often encounters a driver or controller for an inductive load. An understanding of the basic principle of switching inductive load is therefore important. Consider that a transistor is used to drive an inductive load. In referring to the transient characteristics described in Chapter 3, Section II, we know that the current flowing in an inductor cannot change instantaneously. Therefore, the circuit shown in Figure 17(a) requires special consideration. Assume the switch is at position B for a long, long time, and the diode D is not being used. The circuit is then at its steady state or the transistor is in its saturation state. As a result, a steady current I_s is flowing from V_{cc} through the inductor and transistor to ground. When the switch S is turned to position A at t = 0, since the current in an inductor cannot change instantaneously, the operational path on the V-I curves of the transistor will no longer follow the DC load line. Instead, it follows the dotted line horizontally at time t = t⁺ and eventually along the avalanche line of the transistor reaches point A as shown in

Figure 17(b) with a time constant L/R_L. By the same reason, when the switch is suddenly turned to position B again, the turn-on path shown in Figure 17(b) is followed. Although the turn-on path would not cause trouble, the turn-off path which follows the avalanche path is evidently not desirable. Therefore a switching diode should be connected across the inductor as shown in Figure 17(a). The diode is mainly used to pass the current I_s as the transistor is turned off instantly. As a result, the operational path follows the perpendicular dotted line instead of the avalanche line shown in Figure 17(b). In conclusion, it is important to point out that a diode with proper current capacity and reverse voltage rating should always be used shown for any inductive load, such as relay, motor, etc., in any digital system design.

REFERENCES

1. **Altman, L.**, Logic's leap ahead creates new design tools for old and new applications, *Electronics*, Feb. 21, 81, 1974.
2. **Altman, L.**, C-Mos enlarge its territory, *Electronics*, May 15, 1975.
3. **Carr, W. N. and Mize, J. P.**, *MOS/LSI Design and Application*, McGraw-Hill, New York, 1972.
4. **Hart, C. M., Slob, A., and Wulms, H. E. S.**, Bipolar LSI takes a new direction with integrated injection logic, *Electronics*, Oct. 3, 1974.
5. **Horton, R. L., Englade, J., and McGee, G.**, I²L takes bipolar integration a significant step forward, *Electronics*, Feb. 6, 83, 1975.
6. **Pederson, R. D.**, Integrated injection logic: a bipolar LSI technique, *IEEE Trans. Comput.*, Feb. 24, 1976.
7. Special Report on I²L Applications, *IEEE Spectrum*, 28, 1977.
8. **Capece, R. P.**, Faster, lower-power TTL looks for work, *Electronics*, Feb. 1, 88, 1979.
9. *ECL Handbook*, Fairchild Semiconductor, Mountain View, Calif., July 1974.
10. *ECL System Design Handbook*, Motorola Semiconductor, Phoenix, Ariz., 1971.

Chapter 5

BASIC FUNCTIONAL MODULES

In addition to logic gates, a digital system would use many other basic functional modules built on gates or energy storage devices or resistor/capacitor timing elements. They are known as bistable multivibrator or flip-flop, astable multivibrator or free running or clock. Among them the flip-flop is used most frequently, and it is thus presented first. A flip-flop is simply a digital memory device which can memorize one and only one thing at a time. Hence, in a flip-flop, one can store either "yes" or "no", "HIGH" or "LOW", "ONE" or "ZERO". In the logic designer's terms it can memorize or store only one bit (1b) at a time. Actually in our daily life, we do use flip-flops quite often. For instance, the wall switch in our home is one type of flip-flop device. When a switch is turned on, it means that an ON information is stored and the switch memorizes it and keeps the light burning; when the switch is turned off, the negated information is then stored. In fact, the switch called Rockette type on the market is actually a set-reset or R-S flip-flop, since if one presses one end of the switch arm to turn it ON, then he can turn it OFF by pressing the other end. Since most beginners have trouble understanding how a flip-flop works, we shall start with a basic circuit and translate it from circuit designer's language into logic designer's, and then progress to more complicated ones.

I. FLIP-FLOP

A. Basic Flip-Flop Circuit

The circuit shown in Figure 1 is a simple two-stage RTL circuit connected in a special way which yields a very interesting result. The circuit has two inputs, S and R, and two outputs, y_1 and y_2. They are related in a nontrivial way. A flip-flop to a logic designer is like an automobile to a driver, which can be either a useful vehicle or a number one killer depending on how well the driver handles the machine. The reader is urged to follow through this section carefully and be sure that he or she understands everything described in the following sections.

Let us analyze the circuit by the BSA technique. There are two kinds of input signals: (1) DC level inputs, such as 0 or +5 V and (2) positive going pulse inputs, such as a pulse from 0 to 5 and then to zero again. We shall analyze both cases.

DC Level Inputs

Table 1 shows the circuit responses for DC level inputs.

Table 1

S	R	Q_1	y_1	Q_2	y_2
+5	+5	ON	0.2	ON	0.2
0	+5	ON	0.2	OFF	2.67
+5	0	OFF	2.67	ON	0.2
0	0	?	?	?	?

The reader should not have trouble verifying the entries of the first three rows. However, for the fourth row, one should think twice before making any decision.

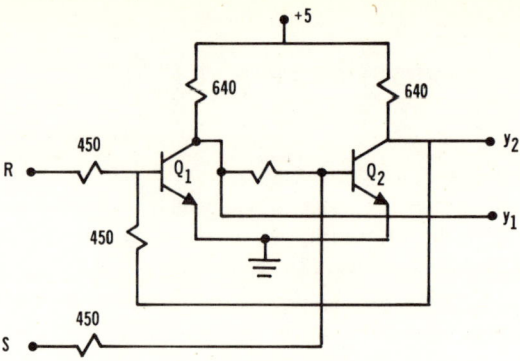

FIGURE 1. Basic flip-flop circuit.

Notice that Q_1 is ON or not only depending on the voltage level at R but also the voltage level at y_2. That is, Q_1 is ON if R or y_2 or both are high. Similarly Q_2 is ON depending on y_2, and S. If y_2 is high, which implies Q_2 is OFF, then Q_1 will be ON. If y_1 is high, which implies Q_1 is OFF, then Q_2 will be ON. In other words, if R = S = 0, then Q_1 can be ON if Q_2 is OFF and vice versa. That is to say if R = S = 0, then there are only two possible cases: either Q_1 = ON, Q_2 = OFF or Q_1 = OFF, Q_2 = ON; but they could not be both ON or both OFF at the same time. One might ask then, which one of these two possible cases it will be if R = S = 0. The answer involves the past history of Q_1 and Q_2. In other words, Q_1 and Q_2 will be the same as what they were just before the R and S both became zero. Analogously, one flipped a coin on the table and walked away. If nobody has touched it since then, what will the state of the coin be, a head or a tail? The answer is obvious that there will be "No Change". If it was a head, it will be a head! One may ask what will happen if Q_1 and Q_2 were both ON or both OFF? In this circuit configuration, it is impossible to have the Both OFF condition, but the Both ON condition occurs if R = S = 5. However, if R = S = 0 follows R = S = 5, the response can be either y_1 = H, y_2 = L, or y_1 = L; y_2 = H depends on which one of the two inputs reaches low first. In order to design a reliable system, this condition should not be allowed. Now we can modify Table 1 to become Table 2.

Table 2

S	R	y_1	y_2
H	H	Not allowed	Not allowed
L	H	L	H
H	L	H	L
L	L	H	L
		L	H

Notice that y_2 is a negative of y_1 if the first row is not allowed, then Table 2 can be simplified further if we define y_1 = Q. Then we have Table 3.

Table 3

S	R	Q	$\overline{Q}$
H	H	Not allowed	Not allowed
L	H	L	H
H	L	H	L
L	L	No change	No change

Let us define Q_n as the present state of the y_1 and Q_{n+1} as the new or the next state of y_1, then we can rewrite Table 3 to become Table 4.

Table 4

S_n	R_n	Q_{n+1}	$\overline{Q}_{n+1}$
H	H	Not allowed	Not allowed
L	H	L	H
H	L	H	L
L	L	Q_n	$\overline{Q}_n$

Positive Going Pulse Inputs

If $Q_{n+1} \triangleq$ the circuit response shortly after the pulse rising edge, $Q_n \triangleq$ the circuit response shortly before the pulse rising edge, and $p \triangleq$ pulse, then we have Table 5.

Table 5

S_n	R_n	Q_{n+1}	$\overline{Q}_{n+1}$
p	p	Not allowed	Not allowed
$\overline{p}$	p	L	H
p	$\overline{p}$	H	L
$\overline{p}$	$\overline{p}$	Q_n	$\overline{Q}_n$

It is important to point out that the circuit actually remembers whether a pulse was applied at the input terminals R or S. Since normally there is not pulse at R or S, during this period if y_1 is HIGH then it means there was a pulse at S, similarly if y_1 is LOW, it implies that a pulse was applied at R. This property provides the circuit the right to claim the glorious name of "Memory Device". In reference to Q, the designer can store a logical 1 in the circuit by applying a pulse at S and a logical 0 by applying a pulse at R. The "S" stands for "Set" and the "R" stands for "Reset". Some designers or manufacturers prefer to use C instead of R; here "C" stands for "Clear", because if a pulse is applied, the [1] and $y_1(Q)$ will be wiped or cleared off. Since there are only two possible steady states, the circuit is also called a bistable circuit or bistable multivibrator or this particular one is known as R-S flip-flop. There are many kinds of flip-flops; this circuit, however, is the heart of all kinds.

Logic Presentation

Notice that the circuit actually contains two resistor-transistor NOR-gates, where one NOR-gate has y_1 as output, and R and y_2 as inputs; while the other NOR-gate has S and y_1 as inputs, y_2 as output. By using MIL-STD-806B symbols, the circuit can be redrawn as shown in Figure 2. Due to its logical configuration, this circuit is also called cross-coupled flip-flop.

We have now moved from the electronic circuit presentation into logical symbolic

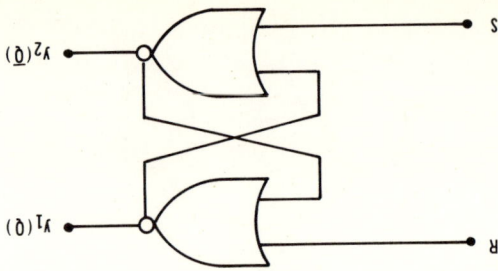

FIGURE 2. Using two NOR gates cross-coupled flip-flop.

presentation. Next, we should try to link the circuit with Boolean presentation which is usually done by direct translation of a truth table. Since the fourth columns in Tables 4 and 5 are a negation of the third columns, they can be ignored and one can write the switching functions from Tables 4 and 5 as follows: in Table 4, since S = R = H is not allowed, we may describe it by the equation, RS = 0. Then, if the subscript n + 1 is defined as next state, the n implies present state. Thus,

$$\begin{aligned}
Q_{n+1} &= S_n \bar{R}_n + \bar{S}_n \bar{R}_n Q_n \\
&= S_n \bar{R}_n + R_n S_n + \bar{S}_n \bar{R}_n Q_n \\
&= S_n (\bar{R}_n + R_n) + \bar{S}_n \bar{R}_n Q_n \\
&= S_n + \bar{S}_n \bar{R}_n Q_n \\
&= (S_n + \bar{S}_n)(S_n + \bar{R}_n Q_n) \\
&= S_n + \bar{R}_n Q_n \quad \text{(For Positive Logic Using NOR-Gates)}
\end{aligned} \quad (1)$$

Similarly, from Table 5, we have

$$Q_{n+1} = p_S + \bar{p}_R \cdot Q_n \quad (2)$$

where p_S denotes the pulse at S terminal, and p_R denotes the pulse at R terminal. Notice that in the above equations, Q_{n+1} and Q_n refer to the same terminal y_1, with Q_{n+1} denotes the logical state or value at the time $t = n + 1$ and Q_n at the time $t = n$. For example, in Equation 2, $Q_{n+1} = [1]$ if there is a pulse occurring at S input. If there is no pulse at R or S, then Q_{n+1} takes the old value, Q_n, or there will be no change of the logical value for Q. Equations 1 and 2 are called input equations of the flip-flops.

State Presentation

The logic diagram presentation, although clear, lacks time sequence information. The switching functions do contain time sequence information but it is somewhat abstract. Since it is found that the *state diagram* gives clearer presentation, it will be considered here.

The diagram shown in Figure 3 gives a clear presentation of the time sequence response. Positive logic is used. The two big circles indicate the two possible steady states of the output terminal y_1. At any instant, y_1 would either be [1] or [0]. Consider $y_1 = [1]$ at this moment, the circuit is staying at the circle on the left-hand side. Now, if RS = 00, or RS = 01, then y_1 will remain unchanged. Therefore a self-circling line is

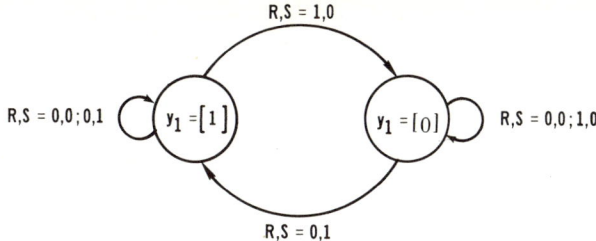

FIGURE 3. State-diagram of NOR-gate cross-coupled flip-flop.

shown. If the input R becomes [1] and S = [0], y_1 will change from [1] to [0], thus the diagram shows an arrow for y_1 = [1] which jumps over to the state of y_1 = [0]. Since RS = 11 is not allowed in this circuit configuration, it is not shown in this diagram. The reader is urged to verify the rest of the operation shown in the diagram.

Conclusion

In this section a basic flip-flop circuit composed of two RTL-NOR gates was used as a vehicle to show how one can derive switching functions, logic diagrams, and state diagrams from a given electronic circuit. The reader is urged to practice this technique on some other electronic circuit so that he will be able to interpret any new electronic circuit into the expression which the logic designers normally use. It is worthwhile to point out that for a low number of inputs or switching variables, the state diagram presentation is a good tool to describe how a sequential circuit behaves dynamically under different input conditions.

B. Basic Integrated Circuit Flip-Flop

In the last section the basic flip-flop circuit made of two discrete NOR-gates was investigated. In this section, the basic flip-flop that contains two I.C. NAND-gates will be described.

Figure 4 shows two 54/74 series I.C. NAND-gates with cross-coupled connection. The dotted lines show the MIL-STD-806B NAND-gate symbols, and the integrated circuits are shown inside the symbol. Although the circuits appear more complicated than the discrete one described in Section I, however, its principle is the same except for the minor detail which will be discussed here.

By applying the BSA technique one can easily obtain the truth table shown in Table 6.

Table 6

R	S	y_1	y_2
L	L	H	H
L	H	H	L
H	L	L	H
H	H	H	L
		L	H

Notice that, with this circuit, it is impossible to have y_1 = y_2 = L. With R = S = H, there are two possible states for the circuit, namely, y_1 = H, y_2 = L, or y_1 = L, y_2 = H. Following the same argument as in the last section, in this circuit R = S = L should not be allowed. Thus the switching function of this circuit for *positive logic* is,

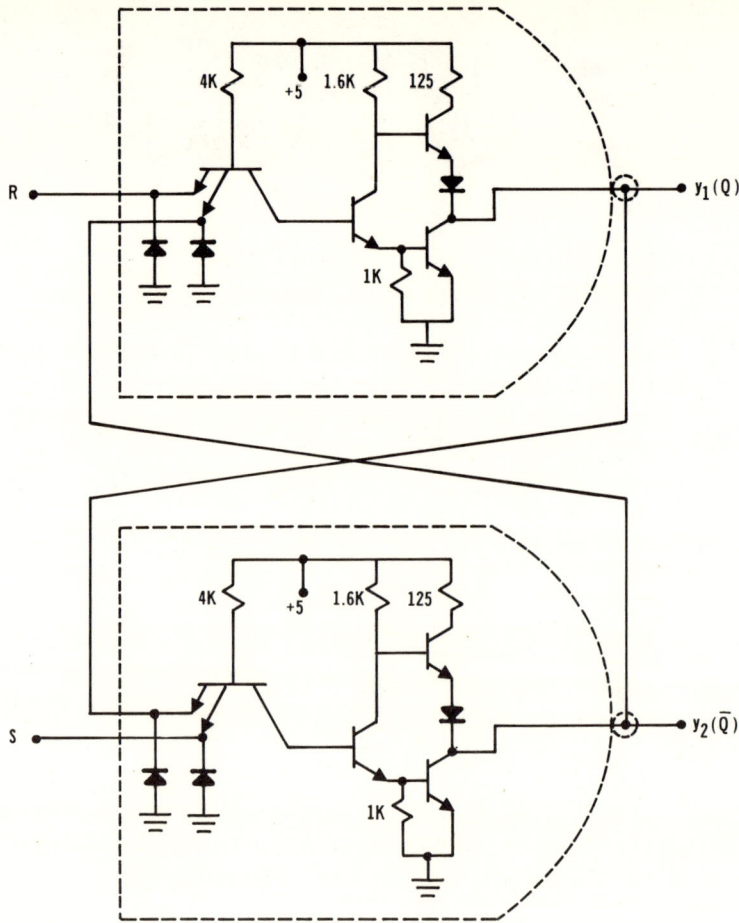

FIGURE 4. Cross-coupled flip-flop using two NAND gates.

$$\overline{RS} = 0$$
$$Q_{n+1} = \overline{R} + SQ_n \text{ (Positive Logic Using NAND-Gates)} \tag{3}$$

and its *state diagram* can be shown in Figure 5.

C. Integrated Circuit Flip-Flop

The reader may find that there are too many kinds of I.C. flip-flops available on the market. Sometimes a logic designer may have a problem choosing the proper one for his or her system. The best way to overcome this kind of difficulty is to analyze a few representative ones. In the foregoing section, the electronic circuits of two basic flip-flops, cross-coupled NOR-gates and NAND gates, were analyzed and their corresponding logic symbols were introduced. In what follows, we shall eventually leave the electronic circuit analysis technique behind and decompose the I.C. flip-flops into basic flip-flops and gates for functional analysis. If the reader does not feel confident to work on the logic presentation level, he can always go back to the foregoing sections for help.

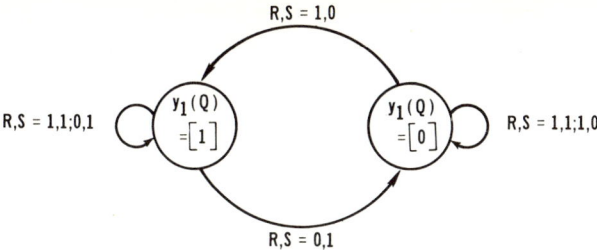

FIGURE 5. State diagram of NAND-gate cross-coupled flip-flop.

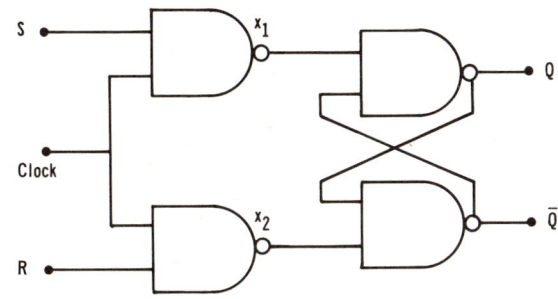

FIGURE 6. Clocked R-S flip-flop.

D. Clocked R-S Flip-Flop

In the basic cross-coupled NAND-gate flip-flop, it is clear that as soon as the set-line has become LOW the output Q-terminal becomes HIGH and a ONE is stored in the flip-flop. However, in some applications the designer wishes to store the information at a specific time. This can be done with the diagram shown in Figure 6.

Note that there are two NAND-gates and one cross-coupled flip-flop, whose electronic circuits and principle of operation were described in the preceding sections. Consider a logical one is to be stored in this device. The S is set to HIGH and the R is set to LOW. The clock line is sitting at LOW. During this time, x_1 and x_2 are both HIGH; therefore, there will be no change in the cross-coupled flip-flop. However, as soon as the clock pulse has arrived, the clock line becomes HIGH and $x_1 = [0]$ and $x_2 = [1]$ which results in Q = [1]. When the clock pulse disappears, the clock line returns to LOW and $x_1 = x_2 = [1]$ and the Q will remain unchanged until the next clock pulse arrives. This circuit is basically the same as the basic R-S flip-flop except that the information is stored at the specific time. Here R = S = HIGH is not allowed since if S = R = HIGH, then x_1 and x_2 will both be LOW while the clock line is HIGH, but $x_1 = x_2 =$ LOW is not allowed for cross-coupled NAND-gate flip-flops. Notice that while the clock line is HIGH, the states of S and R have direct influence on Q; therefore it is important that the state of S and R should be maintained at the desirable state while the clock line is HIGH.

E. J-K Master-Slave Flip-Flop

Principle of Operation — Derivation of Input Equations

Since there is always one state (R = S = [0] in cross-coupled NAND flip-flops), which is not allowed in R-S type flip-flops, the designer should always make sure that in his system the not allowed state will never occur. However, in the J-K flip-flop all possible states are allowed. Figure 7 shows the functional diagram of a J-K master-slave flip-flop. In Figure 7(a), there are two flip-flops, master and slave, and two

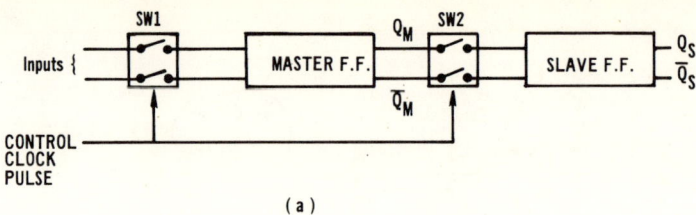

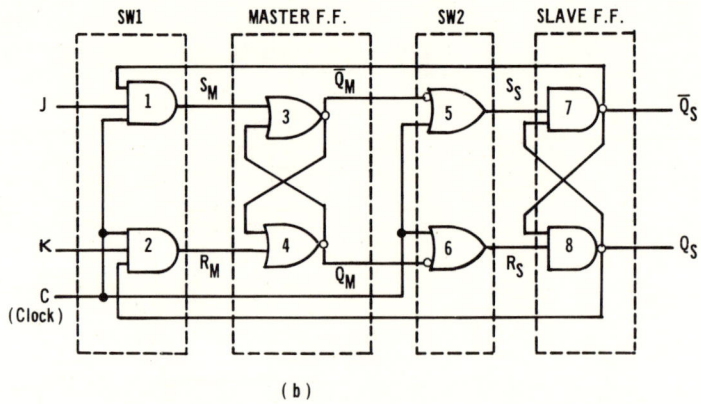

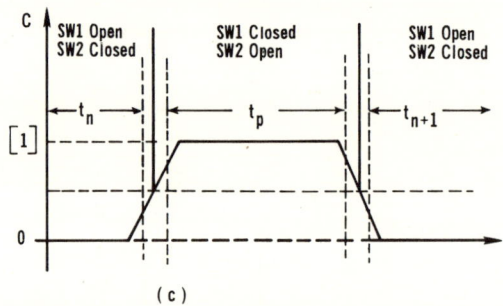

FIGURE 7. J-K master-slave flip-flop.

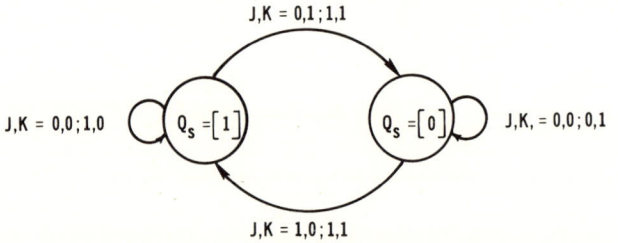

FIGURE 7(d). State-diagram of J-K master-slave flip-flop.

switching blocks, SW1 and SW2. The control clock pulses close and open switches SW1 and SW2 alternately. That is, when the clock line is LOW, SW1 is open and SW2 is closed. The information in the master flip-flop is then transferred to the slave flip-flop. When the clock line is HIGH, SW1 is closed and SW2 is open. Thus the master flip-flop is isolated from the slave flip-flop, and the input information is stored in the

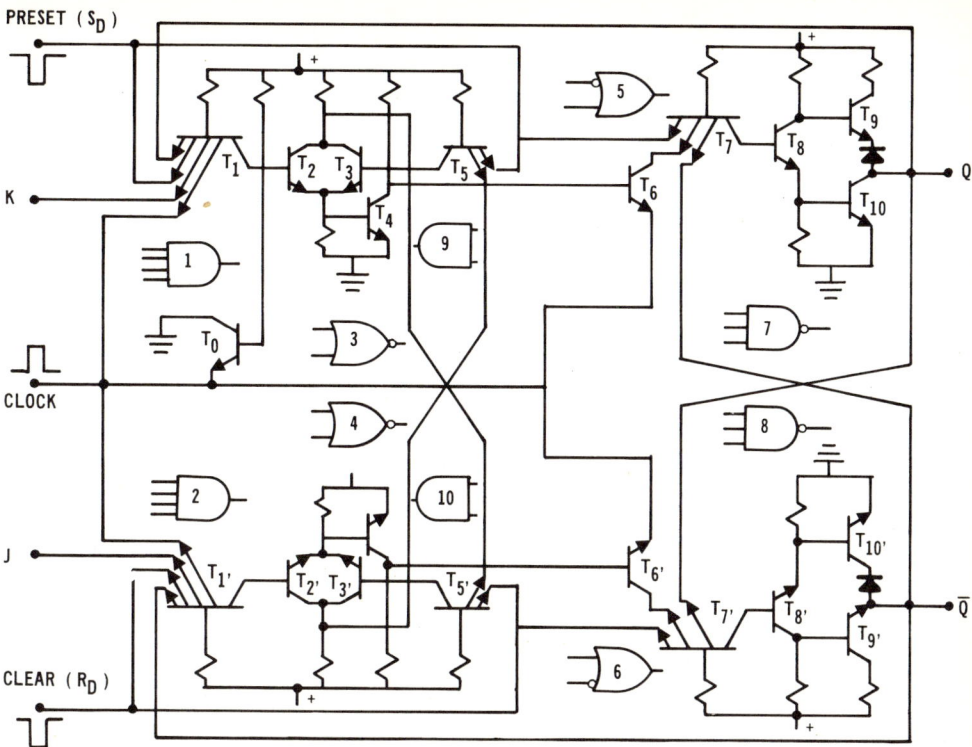

FIGURE 7(e). Circuit diagram of J-K master-slave flip-flop with preset and clear.

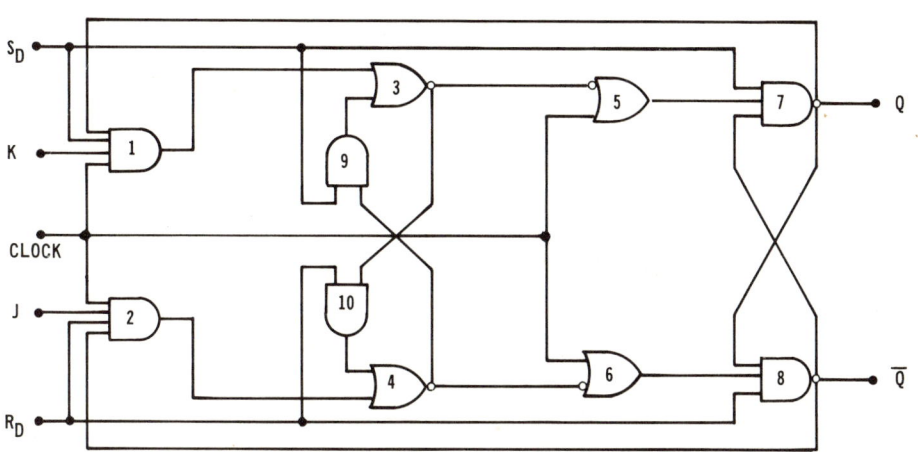

FIGURE 7(f). Logic diagram of J-K master-slave flip-flop with preset and clear.

master flip-flop. Figure 7(b) is logic diagram of the J-K master-slave flip-flop which is a direct translation of the block diagram shown in Figure 7(a). Gates, 1, 2, 5, and 6 are the switches. Gates 3 and 4 are the master flip-flop which is nothing but the simple cross-coupled NOR-gate flip-flop. Gates 7 and 8 are the cross-coupled NAND-gate flip-flop which acts as the slave in this network. There are two feedback lines connecting the output Q_s to the input gate No. 1 and $\bar{Q}_s$ to the gate No. 2. These two feedback ones are used for eliminating the not-allowed state existing in an R-S flip-flop. From Figure 7(b) one may write the switching functions:

108   Handbook of Digital System Design for Scientists and Engineers

$$S_M = C \cdot J \cdot \overline{Q}_S \qquad (4)$$

$$R_M = C \cdot K \cdot \overline{Q}_S \qquad (5)$$

$$S_S = Q_M + C \qquad (6)$$

$$R_S = \overline{Q}_M + C \qquad (7)$$

For the master flip-flop,

$$S_M R_M = 0 \qquad (8)$$

$$(Q_M)_{n+1} = (S_M)_n + (\overline{R}_M)_n (Q_M)_n \qquad (9)$$

For the slave flip-flop

$$\overline{R}_S \cdot \overline{S}_S = 0 \qquad (10)$$

$$(Q_S)_{n+1} = (\overline{R}_S)_n + (S_S)_n (Q_S)_n \qquad (11)$$

Suppose that one wishes to store a [1] into this circuit and Q_S is initially [0]. First, one should set J = [1] and K = [0]. But nothing will happen as long as the clock line is low. Because $S_M = R_M = 0$ from Equations 4 and 5, Q_M and $\overline{Q}_M$ remain unchanged according to Table 4. From Equations 6 and 7, $S_S = Q_M$, $R_S = \overline{Q}_M$ or $\overline{R}_S = S_S$. Since it was assumed that $Q_S = [0]$, then $\overline{Q}_S = [1]$ and $R_S = 1$, hence $S_S = [0]$. Thus $Q_M = [0]$, $\overline{Q}_M = [1]$. Now, as the clock line becomes [1], from Equations 4 through 11, we have

$$S_M = [1] J [1] = [1] \qquad (12)$$

$$R_M = [1] K [0] = [0] \qquad (13)$$

$$S_S = R_S = [1] \qquad (14)$$

$$Q_M = [1] \qquad (15)$$

$$(Q_S)_{n+1} = (Q_S)_n \qquad (16)$$

Effectively, gates 1 and 2 connect S_M to $J \cdot \overline{Q}_S$ and R_M to $K \cdot Q_S$, respectively, which stores the input information in the master flip-flop. However, due to C = [1], $S_S = R_S = [1]$ regardless of the current states of Q_M and $\overline{Q}_M$, which effectively isolates the slave from the master. Since at this time $S_S = R_S = [1]$, the slave flip-flop remains unchanged, i.e., $Q_S = [0]$.

When the clock line becomes LOW again, $S_M = R_M = 0$ and gates 1 and 2 effectively disconnect the master flip-flop from $J \cdot \overline{Q}_S$ and $K \cdot Q_S$: and gates 5 and 6 cause $S_S = Q_M = [1]$ and $R_S = \overline{Q}_M = [0]$ which effectively connects the master flip-flop to the slave flip-flop. Therefore Q_S becomes [1] and the operation of storing [1] is completed. Following the same steps, one can show how a [0] can be stored in the circuit by setting J = [0], K = [1]. Figure 7(c) summarizes the operation sequences:

In t_n: The master flip-flop is connected to the slave flip-flop.
t_p: The period of which the clock is at high level, the master flip-flop is connected to the inputs and is isolated from the slave flip-flop; the input information is stored in the master flip-flop.
t_{n+1}: The master flip-flop is isolated from the inputs and is connected to the slave flip-flop, so that the new information in the Master flip-flop is stored in the mlave flip-flop.

Let us now consider the cases when both inputs are equal. If $J = K = [0]$, then $S_M = R_M = 0$, nothing will change. If $J = K = [1]$, and the circuit is in t_n- period, there is still no change. However, in t_p period, from Equations 4 through 9, we have $S_M = (\overline{Q}_s)_n$, $R_M = (Q_s)_n$, $S_S = R_S = [1]$, and $(Q_M)_{n+1} = (\overline{Q}_s)_n + (\overline{Q}_s)_n(Q_M)_n = (\overline{Q}_s)_n$. In T_{n+1} period, then, $R_S = (\overline{Q}_M)_{n+1} = (Q_s)_n$, and $S_S = (Q_M)_{n+1} = (\overline{Q}_s)_n$. Thus, from Equation 11, $(Q_s)_{n+1} = (\overline{Q}_s)_n + (Q_s)_n(Q_s)_n = (\overline{Q}_s)_n$. Equation 12 shows that, if $J = K = [1]$ in the t_n- period, then in the t_p- period, the negation of the old information in the slave flip-flop is stored in the master flip-flop. Thus, in the $t_{n+1}-$ period, the new information in the master flip-flop, which is nothing but the negation of the old information in the slave flip-flop, is stored in the slave flip-flop. In other words, $J = K = [1]$ in the t_1- period would cause the slave flip-flop to negate its own old information after the clock pulse.

Physically, if one connects a light bulb at Q_s and set $J = K = [1]$, the clock pulse is analogous to one who presses a push-button light switch. One push will turn off the light if the light is ON, and it will turn on the light otherwise.

Table 7 lists the operation of the J-K master-slave flip-flop. Note that there is no forbidden state.

Table 7

J_n	K_n	$(Q_s)_{n+1}$
0	0	$(Q_s)_n$
0	1	[0]
1	0	[1]
1	1	$(\overline{Q}_s)_n$

From the table one can easily find the switching function of this flip-flop, i.e., $(Q_s)_{n+1} = \overline{J}_n\overline{K}_n(Q_s)_n + J_n\overline{K}_n + J_n\overline{K}_n(\overline{Q}_s)_n = \overline{J}_n\overline{K}_n(Q_s)_n + J_n\overline{K}_n[(Q_s)_n + (\overline{Q}_s)_n] + J_nK_n(\overline{Q}_s)_n = \overline{K}_n(Q_s)_n + J_n(\overline{Q}_s)_n$ and the state diagram is shown in Figure 7.

A Practical I.C. Circuit

Figure 7(e) shows a typical circuit diagram of an I.C. J-K master-slave flip-flop with Preset and Clear. The circuit appears to be too complicated for a logic designer to comprehend. However, if the circuit is partitioned in subsections according to their logical functions, it will become quite clear. Thus, in this figure, the circuit components are clustered in several subgroups and each group is labeled with a logic symbol according to its logic operation. Since the Q-side and the Q-side circuits are identical, we shall discuss only the K-side: Transistor T_1 is a four-input AND-gate. T_2, T_3, and T_4 constitute a NOR-gate having two outputs, one at the collectors of T_2 and T_3 and the other at the collector of T_4, so that the two outputs are electronically separated but logically the same. T_5 is a two-input AND-gate. T_7, T_8, T_9, and T_{10} constitute the familiar NAND-gate. It is interesting to note the function of T_6. This single transistor functions as an OR-gate with one output (the collector) and two inputs (the base and the emitter). The base input terminal automatically inverts the switching variable.

110   Handbook of Digital System Design for Scientists and Engineers

Transistor T_0 is used as a clamping device which assures that the clock line would drop to negative no more than a few tenths of a volt if the clock pulse is ringing.

Figure 7(f) shows the logic diagram of the circuit shown in (e), which is basically the same as that shown in Figure 7(b) except that the AND-gates, No. 9 and No. 10, were inserted in the cross-coupled-NOR flip-flop for the additional inputs of Present (S_D) and Clear (R_D). Because of these additions, the flip-flop becomes a versatile one. It can be used as a set-reset flip-flop or a triggered or toggle flip-flop or a J-K flip-flop. It is important to note that J-K and clock terminals are normally at LOW, while (S_D) and Clear (R_D). Because of these additions, the flip-flop becomes a versatile one. It can be used as a set-reset flip-flop or a triggered or toggle flip-flop or a J-K flip-flop. It is important to note that J,K and clock terminals are normally at LOW, while the S_D and R_D are normally at HIGH. From the electronic circuit in (e), it is obvious that a negative pulse going from HIGH to LOW at R_D and S_D, but not both, will cause the action, which will overrule the J-K and clock inputs and force the master and slave flip-flops to stay in the desired state.

F. D Flip-Flop

D flip-flop is sometimes called delay flip-flop or data-latched flip-flop. Basically, it has two outputs, Q and $\overline{Q}$, and two inputs, namely, D, the data input and C, the clock input. The data bit at D will be stored in the flip-flop when the clock line is HIGH. Thus the switching function of this type of flip-flop is simply, $Q_{n+1} = D_n$. The simplest data latch flip-flop is shown in Figure 8. In this diagram, the cross-coupled NAND-gate flip-flop is the memory device. Gates 1 and 2 are so connected that the condition of $\overline{R} \cdot \overline{S} = 0$ is assured, and the following truth table can easily be verified.

D_n	C	S_n	R_n	Q_{n+1}
0	0	1	1	Q_n
1	0	1	1	Q_n
0	1	0	1	[0]
1	1	1	0	[1]

The table simply means that the flip-flop is isolated from the D-input before and after the clock but that the information at D is stored during the clock pulse. Although it is a simple matter to convert R-S flip-flop or J-K flip-flop to a D flip-flop, we shall now analyze a different type of flip-flop called edge-triggered D flip-flop which accepts the information only at the rising or the falling edge of the clock. This feature is desirable for a system whose timing problem is critical. Since the device does not respond to the input during the clock pulse, the pulse-width of the clock is not critical any more.

Figure 9 shows a typical commercially available edge-triggered D flip-flop. The circuit diagram is shown in (a) and its corresponding logic diagram is shown in (b). Notice that transistors T_1, T_2, T_3, T_4; $T_1{'}$, $T_2{'}$; and $T_3{'}$, $T_4{'}$, respectively, constitute the NAND-gates 1, 2, 3, and 4 shown in (b). They share the same diode D as shown in (a). NAND-gates 5 and 6 are made of standard circuitry; it is not necessary to be analyzed here. However, the circuitry for gates 1 through 4 is confusedly simple. Actually, if one imagines that there was a resistor connection from the collector of T_2 to the power supply +V, then T_1 and T_2 constitute a straightforward AND-INVERTED or NAND-gate since T_1 is a simple AND-gate and T_2 is a single transistor inverter. In this circuit, T_1, T_2 and T_3, and T_4 form the cross-coupled NAND-gate fip-flop, and there are cross-connections between the collector of T_2 and emitter of T_3, similarly, the collector of T_4 and emitter of T_1. Thus the imagined resistor in the collector circuit of T_2 would be in parallel with the resistor in the base of T_3, and there is no reason that T_2 cannot share the resistor in the base of T_3. This is why the imagined resistor is a redundant

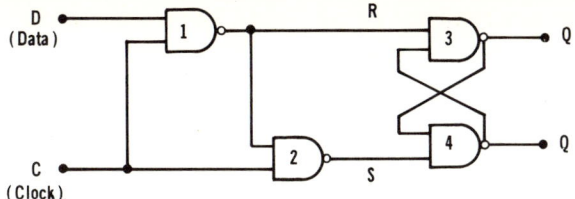

FIGURE 8. Logic diagram of D flip-flop.

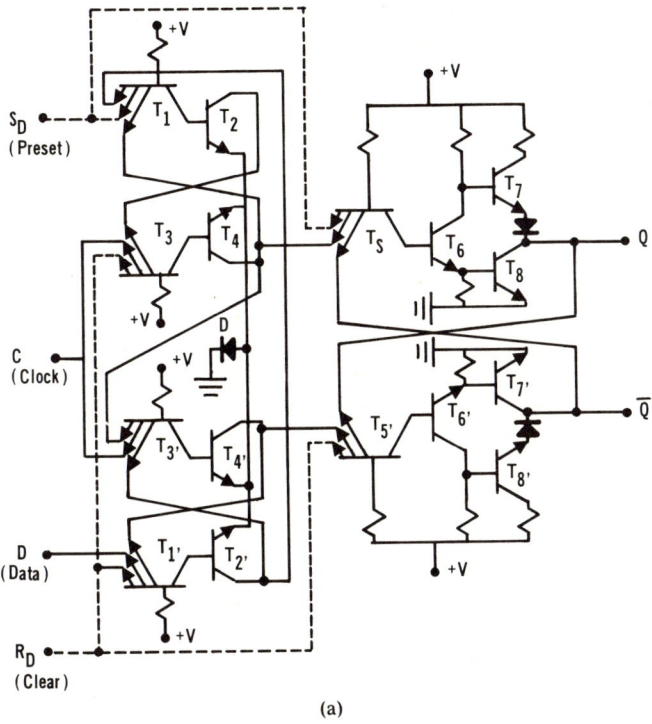

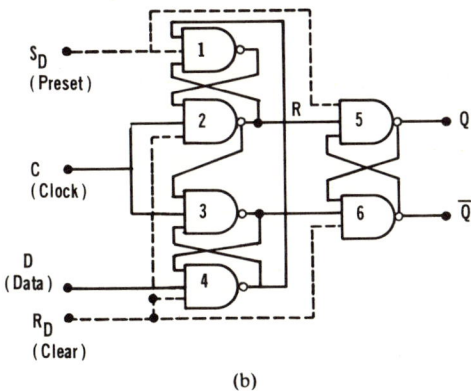

FIGURE 9(a). Edge-triggered flip-flop. Circuit diagram. (b) Logic diagram.

one and is then omitted from the circuit. With the understanding of how the simple circuit constitutes the NAND-gate, we shall now investigate how the logic diagram in (b) functions as a D flip-flop. The solid line portion of the diagram shows the basic D flip-flop and the dotted lines show the additional features of "set" and "reset" which are normally at HIGH state. To understand the D flip-flop only, let us pretend that the dotted lines are nonexistent (since HIGH is equivalent to open circuit); all gates except No. 3 become two-input NAND-gates. The clock line is normally at LOW state which causes R = S = [1] and isolates the 5—6 cross-coupled flip-flop from the inputs; the old information remains undisturbed. Now let $\theta \underline{\Delta}$ the threshold of the gate for changing state, and $y_i \underline{\Delta}$ output of ith-gate; then, let $t = t_n$ as shown in (b) and D = [1]. Since $R_n = S_n = [1]$, we have $y_4 = \overline{D} = [0]$ and $y_1 = [1]$. As C rises to $\geq \theta$, R = [0], S = [1]; thus, Q = [1] = D. Consider, $t = t_p$, or C = [1], but D changes to [0], then, $y_4 = \overline{D} = [1]$.

But R = [0] which locks $y_1 = [1]$, thus S = [1]. Therefore, the output flip-flop remains unchanged during $t = 5_p$ even though D changes to [0]. As C returns to [0], or $t = t_{n+1}$, both R and S return to [1]. The output flip-flop is then isolated from the input. Similarly, consider D = [0] while $t = t_n$. Since $R_n = S_n = [1]$, $y_4 = [1]$, and $y_1 = [0]$. As C rises to $\geq \theta$, S = [0]. But R will still be [1]. Therefore, Q = [0] = D and the information is transferred to the output flip-flop. Consider, D changes to [1] while $t = t_p$, since S = [0], then $y_4 = [1]$, regardless of what state D is in. Hence, it is equivalent to shut-off gate No. 4 and will not respond to the Data line. Now, when C returns to zero or $t = t_{n+1}$, R = S = [1], which isolates the output flip-flop from the input, the stored information remains unchanged until the next clock pulse arrives. Therefore, this D flip-flop only responds to the rising edge of the clock pulse.

G. Conclusion

In this section the BSA technique was used to analyze the basic cross-coupled flip-flops made of NOR-gates and NAND-gates. It was an attempt to link the electronic circuit, logic diagram, and switching functions of the basic flip-flops in a coherent manner. Then, some practical I.C. flip-flops were analyzed by first showing the actual circuit diagram, then partitioning the diagram into cross-coupled flip-flops and gates, and finally translating it into a logic diagram. The basic repeating logic element was analyzed in the electronic circuit point of view and the whole flip-flop was treated as a small digital system or a logic network and its logical characteristic equation or switching function was derived. From now on, the flip-flop can be considered as a black box characterized by its switching function. However, a novice is usually fascinated and confused by the varieties of flip-flops available on the market. Actually, it is just like that on the automobile market. Although there are many varieties of cars, their basic principle of operation is the same, yet they are different only in optional features. According to their characteristic switching functions, one may classify the flip-flops into four main classes, i.e., R-S, J-K, D, and T (toggle) flip-flops. Table 8 shows how they are related to each other. However, according to their triggering methods, one may classify them into *direct, clocked, master-slave,* and *edge-triggered* flip-flops. Although we have shown only the J-K master-slave flip-flop, there is also a D-type master-slave flip-flop, etc. Here the interesting points are not what types of master-alave flip-flops there are, but what the distinction characteristics are for each triggering method, namely, (1) for *direct triggering,* the output changes as the input does; (2) for *clocked* flip-flop, it responds only when the clock line is enabled, otherwise the device is isolated from the input; (3) for *master-slave* flip-flop, the master flip-flop is connected to the input in the rising and the period of the clock pulse, while it is isolated from the inputs and connected to the slave flip-flop in the falling of the clock and the

Table 8
FLIP-FLOP CONVERSION TABLE

FROM \ TO	J-K FLIP-FLOP	D FLIP-FLOP	T (TOGGLE) FLIP-FLOP
RS			
JK			
D			

nonclocked periods; thus the output and the input are at all times separated, which is most desirable when any closed loop in a system is not wanted; and (4) for *edge-triggered* flip-flops, the device only responds to the input at the rising edge (or falling edge) — it keeps the information faithfully unchanged at any other time. For one thing, the pulse width of the clock is not critical for this device, and for a high speed synchronous system, the device would cause less noise problem than others.

II. MONOSTABLE MULTIVIBRATOR (ONE-SHOT)

As we know from the last section, a flip-flop has two stable states, i.e., Q = [1] or $\overline{Q}$ = [0]; it may stay in any one of the two states for an indefinite length of time. However, the one-shot would normally stay in one state when its input is not excited and enters another state if excited. But it cannot stay in the other state forever, although the time for staying in the other state is controllable. The major applications of a one-shot are in waveshaping, pulse-width modification, and time delay. The discrete circuit one-shot is very sensitive to noise. The designer would not use it unless he has no other choice. However, the integrated circuit one-shots now on the market do possess desirable noise immunity.

A. Discrete Circuit One-Shot

Since the discrete one-shot is easier to understand, it has the same principle of operation as others, we shall start with this circuit. Figure 10(a) shows a simplified one-shot circuit for demonstrating its principle of operation. Consider that the switch has been on position H for a long, long time. The capacitor C_x can be treated as an open circuit and the voltage across it is 4.3 V. Let

$$\frac{5 - 0.2}{R_C} \ll \beta \frac{5 - 0.7}{R_x}$$

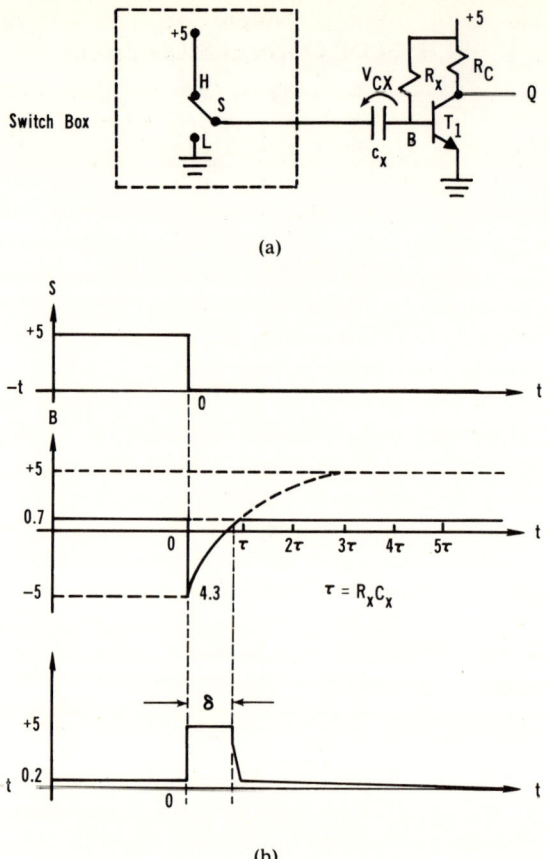

FIGURE 10. Simplified one-shot circuit.

where, β is the current gain of the transistor, then T_1 is saturated, $Q = 0.2$ and $B = 0.7$. Now, at $t = 0$, the switch is closed to L, the B becomes -4.3, which turns off the transistor T_1, and Q jumps to $+5$ V. Effectively, the point B is disconnected from the base, the R_x and C_x become a simple R-C network connecting from $+5$ to ground. The capacitor is being charged toward $+5$ with time constant $\tau = R_x C_x$. Thus we can write $B(t) = B(\infty) + [B(0^+) - B(\infty)]e^{-t/\tau} = 5 + (-4.3 - 5)e^{-t/\tau}$.

But, the transistor will be turned on again as soon as $B(t)$ has reached $+0.7$ V. To determine the time δ at which T_1 conducts, we can solve the Equation, $0.7 = 5 - 9.3e^{-\delta/\tau}$. We have

$$\delta = \tau \ln \frac{9.3}{4.3} = R_x C_x \ln 2.16$$

$$= 0.73 \, R_x C_x$$

Since $\beta \gg R_x/R_c$, the transistor almost enters the saturation region immediately after $t = 0.73 \, R_x C_x$. The waveforms of S, B, and Q are shown in Figure 10(b). It is important to point out that the output pulse starts at $t = 0$ with a pulse width equal to $0.73 \, R_x C_x$ which is a function of R_x and C_x and independent to the input waveform. However, in order to repeat this process, the switch has to be switched back to H for a period of time to allow the capacitor to be charged to 4.3 V again.

Figure 11(a) shows a simple discrete one-shot. By comparing this circuit with Figure 10(a), one can see that the switch S is being replaced by the transistor T_2 and its control circuit R_1 and R_2. The circuit functions as follows. Let $V_i = 0$, since the circuit is at rest, C_x is fully charged and no current flows in or out of the capacitor. The capacitor is effectively an open circuit branch. T_1 is saturated, $Q = 0.2$. For convenience, let $R_1 = R_2$ then the voltage at the base of T_2 is ½ $Q = 0.1$ which will keep T_2 OFF and C_x is connected to $+5$ through R_{C2}. Now, a positive going pulse of $+5$ V is applied to V_i, which will drive T_2 into saturation. Thus S drops from $+5$ to 0.2 V, which is equivalent to turning the switch S (Figure 10(a)) from H to L. The responses of the right-hand side portion of the circuit shown in Figure 11(a) closely follow what has been described in Figure 11(b) except that the switch in Figure 10(a) does not automatically go back to H, but the transistor switch T_2 in Figure 11(a) does. Because the resistor R_1 senses the state of Q, and resistor R_1, R_2, and T_2 constitute an RTL NOR-gate, the collector S of T_2 stays LOW as long as Q or V_i or both are HIGH. As soon as both V_i and Q have become LOW, T_2 is OFF which connects the capacitor C_x back to $+5$ through R_{C2}. The capacitor C_x is then being charged to $5 - 0.7$ with time constant $R_{C2}C_x$; and after approximately $5R_{C2}C_x$, the C_x is fully charged and the circuit is ready for receiving another excitation from V_i to generate another one-shot. The circuit responses to input pulses with different pulse-widths are shown in Figure 11(b), the dotted lines show the response of the input with wider pulse-width. Notice that the output pulse-width is independent of input pulse-width, and is dependent on R_xC_x. The reader is urged to verify the waveforms shown in Figure 11(b). Why is the initial voltage of B = -4.1 instead of -4.3 V? Why is the slow rising-edge of S, and why is the falling-edge of Q much steeper?

B. I.C. One Shot

Figure 12 shows a typical I.C. one-shot. Although the circuit appears quite complicated, for functional analysis it need not be difficult. The key components are (1) R_x and C_x, which are provided externally for determination of the output pulse-width. The manufacturer usually provides an equation which gives the pulse-width in terms of C_x and R_x; (2) transistors T_1 and T_4 which are, respectively, equivalent to T_1 and T_2 in the discrete one-shot shown in Figure 11(a), are functioning as the two switching devices; (3) the trigger input V_{in} and the gate-terminal which allows or prohibits the trigger signal to enter the one-shot; and (4) transistors T_3 and T_7 are the output buffer amplifiers for Q and $\overline{Q}$; T_2 and T_6, respectively, interface T_1, T_3, and T_4, T_7; T_8 is an emitter-follower for recharging C_x after the one-shot; T_5 interfaces the collector of T_1 and the base of T_4, which is equivalent to R_1 in Figure 11(a); T_2, collector of T_1 and the base of T_4, which is equivalent to R_1 in Figure 11(a). T_2, T_5 and T_6 function exactly the same way as the multiemitter transistor at the input of a NAND-gate that we have studied, in which either the emitter-base diode junction is ON or the collection-base junction is ON, but not both. Notice that the triggering scheme is slightly different from the discrete one-shot. In the I.C. one-shot, the triggering input V_{in} is applied to T_1 for turning T_1 off, while in the discrete circuit the trigger is applied to T_2 in Figure 11(a) for turning T_2 ON and in turn, switching T_1 OFF. They serve the same purpose but via different routes. Now, we are ready to see how the I.C. one-shot functions. At rest, T_1 is ON which results in: (1) base-emitter junction of T_2 will be ON which will turn T_3 OFF, (2) base-emitter junction of T_5 will be ON which turns off T_4, and base-collector of T_6 will be ON so that T_7 will be saturated. Thus we have $\overline{Q}$ = [1], Q = [0]. Consider now, the Gate = [0], and a pulse is applied at V_{in}. The rising edge of V_{in} causes a positive spike at S, but the diode D_1 blocks the spike and nothing will happen to the one-shot. However, the falling-edge of V_{in} causes the S to go negative which turns ON D_1 and pulls the base of T_1 to negative, thus T_1 will be OFF and the following chain reaction occurs: (1) base-collector of T_2 will be ON and T_3 becomes

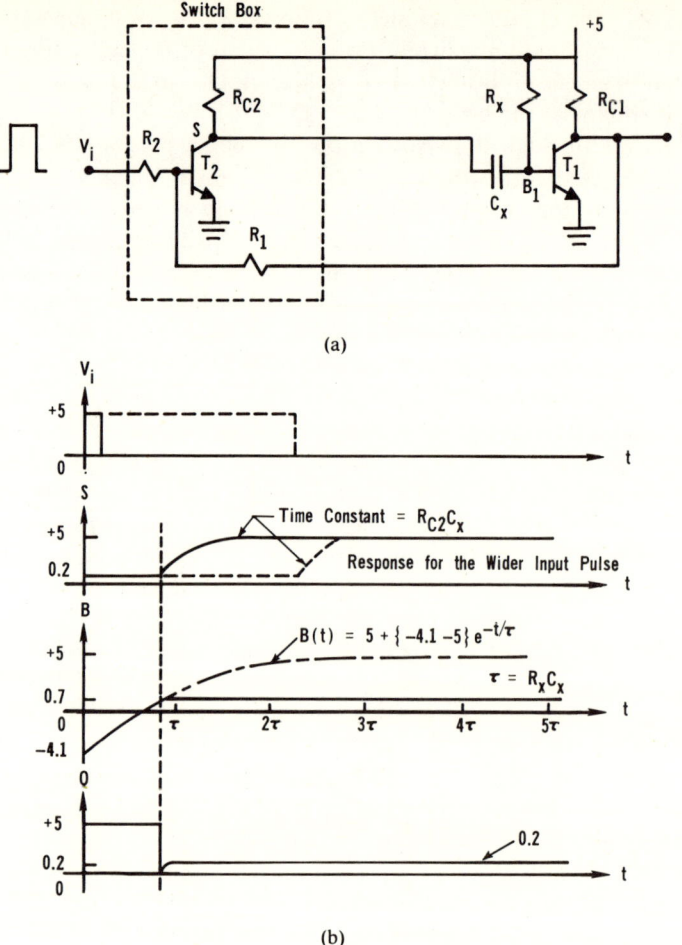

FIGURE 11. Discrete one-shot circuit.

saturated, (2) base-collector of T_5 will be saturated so that the base-emitter of T_6 will be ON and T_7 is OFF, (3) since T_4 is saturated, it effectively ties the left-hand side of C_x to ground; therefore, B becomes negative which keeps T_1 OFF although by this time V_{in} pulse is gone. The circuit remains in this state until C_x is charged through (R_x + 1.5K) and the voltage at B becomes positive enough to turn on T_1. The sequence then is basically the same as that which was described for discrete one-shot. Notice that the provision of the control gate is very desirable since the circuit can be shut off while idling so that the noise will not trigger the one-shot falsely.

C. One-Shot Using I.C. 555 Chip

Figure 13 shows the functional block diagram of a one-shot configuration using I.C. 555 timer; where CMP1 and CMP2 are two analog comparators with $V_1 = \frac{2}{3} V_{cc}$ and $V_2 = \frac{1}{3} V_{cc}$ as one of the two inputs for CMP1 and CMP2, respectively. R_1, R_x, and C_x are the external circuit components. Just like the simple discrete one-shot, R_x and C_x control the pulse-width of the one-shot output. R_1, however, assures that the trigger input pin is at V_{cc} volt in quiescent state. In view of the circuit of the 555 provided by the manufacturer, one can derive the special characteristic of the R-S flip-flop as shown in Table 9.

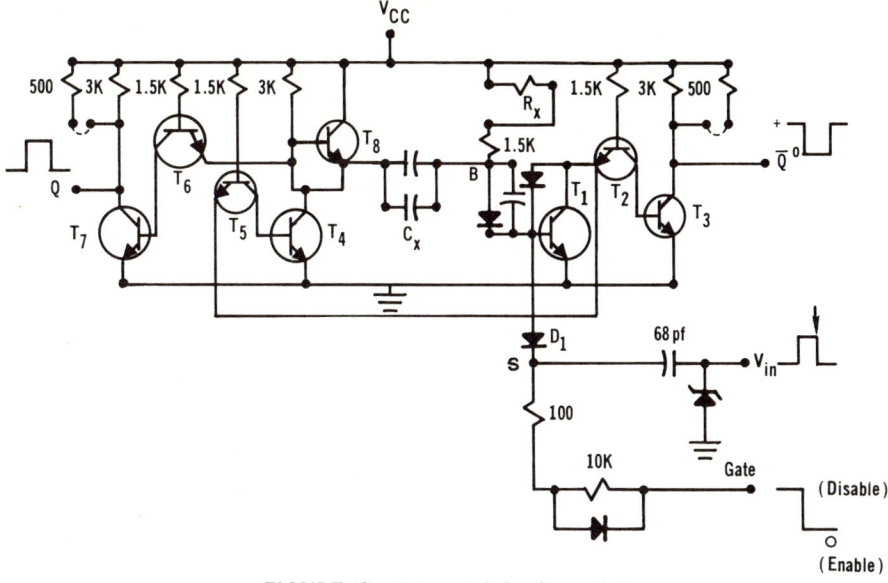

FIGURE 12. Integrated circuit one shot.

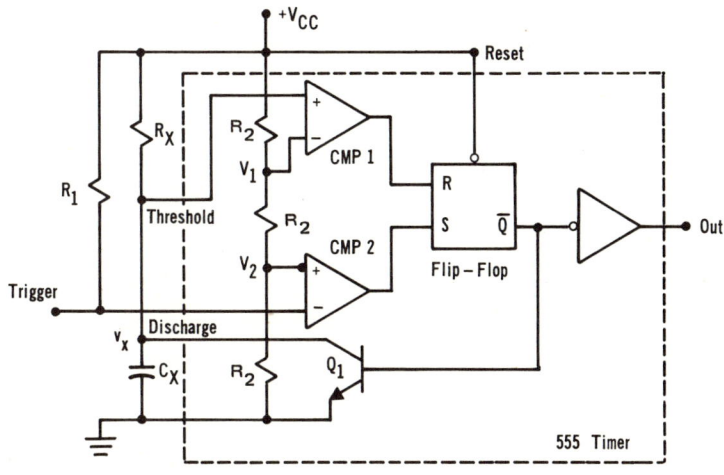

FIGURE 13. Using 555 timer as a one shot.

Table 9

S	R	$\overline{Q}_{n+1}$
0	0	Q_n
1	0	0
0	1	1
1	1	0

Consider the circuit is presently in the following state:

$$\left. \begin{array}{l} \text{Trigger} = \text{High} \rightarrow S = [0] \\ v_x(t) = \text{High} \rightarrow R = [1] \end{array} \right\} \overline{Q} = [1] \rightarrow \text{Out} = [0]$$

where, $\rightarrow$ = imply or result in

118   *Handbook of Digital System Design for Scientists and Engineers*

then Q_1 conducts and discharges C_x with a time constant $\tau_d = C_x R_s$, where, R_s is the saturation resistance of Q_1, which normally is in the order of a fraction of 1 Ω. As a result, the output of CAMP 1 goes LOW, and the circuit enters the following state:

$$\left.\begin{array}{l} \text{Trigger} = \text{High} \rightarrow S = [0] \\ v_x = \text{Low} \rightarrow R = [0] \end{array}\right\} \rightarrow \bar{Q} = 1 \text{ (no change)} \rightarrow \text{out} = [0]$$

The circuit will stay in this state indefinitely until a negative going pulse sets the trigger terminal to LOW at time $t = 0$, or any voltage lower than v_2 which is ⅓ V_{cc}, then

$$\left.\begin{array}{l} \text{Trigger} = \text{Low} \rightarrow S = [1] \\ v_x(0^-) = v_x(0^+) = \text{Low} \rightarrow R = [0] \end{array}\right\} \rightarrow \bar{Q} = [0], \text{ out} = [1]$$

As a result, Q_1 is OFF, capacitor C_x is then being charged by V_{cc} through resistor R_x with a time constant, $\tau_c = R_x C_x$, and

$$v_x(t)0 = v_x(\infty) + [v_x(0^+) - v(\infty)] \epsilon^{-\frac{t}{\tau_c}}$$

where

$$v_x(0^+) = 0, v_x(\infty) = V_{cc}$$

Thus,

$$v_x(t) = V_{cc} \left[1 - e^{\frac{t}{\tau_c}}\right] \tag{17}$$

However, when

$$v_x(t) \geq V_1 = \frac{2}{3} V_{cc}$$

then R = 1.

At this point, if trigger is still low, then we have,

$$\left.\begin{array}{l} S = 1 \\ R = 1 \end{array}\right\} \rightarrow \bar{Q} = 0, \text{ Out} = 1$$

However, if the trigger pulse is gone, then

$$\left.\begin{array}{l} S = 0 \\ R = 1 \end{array}\right\} \rightarrow \bar{Q} = 1, \text{ Out} = 0$$

Here, $\bar{Q} = 1$ causes the transistor Q_1 to be ON and discharging C_x. Let $t = T$ when $v_x(t)$ reaches ⅔ V_{cc}, Equation 7 becomes,

$$v_x(T) = \frac{2}{3} V_{cc} = V_{cc} [1 - e^{\frac{-T}{\tau_c}}]$$

$$T = \tau_c \ln 3 \tag{18}$$

$$= 1.1 R_x C_x$$

In conclusion, the output will remain at HIGH state for a $T = 1.1\ R_x C_x$ period if the trigger pulse-width is equal to or less than T; otherwise the output pulse-width will be equal to that of the trigger pulse. Since C_x requires a finite time to be discharged, this one shot is not a retriggering type. In view of Equation 18 it reveals that the output pulse-width T can be virtually designed as precision as the components R_x and C_x. That is, T is independent of V_{cc} and other circuit variables.

III. ASTABLE (FREE-RUNNING) MULTIVIBRATOR

Astable multivibrator is simply a square wave oscillator whose frequency is determined by two identical pairs of R-C networks. By controlling the R-C network individually, the time period for the HIGH state and LOW state of the circuit can be unsymmetrical. The essential application of this circuit in a digital system is being used as a CLOCK. There are many kinds of astable multivibrators available on the market. For application which requires accurate frequency reference, one may select crystal-controlled and temperature-compensated astable multivibrators; however, if the designer is only concerned with the proper sequencing in his system, he may use the basic circuit described in the following section.

A. Basic Astable Multivibrator

The circuit shown in Figure 14 is a basic astable circuit with two pairs of R-C networks, $R_x C_x$ and $R_x' C_x'$, for frequency control. The circuit can be reduced to the one-shot circuit shown in Figure 11(a) by (1) eliminating all diodes, R_2' and R_2; (2) shortening $\overline{Q}'$ to $\overline{Q}$, Q' to Q; (3) replacing C_x' by a resistor; (4) disconnecting R_x' from +5 and using it as the input resistor as in the one-shot circuit. Thus T_2 becomes the switch box of the one-shot. Similarly, one may use T_1 as the switch box, and R_x as the input resistor; then one has another one-shot with $\overline{Q}$ as its output. Therefore the astable can be treated as two one-shot circuits which are cross-connected such that they will be self-exciting to maintain the free-running requirement for the astable circuit. The reader may recall that in Figure 11, the capacitor C_x should be fully charged to 4.3 V before the next action is initiated which causes the slow rise voltage waveform at S. This would not be desirable for an astable circuit, since S will be the $\overline{Q}$ in the astable circuit, which happens to be one of the outputs. Therefore, R_2, R_2', D_1 and D_1' are added to the circuit shown in Figure 14(a). With this modification, the C_x and C_x' are, respectively, recharged through R_2' and R_2 so that the waveforms at $\overline{Q}'$ and Q' have slow rise time. To be more specific, consider the points at Q' and Q. While C_x' is charging T_2 is ON and T_1 is OFF; Q' rises slowly but Q tends to jump to +5 immediately. Of course, if Q and Q' were shorted, Q would rise slowly as Q'. However, the diode D_1 is isolating Q from Q' since Q > Q' while C_x' is being charged and Q = Q' when C_x' is fully charged, thus D_1 is always cut off during this period. Now, as T_1 is being turned on, D_1 will always be ON since Q will be 0.2 V and Q' > Q or Q' ≃ 0.9 which effectively ties the right-hand side of C_x' to ground so that T_2 will be kept OFF until it is charged to +0.7 V. The operation is the same as that of the one-shot. It will not be repeated here. The waveforms of the points that are interesting to us are shown in Figure 14(b). The output waveform at Q has a fast rise time.

In practice this circuit has one drawback. It is possible that both T_1 and T_2 stay saturated at the same time. If so, the circuit will not oscillate until a disturbance is applied to the circuit. The simplest way to get the circuit to start oscillating is to momentarily short the base of either one of the transistors. In order to insure that the circuit will self-start, the D_2 and D_2' are added in the circuit. As long as the parallel value of R_2 and R_c is slightly less than that of R_x, the transistor will be kept in the

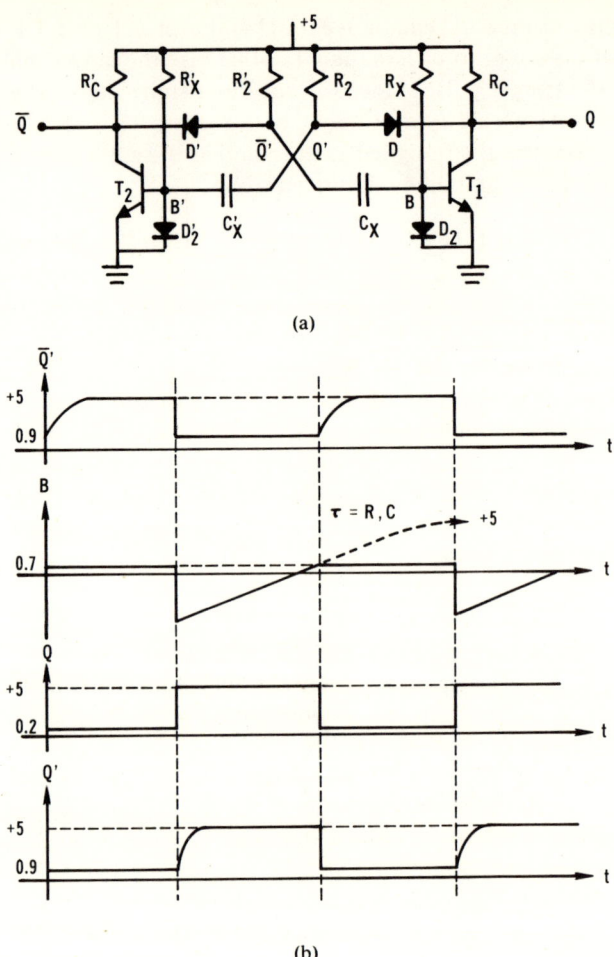

FIGURE 14. Basic astable multivibrator.

active region while the capacitors are fully charged. Since the circuit is connected with positive feedback, any natural noise from anywhere can initiate the oscillator; thus the circuit shown is considered a self-started astable multivibrator.

B. Astable Multivibrator Using 555 Chip

Figure 15(a) shows the circuit configuration for astable multivibrator using 555 timer. The capacitor C_x is being charged through $R_1 + R_2$ with a time constant $\tau_c = C_x(R_1 + R_2)$ and discharge through R_2 with a time constant $\tau_d = C_x R_2$. Consider the following three possible cases:

1. For $v_x(t) < \frac{1}{3} V_{cc}$, the flip-flop will have its inputs, $R = 0$, $S = 1$ and according to Table 9, $\overline{Q} = 0$, transistor Q_1 will be OFF, C_x will be charged through $R_1 + R_2$.
2. For $\frac{1}{3} V_{cc} < v_x(t) < \frac{2}{3} V_{cc}$, we have, $R = 0$, $S = 0$; capacitor C_x will continuously be charged.
3. For $v_x(t) > \frac{2}{3} V_{cc}$, we have, $R = 1$, $S = 0$ and Q_1 will be ON and C_x will be discharged until $v_x(t) < \frac{1}{3} V_{cc}$. The operation repeats itself and the output yields a square wave clock. It is in HIGH state when C_x is being charged and LOW

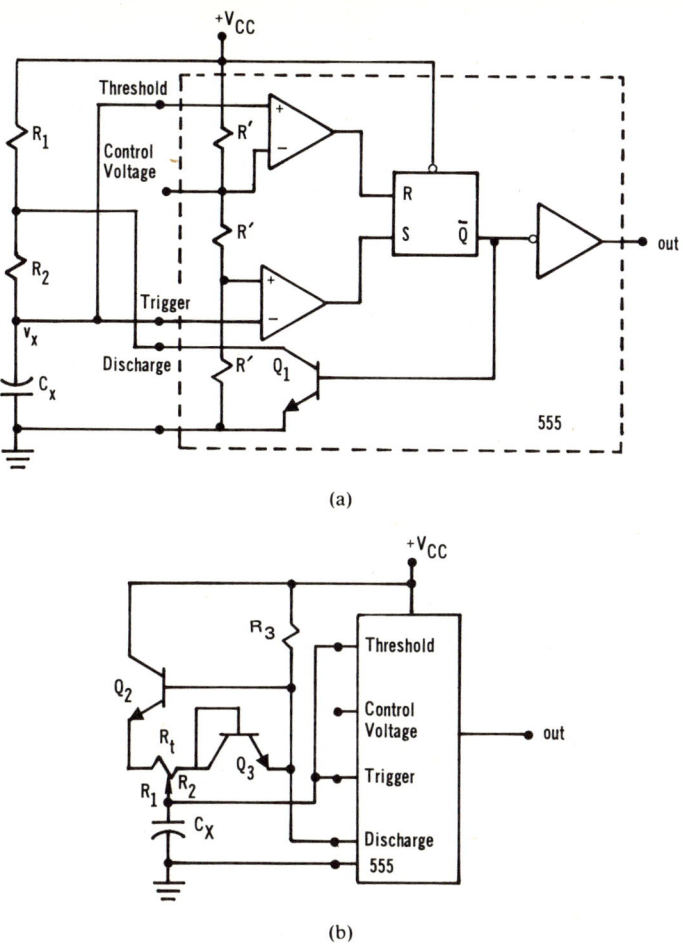

FIGURE 15. Astable multivibrator using 555 timer.

state when C_x is being discharged. Follow the derivation similar to that described in Section II.C, we have, t_H = time while output is HIGH = $0.693 (R_1 + R_2)C_x$; t_L = time while output is LOW = $0.693 R_2 C_x$.

The frequency of the clock,

$$f = \frac{1}{t_H + t_L} = \frac{1.44}{(R_1 + 2R_2)C_x} \tag{19}$$

In some applications, one would like to adjust the duty-cycle or the ratio of the HIGH and LOW periods of the clock without changing its frequency. To do this, one can use the circuit shown in Figure 15(b). Here, the frequency of the clock,

$$f = \frac{1.44}{R_t C_x}$$

where $R_t = R_1 + R_2$, the total resistance of the potentiometer; thus the frequency will remain constant as the charging and discharging time is being controlled by the setting of the potentiometer. Here, transistor Q_2 is turned OFF during discharging period and turned ON during charging. Q_3 is used as a diode which isolates resistance R_2 from the charging loop, therefore we have, $t_H = 0.693\ R_1C_x$; $t_L = 0.693\ R_2C_x$; and

$$f = \frac{1.44}{(R_1 + R_2)C_x} = \frac{1.44}{R_tC_x}$$

IV. SCHMITT TRIGGER CIRCUIT

The Schmitt trigger circuit is another kind of regenerative circuit. Its application is usually as a pulse shaper for improving the rise and fall time of a signal or a level detector. The circuit has a predetermined threshold voltage V_H, if the input signal is less than V_H, the output will stay LOW. As soon as the input voltage has reached to V_H, the output voltage jumps to HIGH. Thus the circuit has the desirable noise immunity property.

Figure 16 shows a basic Schmitt trigger circuit. For BSA purposes, a transistor with high current gain is assumed so that the effect of I_b, the base current, can be neglected. Consider $V_i = 0$, then T_1 will be OFF, T_2 will be ON. We have,

$$B = \frac{R_2}{R_{C_1} + R_1 + R_2} \cdot 6 = \frac{5}{10} \cdot 6 = 3 \tag{20}$$

$$E = B - 0.7 = 2.3$$

Now, as V_i reaches $E + 0.7 = 3$ V, T_1 begins to conduct and I_{C1} causes a voltage drop across R_{C1} which results in decreasing B to less than 3 V. As soon as B has dropped below 3 V, T_2 will be OFF. Under this condition, B can be determined by applying the Thévenin theorem.

Let $Q_{th}\ \Delta$ Thévenin voltage at $\overline{Q}$, $R_{th}\ \Delta$ Thévenin resistance at $\overline{Q}$, and T_1 is actively nonsaturated, then $Q_{th} = 6 - I_{C1}R_{C1}$, $R_{th} = R_{C1} = 0.5K$, and

$$B = \frac{R_2}{R_{th} + R_1 + R_2} \cdot \overline{Q}_{th} = \frac{5}{10}[6 - I_{C_1}0.5] \tag{21}$$

$$= 3 - 0.25 I_{C_1}$$

when $V_i = 3$ V, $I_{C1} = (3 - 0.7)/R_E = 2.3/1K = 2.3$ mA, when $V_i = 4$, $I_{C1} = 3.3$ mA. Hence, B(at $V_i = 3$ V) $= 3 - 0.25 \times 2.3 = 2.43$, and B(at $V_i = 4$ V) $= 3 - 0.25 \times 3.3 = 2.17$.

From the above calculation, one can conclude that B = 3 V for $V_i = 3^-$ V; and B = 2.43 V for $V_i = 3^+$ V. However, T_1 will be ON, T_2 will be OFF if $V_i = B + \varepsilon$, and T_1 will be OFF, T_2 will be ON, if $V_i = B - \varepsilon$, where $\varepsilon\ \Delta$ infinitesimal value.

Although B is a function of I_{C1} according to Equation 17, it has only two critical values, i.e., $B_H = 3.0$ V for $V_i = 3^-$, and $B_L = 2.43$ v for $V_i = 3^+$. In other words, B = $B_H = 3$ V, if V_i is approaching 3 V from the low value; and B = $B_L = 2.43$, if V_i is approaching 3 V from the high value. Therefore, (1) if V_i is approaching from low value, T_2 will be turned OFF when $V_i = 3 + \varepsilon$, and (2) if V_i is approaching from high value, T_2 will be turned ON when $V_i = 2.43 - \varepsilon$. The difference of B_H and B_L is known as the hysteresis of the Schmitt trigger. Figure 16(b) depicts the corresponding waveforms at different points of the circuit. Figure 16(c) depicts its characteristic transfer curve.

The integrated Schmitt trigger circuit is now available on the market. Texas Instru-

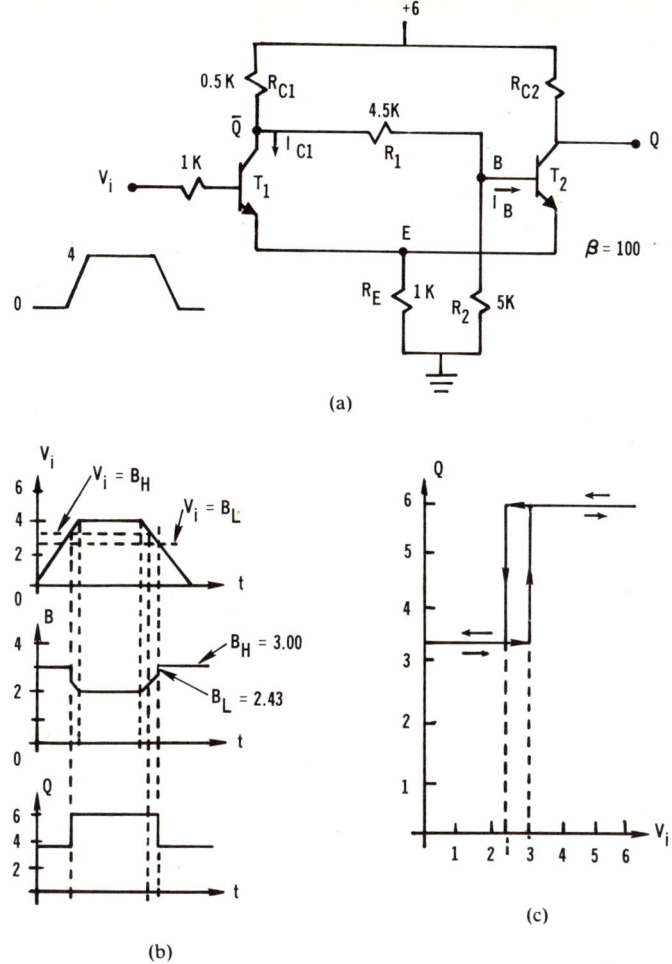

FIGURE 16. Basic Schmitt trigger circuit.

ments for example, provides an I.C. Schmitt trigger circuit registered as SN7413 whose input stage is a four multiemitter transistor, and its input-output transfer curve is shown in Figure 17. Figure 18(a), (b), and (c) shows a few typical applications of the circuit.

In Figure 18(a), the circuit functions as a pulse stretcher. The waveforms at different points are shown accordingly. Note than as x is HIGH, V_{in} is LOW and when x drops to low, the capacitor C is being charged to HIGH through the output of the NAND gate. V_{in} exponentially rises and $\overline{Q}$ stays at Q_H = 5 V until V_{in} reaches V_H, the upper threshold of the Schmitt trigger, at which the Q drops to LOW. The stretched time is determined by the time constant τ which is the product of C and the effective parallel resistance of the output and the input resistance of the NAND gate and Schmitt trigger, respectively. Knowing the input waveform V_{in}, one can plot the output $\overline{Q}$ through the input-output transfer or hysteresis curve shown in Figure 17.

Figure 18(b) is an astable multivibrator using a Schmitt trigger. Consider the instant at which V_{in} is rising from V_L the lower threshold, according to the hysteresis curve, $\overline{Q}$ stays at HIGH but is moving toward the right as the arrow shows. During this time the C is being charged through R and the base resistor (4K) of the multiemitter transistor at the input Schmitt trigger with a time constant

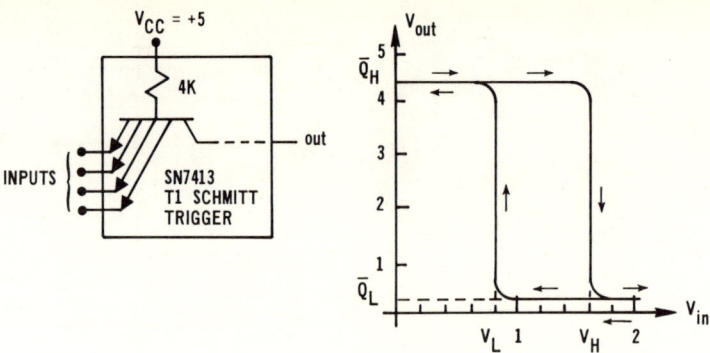

FIGURE 17. Schmitt trigger SN7413.

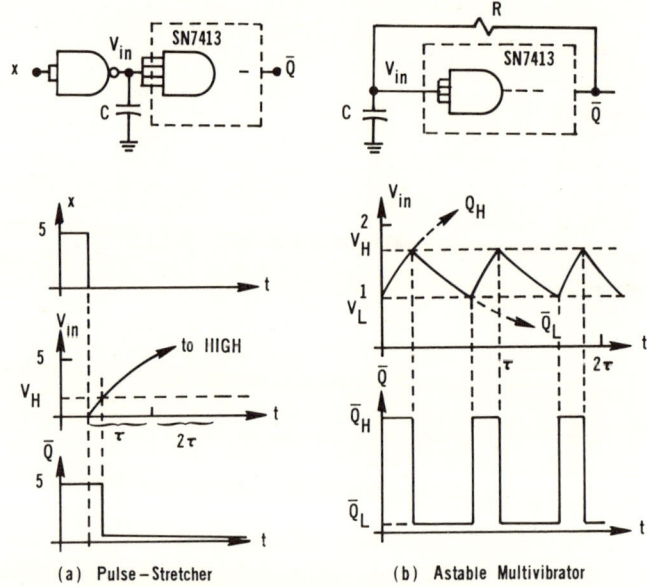

(a) Pulse-Stretcher   (b) Astable Multivibrator

FIGURE 18. Some applications of Schmitt triggers. (Courtesy of Texas Instruments, Inc.)

$$\tau \simeq C\left(\frac{R \cdot 4K}{R + 4K}\right)$$

toward $\overline{Q}_H$. As soon as V_{in} has reached the upper threshold V_H, $\overline{Q}$ drops to LOW ($\overline{Q}_L$) and V_{in} discharges exponentially toward $\overline{Q}_L$ with a time constant.

$$\tau' \simeq C\left(\frac{R \cdot R_{in}}{R + R_{in}}\right)$$

where $R_{in} \triangleq$ the input resistance of the Schmitt trigger, which need not be 4K this time. During this time $\overline{Q}$ stays LOW until V_{in} reaches V_L from V_L^+ at which time the $\overline{Q}$ jumps to $\overline{Q}_H$ according to the hysteresis. The cycle repeats and thus the circuit functions as an astable multivibrator. The waveforms at the points of interest are shown in Figure 18(b).

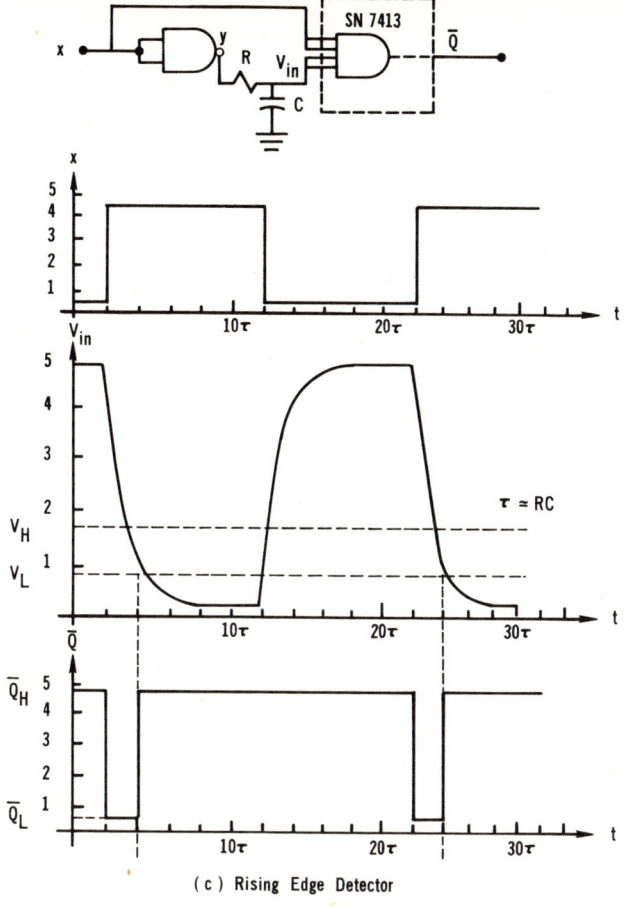

(c) Rising Edge Detector

FIGURE 18(C).

Figure 18(c) shows another application of the Schmitt trigger circuit. The circuit functions as a rising-edge-detector. The four input terminals are divided into two groups and connected as shown. Consider x is at LOW, and V_{in} is HIGH, then $\overline{Q}$ will be HIGH. Now, as x jumps to HIGH, y drops to LOW immediately but V_{in} is being discharged through R exponentially toward LOW. Before V_{in} reaches V_L, all four input terminals of the Schmitt are greater than V_L, thus $\overline{Q}$ will stay LOW until V_{in} falls to V_L. At this time, according to the hysteresis the $\overline{Q}$ jumps to HIGH. Now, as x drops to LOW again, the y responds to HIGH, which will charge the capacitor C through R. Thus V_{in} increases exponentially toward HIGH. Although V_{in} passes through V_H, at which $\overline{Q}$ would normally drop to LOW, yet $\overline{Q}$ will stay HIGH, in this case, due to the fact that x is LOW and two of the four inputs of the Schmitt trigger stay LOW which is lower than V_H. The waveforms at the points of interest are shown in Figure 18(c).

V. REGISTERS AND COUNTERS

A. Registers

A register is basically a collection of a set of flip-flops being logically connected together to perform one or more specific functions, such as latch, buffer, right or left-shift, parallel-in/parallel-out, serial-in/parallel-out, etc. Fortunately, registers with

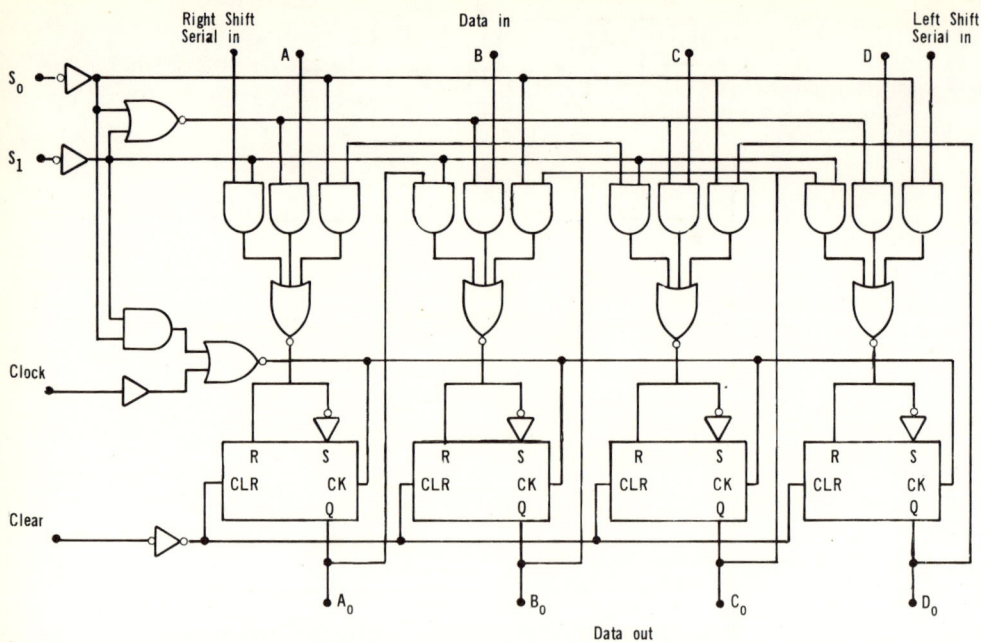

FIGURE 19. Logic diagram of a universal shift register.

specific functions are available on the market. As a designer one would only need to be familiar with the functions and make proper selections. As an example for familiarization, Figure 19 shows the logic diagram of a 4-b bidirectional universal shift register, 74194. Note, that there are three distinct subsets of terminals, i.e., the controls, input data, and output data. The control lines contain the function or mode controls, s_0, s_1, clear, and clock. The following table shows the code of the mode controls.

S_0	S_1	Mode
0	0	Clocking of the flip-flop is inhibited
0	1	Shift-left synchronous with rising edge of the clock
1	0	Shift-right synchronous with rising edge of the clock
1	1	Parallel loading of input data

By setting the control mode, one can use this register as a serail-in/serial-out, parallel-in/parallel-out, serial-in/parallel-out and parallel-in/serial-out data manipulator. It is important to point out that the designer should carefully study the timing diagram provided on the data sheet before designing the control signals.

B. Counters

A counter is also a collection of a set of flip-flops, which is used for counting events or sequencing some controls. Again, there are many kinds of counters available on the market in I.C. packages. For example, one can purchase some basic counters, such as, up/down binary counters, ring counter, decode counter, etc., at very reasonable cost. More sophisticated ones which offer different kinds of input/output options are also available. Although more sophisticated, they are all designed with the basic counters, or flip-flops with gates. To illustrate, a design procedure for a Modulo-6 counter using D flip-flops is described in the following example.

Since a count of six is desirable, we need three D flip-flops. The design procedure is presented as follows.

Transition Table

```
       D-FF
   X   Y   Z    Sequence
   0   0   0    0 ←----┐
   0   0   1    1      │
   0   1   0    2      │
   0   1   1    3      │ For Modulo-6
   1   0   0    4      │
   1   0   1    5 ←----┘
   1   1   0    6
   1   1   1    7
```

Derivatation of Input Equations

Since the sequence for Modulo-6 is,

State 0 → State 1 → ... → State 5

and the input equation for D flip-flop is: Q_{n+1} = (clock) · D_n, the input equation for each flip-flop can be derived based on the Transition Table and its Karnaugh map. Let the D-input for X, Y, and Z flip-flop be D_x, D_y, and D_z, respectively, then for X flip-flop:

X \ YZ	00	01	11	10
0	0	0	1	0
1	1	0	—	—

X_{n+1} ≜ Next sequence output of X flip-flop

Input Equation → $D_x = XZ' + YZ$

For Y flip-flop:

X \ YZ	00	01	11	10
0	0	1	0	1
1	0	0	—	—

Y_{n+1} ≜ Next sequence output of Y flip-flop

Input Equation → $D_y = X'Y'Z + YZ'$

For Z flip-flop:

X \ YZ	00	01	11	10
0	1	0	0	1
1	1	0	—	—

Z_{n+1} ≜ Next sequence output of Z flip-flop

Input Equation → $D_z = Z'$

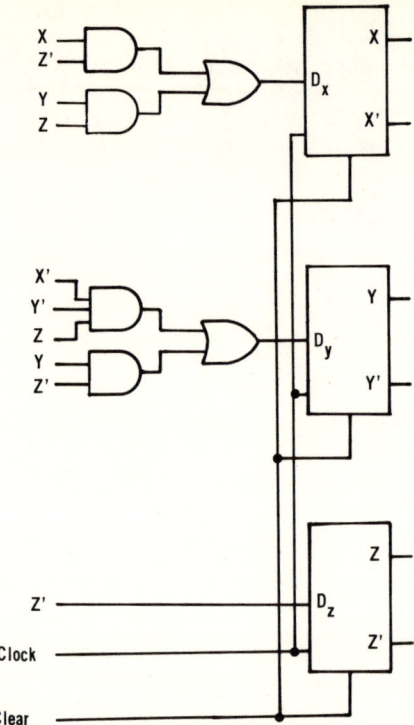

FIGURE 20. Logic diagram of Modulo-6 counter.

Logic Diagram

Figure 20 shows the logic diagram of the Modulo-6 counter; here the edge-triggered D flip-flop is used. This design technique can be used to design counter of Modulo-x where x is any positive integer and is less or equal to 2^n, with n as total number of flip-flops required.

VI. ENCODERS, DECODERS, MULTIPLEXERS, AND DEMULTIPLEXERS

A. Encoder

An encoder is a logic network which encodes a specific input-line by a set of binary output lines. Keyboard encoding or range selection, for examples, are typical applications. Figure 21 shows the block diagram and the truth table of the ten-line to four-line priority encoder 47147. Note that there are only nine input-lines. The zero-line of the ten input-lines is economically omitted, because logically, when all nine lines are HIGH, it implies that the zero-line is LOW.

B. Decoder

A decoder as its name implies, is a logic network which functions as the inverse of the encoder. Address decoding network of a memory is one of the typical applications. For example, 74154 is a device which decodes four input lines into one of 16 mutually exclusive output lines, which performs the inverse function of the 74147 encoder.

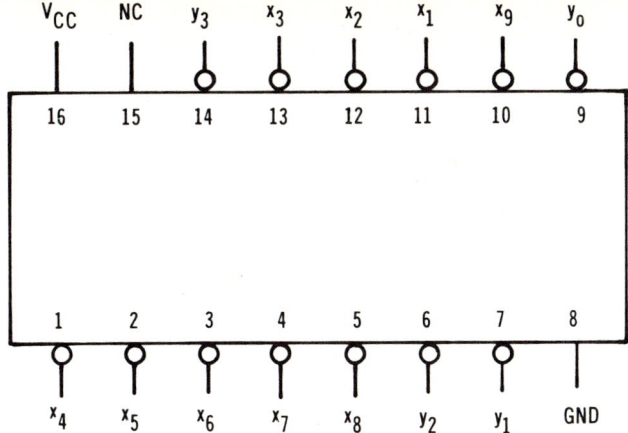

Truth Table

INPUTS									OUTPUTS			
x_1	x_2	x_3	x_4	x_5	x_6	x_7	x_8	x_9	y_3	y_2	y_1	y_0
1	1	1	1	1	1	1	1	1	1	1	1	1
-	-	-	-	-	-	-	-	0	0	1	1	0
-	-	-	-	-	-	-	0	1	0	1	1	1
-	-	-	-	-	-	0	1	1	1	0	0	0
-	-	-	-	-	0	1	1	1	1	0	0	1
-	-	-	-	0	1	1	1	1	1	0	1	0
-	-	-	0	1	1	1	1	1	1	0	1	1
-	-	0	1	1	1	1	1	1	1	1	0	0
-	0	1	1	1	1	1	1	1	1	1	0	1
0	1	1	1	1	1	1	1	1	1	1	1	0

FIGURE 21. 74147 10-to-4 encoder.

C. Multiplexer

A multiplexer is another type of MSI package. Logically, its time multiplexes an output line to a set of input lines. Electrically, it functions as a rotary switch with the rotating arm as its output. A device such as 74153 is a dual four-line to one-line multiplexer. Two control-lines are used to selectively connect one of the four input lines to the output. Figure 22 depicts the logic diagram of 74153 and its rotary switch equivalent.

D. Demultiplexer

A demultiplexer is a device which performs the inverse function of a multiplexer. It electrically distributes an input data line to a set of output data lines. It is similar to a rotary switch whose central arm, however, is now used as an input line. It is therefore a one-to-many data distribution.

E. Decoder/Demultiplexer

Since the logical diagram for both decoder and demultiplexer are identical, except the way of defining the inputs, the device such as 74155 can be used either as a decoder

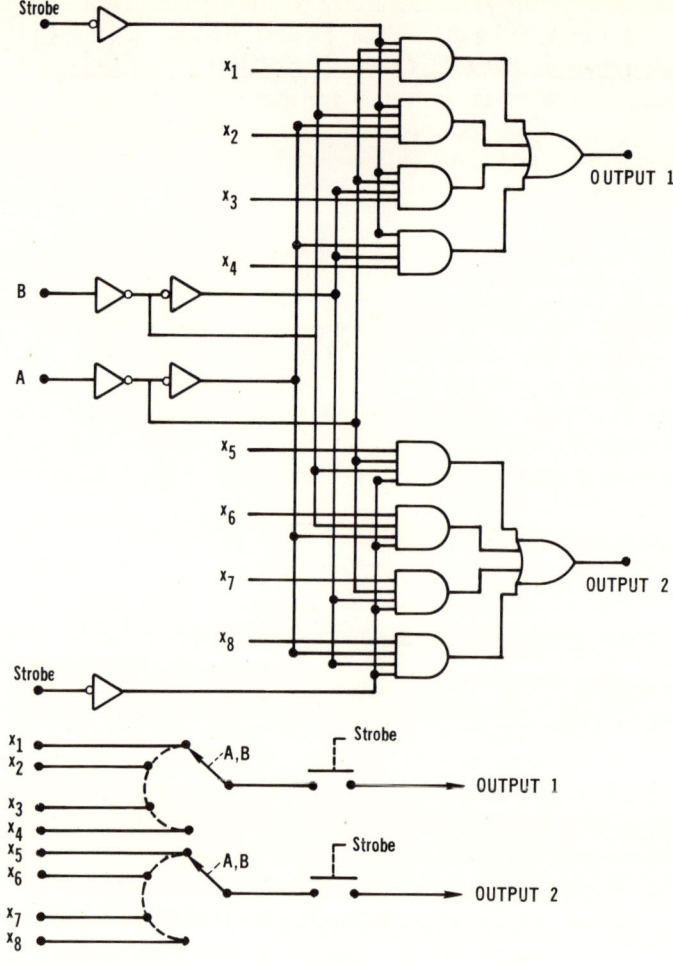

FIGURE 22. Logic diagram of multiplexer 74153.

or as a demultiplexer. By properly defining or labeling the inputs, 74155 can be either used as a dual two-line to four-line decoder, or as a dual one-line to four-line demultiplexer.

VII. LINE RECEIVERS AND DRIVERS

What we have presented in the preceding sections are basically logic elements used within a system. For TTL logic, the binary voltage levels are: logic HIGH $\geq$ 2.4 V, and logic LOW $\leq$ 0.5 V. In many cases, however, we have to deal with data communication, or sending/receiving data through transmission lines. Special devices are therefore required for these applications. There are basically three problems involved, namely, (1) logic level conversions between different systems, (2) impedance matching, and (3) noise immunity and common-mode-rejection. For example, Signetics Corporation provides EIA/MIA Line Driver (8T15) and Receiver (8T16) which will satisfy the EIA (Electrical Industry Association) standard RS-232B and C. For these the binary logic levels are ±6 V; therefore for 8T15 the nominal power supplies for V_{cc} and V_{EE} are, respectively ± 12 V. Receiver 8T16 has a desirable hysteresis feature which will receive ± 6 $\sim$ ±12 V signal and convert it into a steady TTL logic level signal. As another example, most of the manufacturers provide devices called transceivers which receive and transmit conventional logic levels with tri-state logic gate. Some of the receivers provide hysteresis property; others have differential inputs for common mode rejection. Since these devices are normally driving or terminating a transmission line which has low characteristic impedance, the driver should have high input impedance so that the characteristic impedance matching can be achieved by shunting external resistors with appropriate values at its output.

VIII. DEBOUNCER

In many cases, we have to use a push-button switch to directly excite logic elements. Unfortunately the push-button switch would never produce a clean pulse; instead, it would generate a series of noisy pulses which are most unwelcome to the multivibrators and cause errors. It is thus mandatory that all the push-button switches be debounced with special circuitry. Figure 23(a) shows a simple debouncer which uses two resistors and a cross-coupled flip-flop. Although not absolutely free of bounce, it does serve the purpose in most cases. Figure 23(b) shows the circuit of one of the tri-state quad switch debouncers. Note that a strobe and a tri-state control are provided, which will, of course, assure the bounce-free action, if proper timing of the strobe is employed. With the provision of tri-state control, the switch can be disconnected from the output by electric signal or software if it's controlled by a computer.

IX. CONSIDERATION OF INPUT/OUTPUT OF A MODULE

Due to the rapid progress in the MSI/LSI technology, the system designer apparently could consider logic module as a blackbox and design the whole system based on this blackbox concept. In practice however, one would need the information on input/output characteristics of a module and consider the compatibility between modules. Fortunately, most of the data sheets do provide partial input/output equivalent circuits so that the designer can use this information for interfacing modules, or if necessary, one could design the interface circuit for the incompatible ones. To clarify this concept, let us consider a typical example. Figure 24 shows the input/output equivalent circuits of a typical 74S194, 4-b bidirectional universal shift register by Texas Instruments. In view of the input equivalent circuit, it reveals that at input HIGH state, the device has very high input impedance and that at input low state, its input impedance is approximately equal to the equivalent resistance R_{eq} with a current $I = V_{cc}/R_{eq}$ flowing out of the device. The clamping Schottky diode at the input would clamp the negative pulse and if the driving source resistance is approximately equal to R_{eq}, the device would not operate properly in input LOW state. Consider the output equivalent circuit. It reveals that the device has a totem-pole configuration. The upper transistor operates as an emitter follower which can provide high drain current in output HIGH state; the 50 Ω resistor in the collector circuit serves as a current limiter if the output is temporarily shorted to gound. The lower Schottky transistor will sink current to its specification when the device is in its output LOW state.

X. A SUMMARY OF SYMBOLS OF LOGIC ELEMENTS AND MODULES

Logic elements and modules have been described in the preceding sections. Although we have been using the most popular symbols that appear in the literature, it is worthwhile to present a summary of the symbolic tables at this point for convenience. Table 10(a) and (b) show the most popular logic symbols and their definitions.

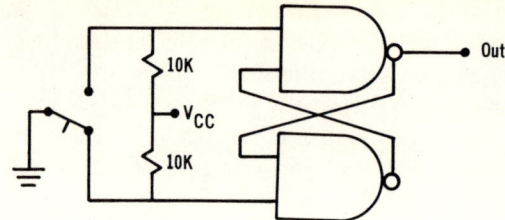

(a) Simple Debouncer Circuit

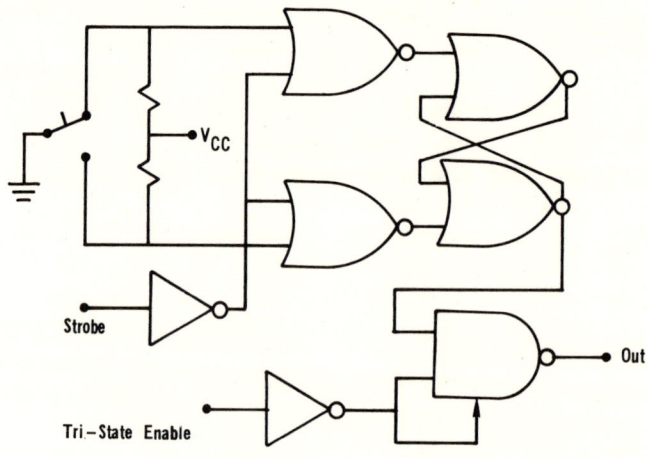

(b) DM75/DM85 44 (Courtesy of National Semiconductor)

FIGURE 23. Switch debouncer circuits.

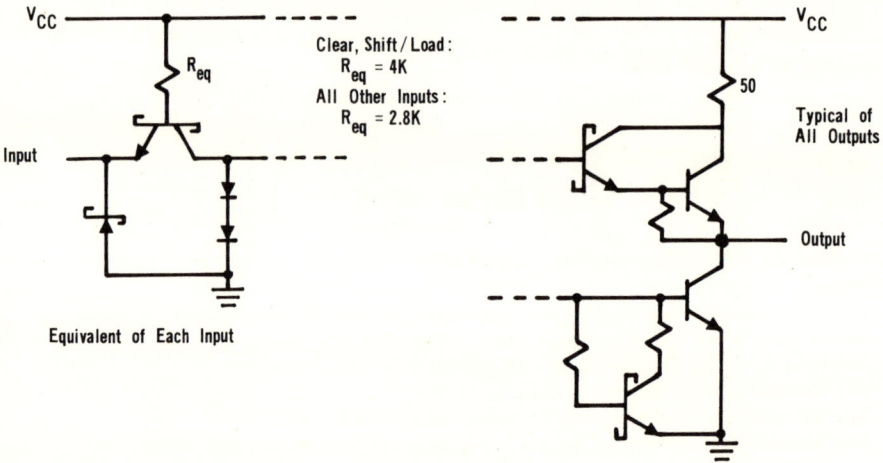

FIGURE 24. 74S194 4-b bidirectional universal shift register. (Courtesy of Texas Instruments, Inc.)

Table 10(a)
SYMBOLS OF LOGIC ELEMENTS AND MODULES

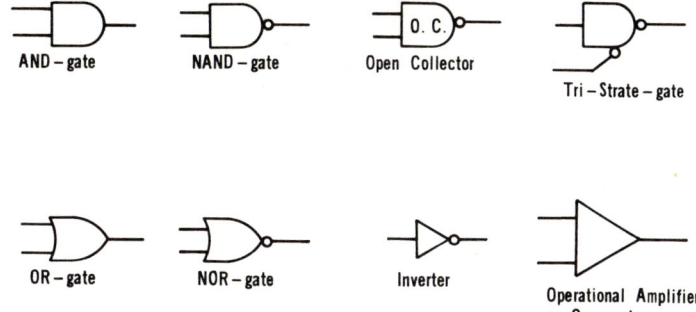

Table 10(b)
SYMBOLS OF LOGIC ELEMENTS AND MODULES

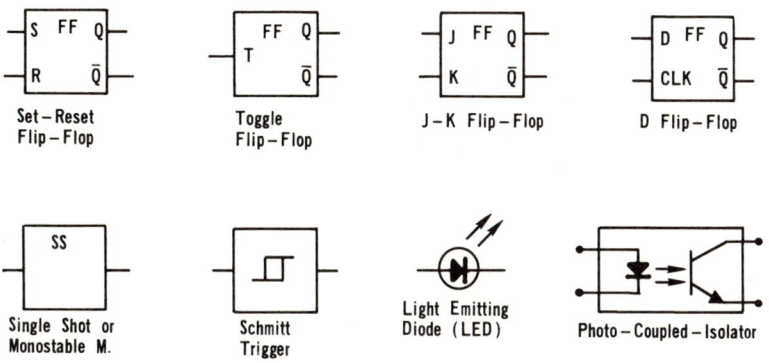

REFERENCES

1. **Millman, J.**, *Microelectronics: Digital and Analog; Circuits and Systems,* McGraw-Hill, New York, 1979.
2. Updating Circuit Symbols, the Graphic Language of Electronics, *Electronics,* April 3, 1975.
3. Analog Data Manual, Signetics, Sunnyvale, Calif., 1977.
4. TTL Data Book, National Semiconductor, Santa Clara, Calif., 1976.
5. TTL Data Book, Texas Instruments, Dallas, 1976.
6. Supplement to the TTL Data Book, Texas Instruments, Dallas, 1977.
7. Data Manual: Logic, Memories, Interface, etc., Signetics, Sunnyvale, Calif., 1976.

Chapter 6

MEMORY SYSTEMS

I. INTRODUCTION

In the early fifties, memory to a logic designer meant flip-flops and registers, whereas memory systems are only digital computer designers' concern. As the technology progresses, the cost of the basic memory devices decreases to the level that a logic designer can now afford to use them to design more sophisticated digital systems. In this chapter we shall present the memory cells/systems for core and semiconductor memories. First, clarifications of the confusing terminologies are in order. Basically, a memory system is a collection of memory cells similar to flip-flops, where one bit information, either ONE or ZERO, can be stored or retrieved. The process of storing digital information is defined as "WRITE", while retrieving is defined as "READ". A set of bits which forms a group is called a digital "WORD", the width of a WORD is defined by the number of bits in that word. Just like an English word which is composed of a set of alphabets, a digital word is composed of a set of digital alphabets, i.e., [1] and [0]. The width of an English word however, varies from one alphabet to many, while the width of digital word in number of bits is normally a constant for each system. A memory system then, may contain tens, hundreds, or thousands of cells organized in word-spaces, where digital information can be stored into (write-process) and retrieved from (read-process). Specifically, in write-process, one should first specify which word-space the information is to be written into, and similarly, in read-process, from which word-space the information is to be retrieved. The location of a specific word is called the "ADDRESS" of that word, and the information stored is defined as the "CONTENT" of that word. The major operations for a memory system are then (1) specifying write/read command, and (2) addressing the location of the word-space. One could imagine a memory system is analogous to the mail boxes in a post office. Each mail box has its unique address, and the mailman places (WRITE) the specific mails into the specific boxes and an addressee takes (READ) the mails out of the specific box.

According to their store/retrieve technologies, memory systems have been classified into the following categories.

A. RAM

A RAM is a Random-Access-Memory system. In this system, one can store or retrieve data at any location or address in random within a specific time. In other words, the data access time for this system is independent of data location.

B. CAM

A CAM is a Content-Addressable-Memory system, also known as Associative Memory system. In this system, memory word has no apparent address. Instead, a segment of the content of a memory word is used as the address of that word. As a typical example in applications, the content of a memory word for an air-flight computer, may contain several segments, i.e., time, date, name of the airline, destination, price, etc. One could use the time and destination segments as the addresses for all memory words which match the desired time and destination, and then one could accordingly choose the preferred date and airline, etc.

C. ROM

A ROM is a Read-Only-Memory system. It is a system containing prestored information by the designer so that the content of it cannot be altered later. Therefore, the content can only be read from it as the name implies. Actually, in our daily life, we see a lot of ROM systems, for example, the street signs, road signs and some kinds of name plates. Their information has been precarved and then placed in specific locations for people to READ-ONLY and they are not supposed to be altered.

D. RMM

A RMM is a Read-Mostly-Memory system. In this system, the content of a memory word, although it can be altered, is mostly to be read. In this kind of memory system, the writing process normally requires a much longer time than the reading process.

II. RAM-MAGNETIC CORE MEMORY

A. Memory Cell
Principle of Operation

A memory cell of a magnet core memory system contains a doughnut-shaped tiny magnetic core with several wires going through the hole as shown in Figure 1(a). The magnetic core has a desirable hysteresis characteristic curve for current (i) vs. magnetic flux (ϕ) shown in Figure 1(b). In Figure 1(a), there are four wires, where read, write, and inhibit wires are used for pulse-excitation purposes. When excited, a signal will be induced in the sense wire with a magnitude proportional to $d\phi/dt$. Figure 1(c) shows the timing diagram of the read/write processes. The principle of operation follows.

At quiescent point, no current flows in any of the wires. The core is in one of the two steady states shown in Figure 1(b), i.e., [1] or [0]. Consider the core is currently in [1] state. The read cycle starts with a current pulse of $(1/2)I_\theta$ in both x_i-wire and y_j-wires (Figure 1(c)). The net current flows through the core-hole will be I_θ which is the threshold of the excitation current (Figure 1(b)) so that it will reverse the direction of the magnetic flux in the core. Due to the hysteresis characteristic shown in Figure 1(b) by the arrows, the core will not return to [1] but stay in [0] state as the current in the wire becomes zero. This flux reversal will then yield a voltage pulse with a detectable magnitude in the sense wire. Now, should the core be in [0] state, although the total excitation (read) current of x_i and y_j wires is still I_θ, the core will however return to [0] state after the current pulse is gone. There will thus be no significant flux change, and it would induce considerable low voltage in the sense wire. Therefore, based on the sense wire output, the x_i-y_i read currents can be used to determine the original state of the core. Consider the write-process: it is important to point out that the core memory is designed in such a way that write-cycle is always preceded by a read-cycle. Therefore, it is evident that the core is always in [0] state before write-cycle. In reference to Figure 1(a) and (c), the write currents in x_i and y_i are always equal in magnitude but opposite in direction with respect to read currents. Depending on what information is to be stored in the core, one could accordingly set the inhibit current to be logic ONE or ZERO as shown in Figure 1(c). By setting an inhibit line with a positive current pulse of $1/2\ I_\theta$, the net write-current through the hole of the core will be $-1/2\ I_\theta$ which would not be able to reverse the flux direction. As a result, the core will remain in [0] state. For writing logic 1 in the core, one just simply disables the inhibit line and the core will then be driven to [1] state by the write-current pulses.

Nonvolatile and Destructive Read-Out

In view of its principle of operation, the magnetic core memory has two important

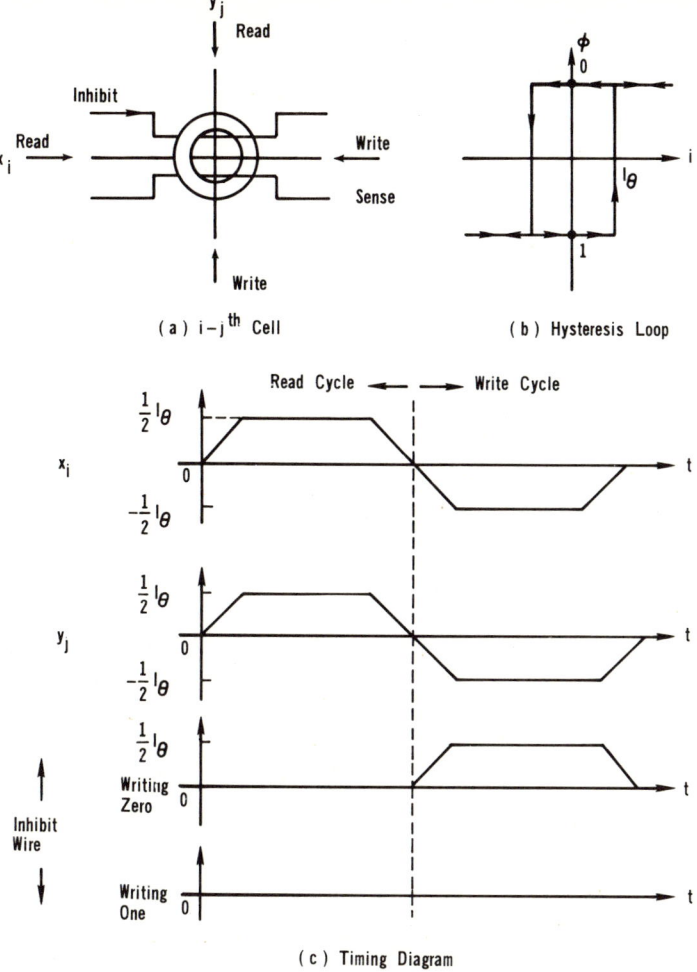

FIGURE 1. Magnetic core memory cell.

characteristics worthwhile to mention. First, since the core does not require standby power for quiescent state and the information will remain unchanged even without power, it is a nonvolatile memory device or a device which is power-fail-proof. Second, as the core always returns to [0] state after being read, it is called destructive read-out device. That is, the original information is destroyed after it is read. This of course is not a desirable feature, therefore a rewrite mechanism after read-out is always required in a core memory system.

B. Configuration of A 4 × 1 Memory Plane

For clarification, Figure 2(a) shows the configuration of a memory plane which has four memory words and each word is only 1-bit wide. The diagram depicts the concept of operation. Figure 2(b) shows the concept of a memory map where the address defined the location of each memory word and the content, either [1] or [0] is the data stored after write-cycle.

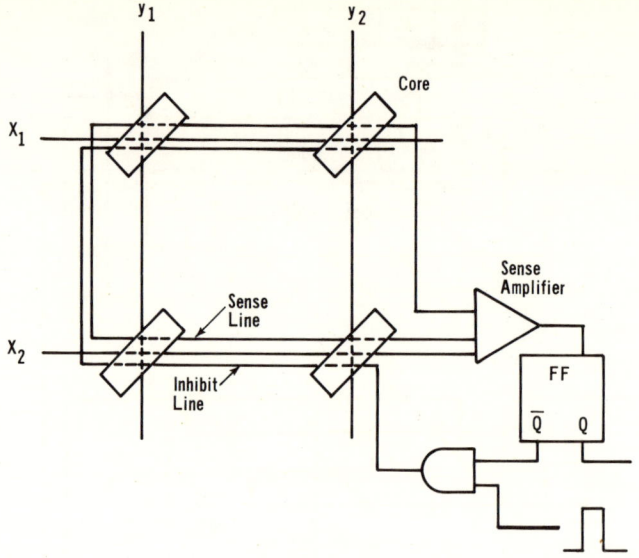

(a) 4 X 1 Memory Plane

ADDRESS				CONTENT
x_1	x_2	y_1	y_2	
1	0	1	0	1/0
1	0	0	1	1/0
0	1	1	0	1/0
0	1	0	1	1/0

(b) Memory Map

FIGURE 2. Configuration of a 4 × 1 memory plane.

C. System Configuration
3-D System

Figure 3(a) shows the configuration of a 16 × 3 words 3-D core memory system. Here, we have 16 memory words and each word is 3-b wide. In this diagram, we introduce the terminology of Memory-Address-Register (MAR) and Memory-Data-Register (MDR). Here, MAR is a 4-b register, A_0, A_1, A_2, A_3, and A_0, A_1 define the y-components and A_2, A_3 define the x-components of the address. For example, if the MAR has the value, $A_0A_1A_2A_3 = 0011$, the right-lower-corner word will be addressed and its content will be loaded into MDR, then the rewrite-cycle follows. As the figure shows, this is obviously a 3-D (three dimensional) system.

Figure 3(b) is a general block diagram of a 3-D memory system. Note that there are four blocks function as read-write, x-select, y-select, and inhibit drivers. Those drivers are basically current amplifiers; in most systems they are made of larger magnetic cores capable of delivering sufficient current to drive through the wires of the memory cells. Although the diagram depicts only the organization of a 3-D core memory system, it also shows the major components for other systems such as 2-D and 2-1/2-D described in the next two sections. Furthermore, it illustrates that any memory system including the semiconductor memory system can be considered as a black-box which interfaces with the outside world with two registers, namely, MAR and MDR, as well as three groups of lines called "Buses" which are functionally known as address-bus, data-bus, and control-bus. For a N × M memory system, there will be N address lines (after

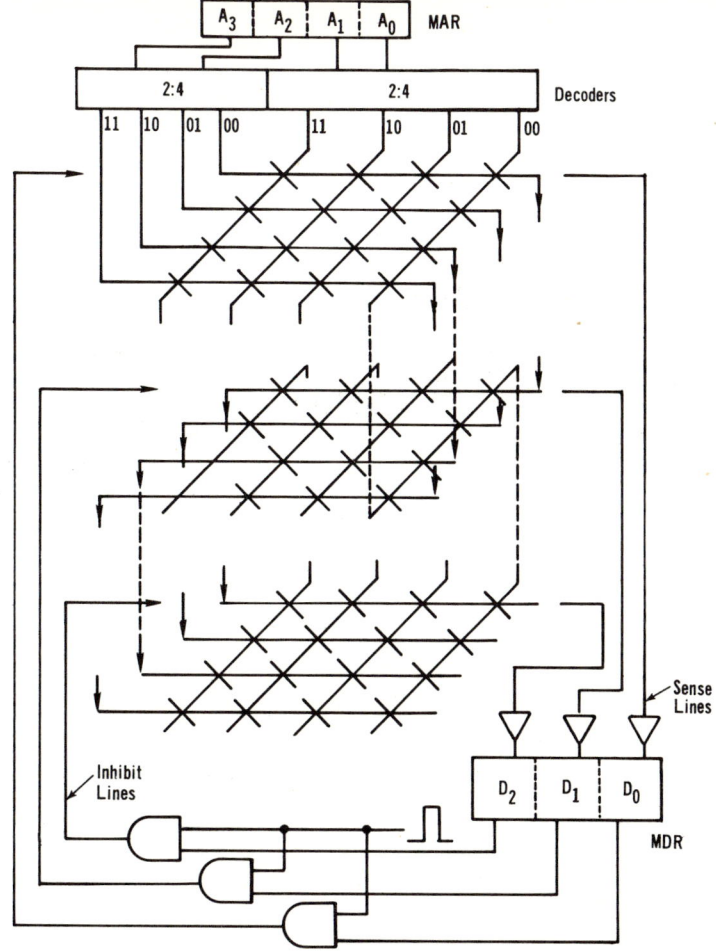

FIGURE 3(a). Configuration of a 16 × 3 words 3-D memory system.

decoded) and M data lines. Accordingly, the width of the MAR would be $\log_2 N$ and that of MDR would be M. The number of lines for control-bus varies from one system to the other. It would contain read/write commands, clock, inhibit, and others.

2-D System

Figure 4 is a simplified diagram which shows the configuration of a 16 × 3 words, 2-D system. For simplicity, the inhibit circuitry is not shown. Note that the address has no x-y components, the 16 words, $W_0, W_2, \ldots W_{15}$, can be randomly selected by one of the 16 decoded address lines. Unlike the 3-D system shown in Figure 3 where each address line drives 12 cores, in this system, each address line drives only 3 cores.

2-1/2-D System

This system as its name implies, stays in the middle of the 3-D and 2-D systems based on its cost and performance. Figure 5 shows how the 3-D system is evolved into 2-1/2-D system. For convenience, the 16 × 3 words 3-D system shown in Figure 3 is reproduced in Figure 5(a). Conceptually, one could cut the x-address lines at the Xs and then re-configurate it in the way as shown in Figure 5(b). Note that the number of y-address lines have reduced to one half of the original, and the number of x-address

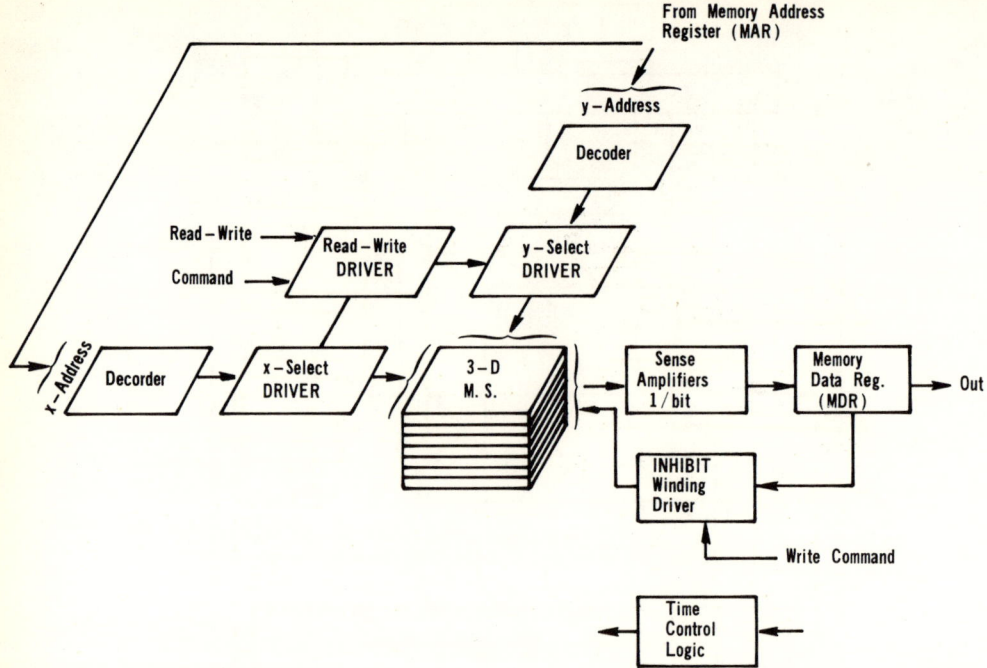

FIGURE 3(b). A general block diagram of a 3-D memory system.

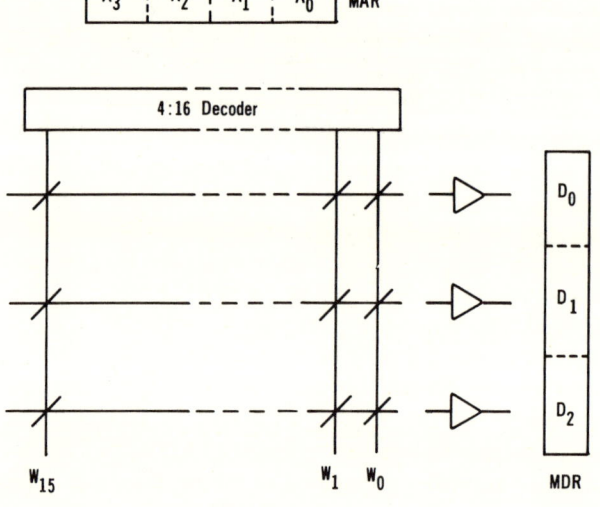

FIGURE 4. Configuration of a 16 × 3 words 2-D memory system.

lines has been doubled, so that the total number of memory words remains unchanged. Note that in this example, while each of the y-address lines are still driving 12 cores, the x-address is driving six cores with three parallel lines, two cores per line.

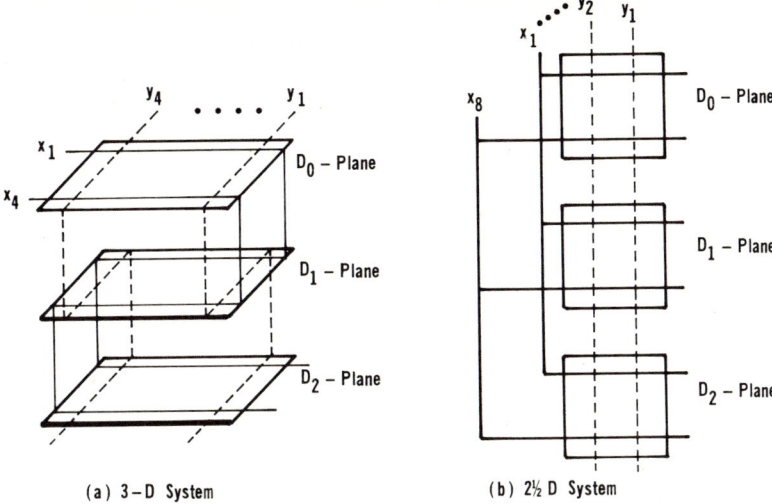

FIGURE 5. Evolution of 2-1/2-D system from 3-D system.

III. RAM-SEMICONDUCTOR MEMORY

A. Memory Cell

Semiconductor memory, as one may easily imagine, is a collection of registers which in turn is composed of a set of semiconductor flip-flops. Therefore, a semiconductor-memory cell is basically a flip-flop plus the minimum necessary peripheral circuitry for address and data write-in and read-out. To understand how a semiconductor memory system is organized and how it functions, let us examine the circuitry of a few different kinds of memory cells.

Static Memory Cll

Figure 6(a) through (d) shows the circuitries of memory cells for different kinds of solid state technologies. Each of the cells basically contains a flip-flop, a word (address) line, and a pair of bit (data) lines. Therefore, their operations are similar. For example, consider the flip-flop shown in Figure 6(a), for Q_1 ON, and Q_2 OFF. The cell is in steady state ad the word line is LOW. In WRITE process, the Q-bit line is set HIGH and $\overline{Q}$-bit line, LOW. When the world-line goes HIGH, the cell is selected; the B-E junctions connected to word-line of both Q_1 and Q_2 will be OFF. Since $\overline{Q}$-bit line is LOW, Q_2 conducts and its emitter current flows into $\overline{Q}$-bit line. In the meantime, Q_1 will be OFF because of the fact that Q-bit line is set HIGH. As a result, a [1] is stored in Q_2 as soon as the world-line has returned to LOW and bit-lines to HIGH. In READ process, the sense amplifier connected to the bit-lines will detect the current if Q_2 has been ON. In summary, the word-line HIGH implies cell selected and the bit-lines carry SENSE signal for READ and DRIVE signal for WRITE. By the similar analysis, one can describe the READ/WRITE process for the ECL cell shown in Figure 6(b). Here, in steady-state, $I_c = I_E$. During the READ/WRITE process, the word-line goes HIGH and $I_c = I_E + I_Q$, where, I_Q flows in the bit-lines. Thus we have again "SENSE for READ" and "DRIVE for WRITE" as described.

Figures 6(c) and (d) are MOS memory cells. Since MOS-FETS are voltage control devices, their configurations are quite straightforward. Q_5 and Q_6 in both circuits are used as switches. They connect the flip-flop to its corresponding bit lines, we have SENSE for READ and DRIVE for WRITE. It is evident that the C-MOS memory cell consumes extremely low standby power.

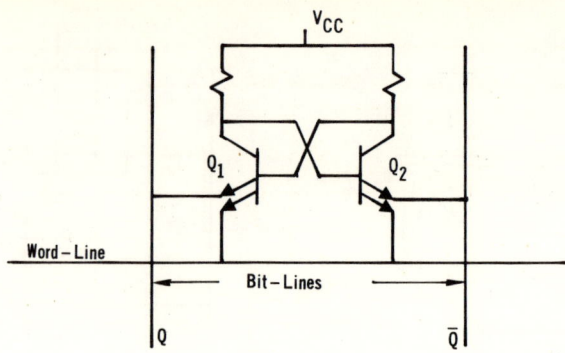

FIGURE 6(a).   TTL memory cell.

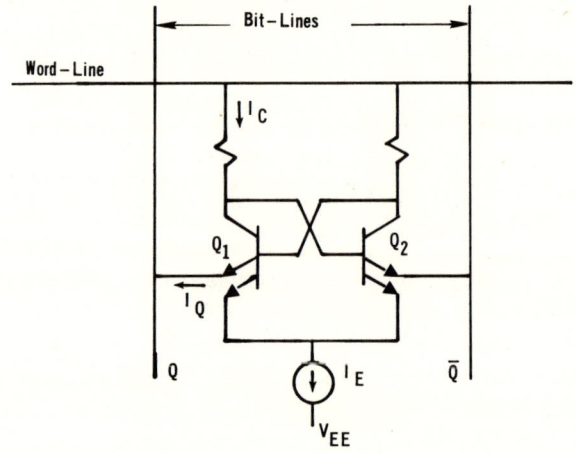

FIGURE 6(b).   ECL memory cell.

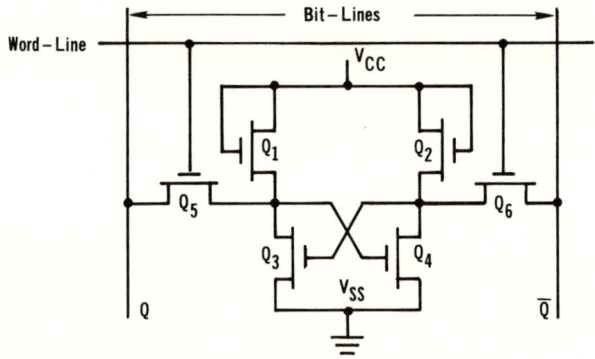

FIGURE 6(c).   MOS memory cell.

Dynamic Memory Cell

It is important to point out that the static memory cell just described requires more space in comparison with the dynamic memory cell. For the static memory cell, information is stored in a flip-flop and thus it can be detained indefinitely as long as the power is ON. For the dynamic cell however, the information is retained in a capacitor which may leak away as time elapses. Therefore it requires a refresh-circuit to reinforce

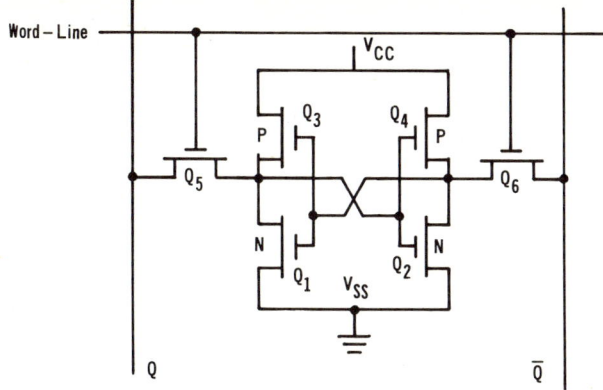

FIGURE 6(d).   CMOS memory cell.

the information every 2 msec. Figure 7(a) shows an older type of three-transistor dynamic memory cell. Here, data is stored in the capacitor through the switch Q_1 and read through Q_3. Due to the circuit simplicity of the memory cell, a more sophisticated peripheral circuitry is required in addition to the standard sense/drive circuits for bit-lines. A brief description of the READ/WRITE process follows. In READ process, the data-out line is first precharged to logic [1], then Q_3 is turned ON by Read-enable (word) line, and the data-out line will be discharged or not discharged through Q_1 depending on the information stored in capacitor C. The sense amplifier connected to data-out line would sense the information stored. For writing, data-in line is activated with the desired information by the DRIVE circuitry connected to it. Q_1 turns ON when the write-enable (word) line is activated and the information in the data-in line is then stored in the capacitor. For refreshing process, the information in the capacitor is first read to the data-out line and then fed back to the data-in line for being written back to the capacitor. This process is repeated every 2 msec.

Figure 7(b) shows a more advanced memory cell of one-transistor. Here, the capacitor again is used as a data storage element. The transistor is selected and activated by the word-line, and it functions as a switch, transferring the information from the capacitor to the data line. Due to its simplicity in circuitry, 16,384 memory cells of this kind can be packed in a 16-pins chip.

Figure 7(c) shows another kind of dynamic memory cell, where the I²L technology is employed. The capacitor, used as the information storage device of the cell, has an effective capacitance of βC μF, where C is the face value of the capacitor and β is the current gain of transistor Q_2. The effective capacitance is also known as the Miller Effect of the transistor. The access time of this cell is much faster than that of the dynamic memory cell using MOS technology. The READ/WRITE operation follows. Let us define that the cell is in logic [1] state, if the capacitor is uncharged or the voltage across the capacitor, $V_c = 0$; and in logic [0] state, if the capacitor is charged. When the memory is in steady state, the word-lines W_p is LOW while W_n and bit-line stay HIGH. Q_1 and Q_2 are then OFF. Just like the other types of memory cells, for bit-line, we have SENSE for READ and DRIVE for WRITE. In READ process, W_n goes to LOW, Now, if $V_c = 0$, Q_2 conducts and then bit-line senses signal, a logic [1] is detected. If V_c = HIGH, however, Q_2 will remain OFF, no signal is detected in bit-line, a logic [0] is assumed. In WRITE process, W_n also goes LOW and the bit-line is set HIGH if the desired information to be stored is [0]. Otherwise, it is set LOW. Consider writing [0]. Since the bit-line is HIGH and W_n is LOW, the capacitor will be charged and logic [0] is then stored. For writing logic [1], the process is more compli-

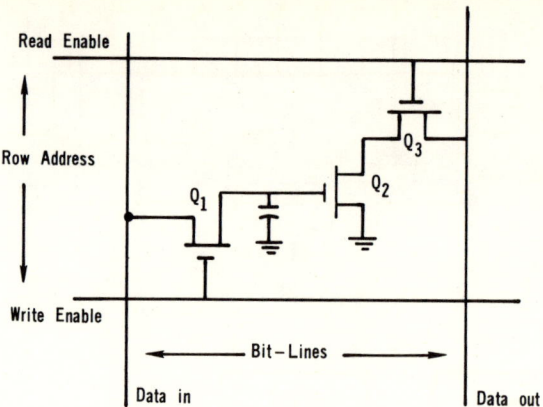

FIGURE 7(a).   3-Transistor memory cell.

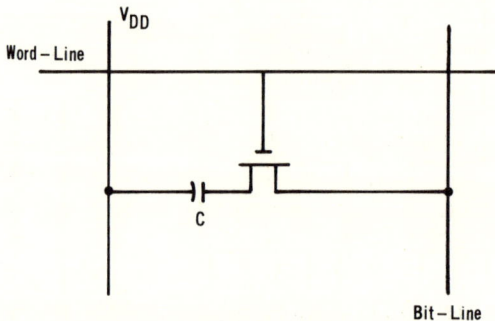

FIGURE 7(b).   1-Transistor memory cell.

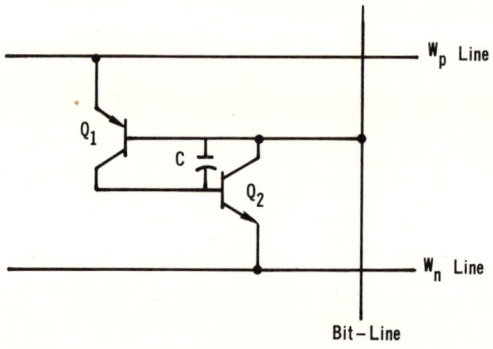

FIGURE 7(c).   I²L memory cell.

cated. First, the bit line is set LOW, then W_n and W_p become HIGH; thus Q_2 stays OFF and Q_1 conducts, which would discharge the capacitor, and logic [1] is then stored. Just like other types of dynamic memory cells, this cell requires refreshing every 2 msec. Due to the circuit configuration, the logic [0] information is automatically refreshed during the READ cycle. Refreshing logic [1] takes place during the writing logic [1] cycle, except that it writes back the current information in the bit-line.

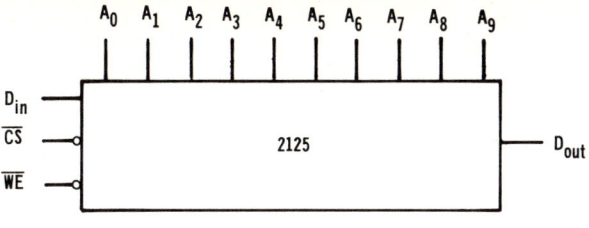

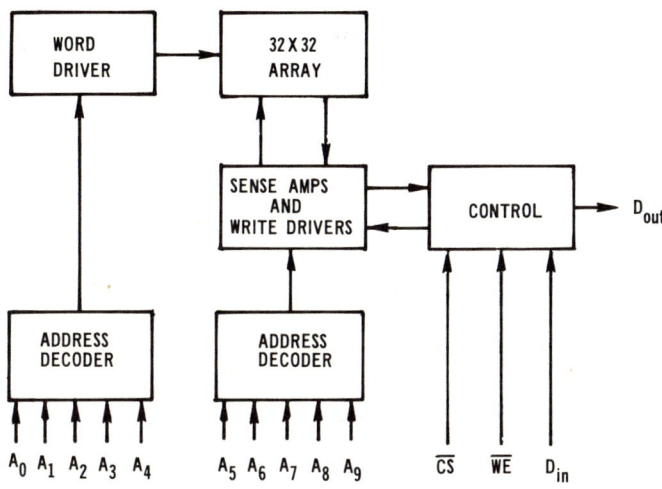

FIGURE 8. 2125 memory chip.

Volatile and Nondestructive Read-Out

In contrast with the core memory, the semiconductor memory cell has a volatile but nondestructive read-out. While nondestructive read-out is desirable, the volatile is not, for an obvious reason, because the information will be lost when power supplies fail. A memory cell of the CMOS type, however, requires very low standby power; therefore CMOS memory with a battery back-up circuitry can be used for systems which require nonvolatile memory.

B. System Configuration

Static RAM Memory System

The semiconductor memory systems are normally organized on chip levels. Although in principle the configuration of a semiconductor memory system is similar to a 3-D core memory system described in the preceding section, it is worthwhile to examine some typical samples. Figure 8 shows the logic symbol and the block diagram of 2125 static ram chip by Intel. Each chip has 1024 words and each word is 1-b wide. Note that there are ten address-bits, $A_0, A_1, \ldots A_9$, which are subdivided into equivalent x-address and y-address and sometimes called ROW-ADDRESS and COLUMN-ADDRESS, respectively. Since the memory word is 1-b wide, there is only 1-b data-in (D_{in}) and 1-b data-out (D_{out}). $\overline{CS}$ stands for chip-select, negative true. That is, when this line is LOW, the chip is selected. $\overline{WE}$ stands for write-enable, negative true. Therefore, when this line is HIGH, the chip is in READ-mode. This memory chip has a tri-state output, therefore the output is in high impedance state when the chip is not in

146   Handbook of Digital System Design for Scientists and Engineers

READ mode. $\overline{CS}$ is actually an address-line when multiple chips are used in a system. For example, Figures 9(a) and (b) show the 3-D configuration of a 64K × 8 or 64 K-byte memory system using 512 INTEL2125 static memory chips. In Figure 9(a), there are 64 subgroups and each group contains eight chips because the memory word is 8-b wide. Thus, each subgroup contains 1024 × 8 words. All the address pins of each chip with the same labels are tied together. That is, all A_0's, A_1's, . . . A_7's and $\overline{CS}$s, respectively, are tied together within each subgroup. However, for simplicity in the figure, only the tie-points of $\overline{CS}$ are shown. Figure 9(b) shows the 16-b MAR (Memory-Address-Register). A_{10}, . . . A_{15} are connected to the 3:8 decoders as shown, whose outputs yield the matrix. Each intersection of the matrix represents an AND gate whose output drives the corresponding $\overline{CS}$. A_0, . . . A_9, respectively, are tied to the corresponding chip-pin with the same label. In this way, each memory word can be addressed by specifying the value of the MAR.

Dynamic RAM Memory System

In comparison with the cell of a static RAM, the dynamic memory cell is far more simple and the packing density of the latter is thus much higher. Since the dynamic cell uses capacitor as the information storage element, it requires refreshing process. Therefore the peripheral circuitry of a dynamic RAM chip is more complicated. Let us examine a typical three-transistor dynamic RAM chip shown in Figure 10. The circuitry of a three-transistor cell is shown in Figure 7(a). In WRITE process, x_i and y_i address lines select the i-jth cell; write command activates the write-enable line. Data provided by the drive-circuit, in the write-column is stored. In READ process, the READ COL is precharged to logic [1]. Selected address lines and read command activate the desired read enable line, the capacitor is charged or discharged by the precharged read column, and thus the sense circuit detects the information. In REFRESH process, the information on each row is read and fed back to the capacitor through the writecolumn. Therefore, the dynamic system is refreshed in a special interval on row-by-row basis. It is evident that a sophisticated timing circuitry is required in a dynamic RAM system. Although straightforward, the peripheral circuitry for a one-transistor-cell, high density RAM would require even more complicated circuitry for addressing and reading processes — the 16K × 1 and 64K × 1 dynamic RAM chip, for example. ROW/COLUMN address information is time multiplexed and differential amplifiers (flip-flop based) are used for detection of the considerably weak data signal from storage capacitors.

IV. CHARGE-COUPLE-DEVICE (CCD)

Charge-Couple-Device is another type of semiconductor memory. The device is basically used as a shift register of an extremely high number of bits. It can be used to fill the gap between the mass memory devices such as magnetic disk or tape, and the RAM. Figure 11 shows its principle of operation. Figure 11(a) shows a segment of the device. Assume that the voltages at each of the electrodes be v_1, v_2, v_3, and v_4 and that $v_3 < v_4 = v_2 = v_1 < \theta < 0$, where θ is the threshold voltage. When the electrode voltage is lower than the threshold of the device, it yields the depletion region. The depth of the region is a function of the applied voltage. When the voltage is sufficiently more negative than the threshold voltage, a POTENTIAL WELL is generated which can receive or store the positively charged holes. To illustrate the data shifting action, Figures 11(b), (c), (d), and (e) show the data shifting process of a three-phase clocked CCD. Figure 11(b) depicts the time waveforms of the voltages at the electrodes for storage and transfer periods. Figures 11(c), (d), and (e) show the shifting of the digital

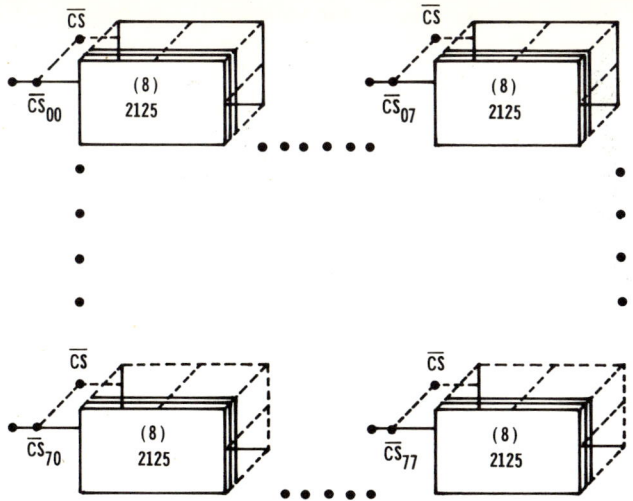

FIGURE 9(a).   A 64 K × 8 memory system block diagram.

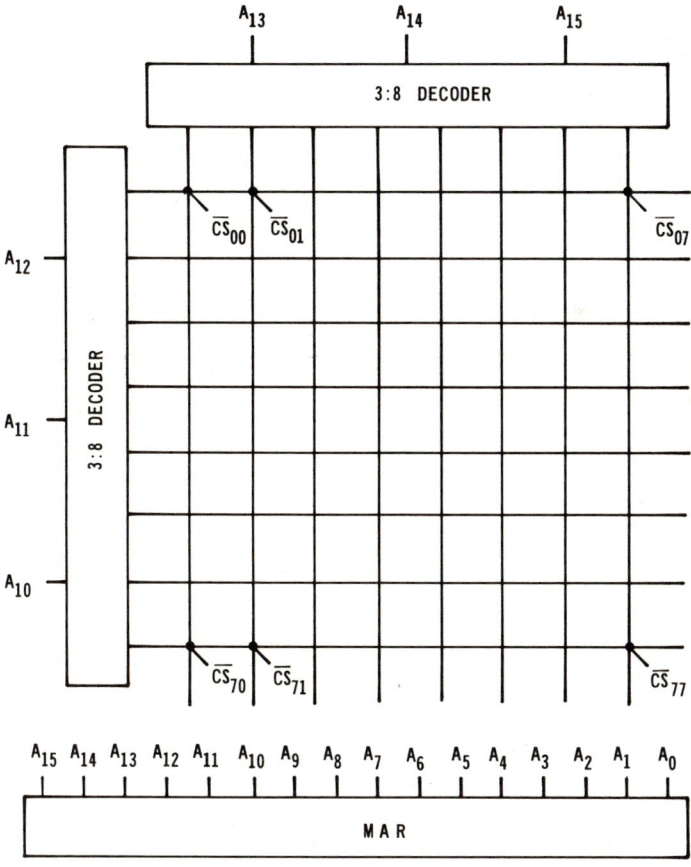

FIGURE 9(b).   Address map for a 64 K × 8 memory system.

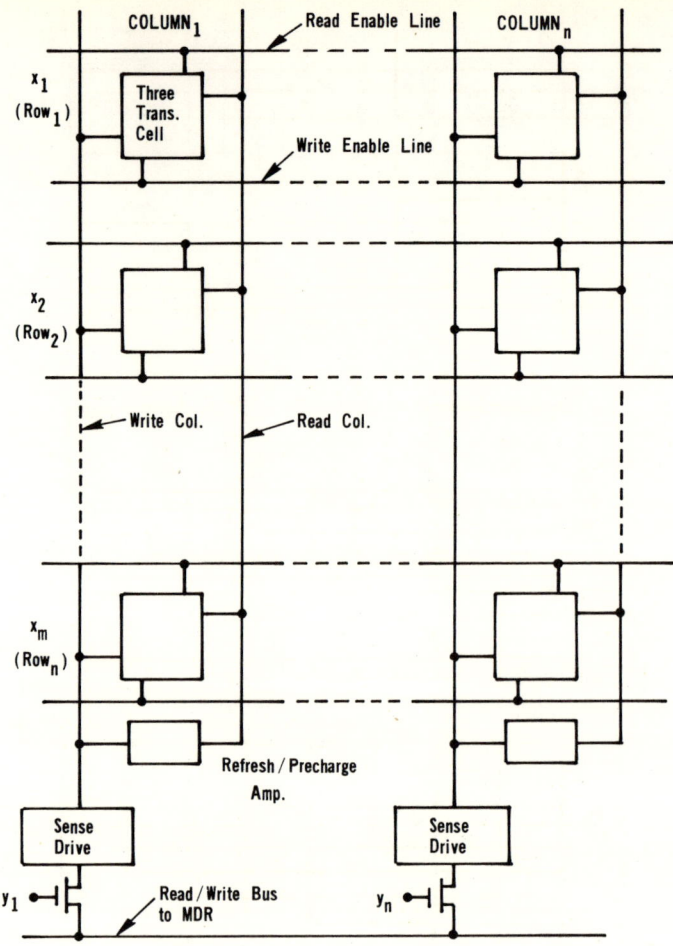

FIGURE 10. 3-Transistor dynamic RAM chip.

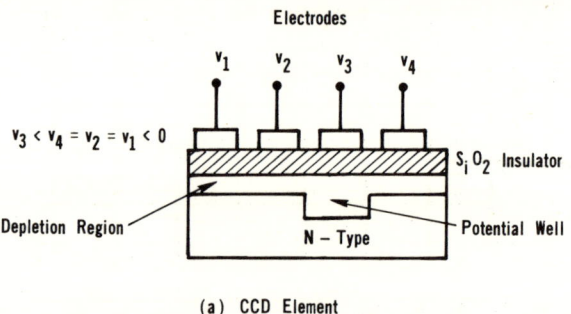

(a) CCD Element

FIGURE 11(a). CCD element.

data. In Figure 11(c), electrodes tied to ϕ_1 generate the corresponding potential wells. Two of the four wells are holding positive charges; thus the device has the digital information as shown accordingly. Entering the shifting mode we have $\phi_2 < \phi_1 < \phi_3$ as shown in Figure 11(b). The corresponding depletion diagram is shown in Figure 11(d). Here, the potential wells of the electrodes tied to ϕ_2 are deeper than their left-hand-

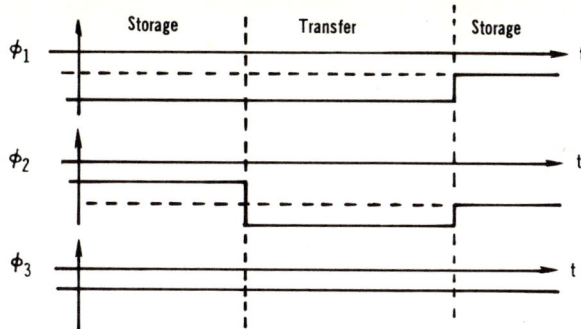

FIGURE 11(b). Voltage waveforms of the electrodes.

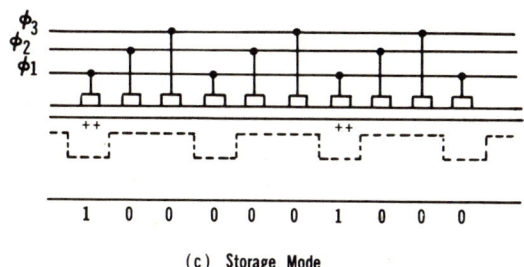

(c) Storage Mode

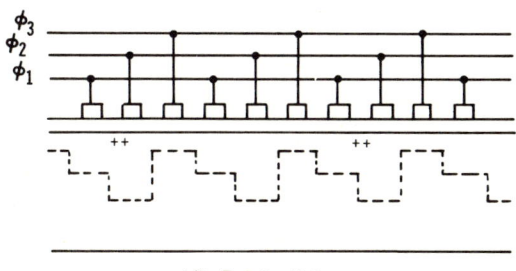

(d) Transfer Mode

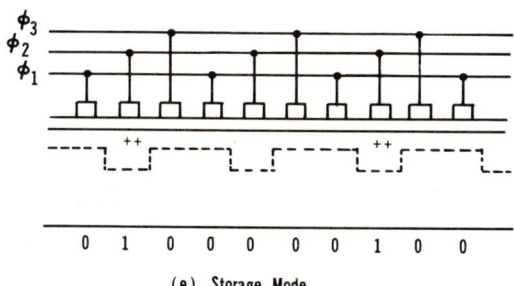

(e) Storage Mode

FIGURE 11(c)(d)(e). CCD data shifting operation.

side neighbors; as a result, the positive charges are transferred to the right as shown. Figure 11(e) shows that the status of the device returns to the storage mode after the transfer and that the corresponding digital data shifted to the right in comparison with Figure 11(c). As one may see for CCD, the fabrication process is simple and its packing density can be very high. The major applications of these devices are digital memory, shift register, image processing, analog, and digital signal processing.

V. ASSOCIATIVE OR CONTENT ADDRESSABLE MEMORY (CAM)

Figure 12(a) shows the operation of a CAM. Assume that the system contains a 4 × 8 memory word with the content as shown. It is to search for a match word whose three most significant bits are 101. Since we are interested in only the three most significant bits, the MASK REGISTER is thus set as shown in Figure 12(a). The matching process is carried out on a bit-by-bit basis; there are three search/match cycles. The search results are chronologically shown on the right in Figure 12(a). Note that once a bit of a certain word is not matched, it would then be eliminated by setting its result bit to zero. In Figure 12(a), we found that the third memory word is selected and its entire content would then be read to MDR. There are cases in which more than one memory word may match the content of the search register after masked operation. In that case, a multiple match resolver is needed. Figure 12(b) shows a typical 16 × 4 CAM chip by Fairchild Camera and Instrument Corporation, whose outputs labeled as M_0, M_1, M_2, M_3, and $\overline{M}_0$ show the search results, and the four enable inputs are for loading desired data into the mask register.

VI. READ-ONLY-MEMORY (ROM)

Read-Only-Memory is a special purpose random-access-memory. It takes a lot longer time to write data into a memory cell than to read data from it. Data can be rewritten into some of them while for others it cannot. One could easily confuse ROM with the terminology RAM. Strictly speaking, one could logically call ROM as a RAM, because a ROM is also a random-access-memory. However, the RAM is traditionally now reserved as Read/Write random-access-memory. Just like RAM described in preceding sections, there are many types of ROMs. Depending on the type of the material being used to make up the memory cell, one can classify them as capacitor ROM, core ROM, and semiconductor ROM. Figures 13(a), (b), and (c), respectively, show the conceptual diagrams of the three different types. Due to the recent rapid progress in solid state technology, semiconductor ROMS almost dominate all fields of ROM applications. Generally speaking, semiconductor ROMs are inexpensive, and have faster access speed. Above all, ROM has the two most desirable properties: namely (1) nonvolatile, and (2) nondestructive read-out. They are widely used for logic realization, table look-up, character generators, etc.

Depending on the method of writing process, the semiconductor ROM can be further classified into the following classes.

A. Nonprogrammable ROM

This type of ROM is permanently programmed during the fabrication process by using a custom-designed mask through a ROM manufacturer. The mask is designed according to the system designer's specifications. It is evident that this kind of ROM is suitable for application where high volume mass production is expected. Once the mask is made, it is difficult or expensive to change the design.

B. Field-Programmable ROM or FPROM

Figure 14 shows a field-programmable ROM. To this device, data can be permanently stored by a burn-in process. That is, to write logic [0] to a cell, an excess current can be applied to that cell so the fuse shown in the emitter circuit can be blown out and logic [0] is then stored. Otherwise a logic [1] remains. Clearly this type of ROM is not reprogrammable.

Search Reg.	1	0	1	1	1	0	1	0

Mask Reg.	1	1	1	0	0	0	0	0

									S_1	S_2	S_3
WORD 1	1	0	0	x	x	x	x	x	1	1	0
WORD 2	0	1	1	x	x	x	x	x	0	0	0
WORD 3	1	0	1	x	x	x	x	x	1	1	1
WORD 4	0	0	0	x	x	x	x	x	0	0	0

Search Results

FIGURE 12(a). Matching operation of CAM.

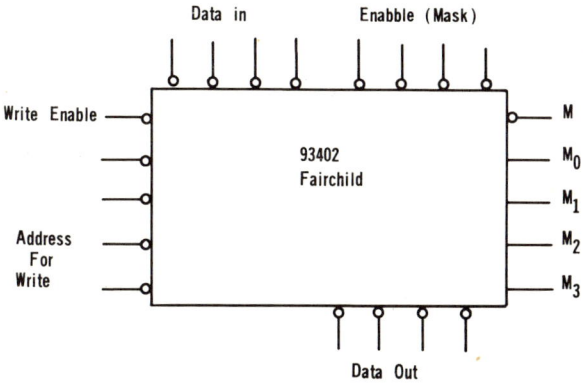

FIGURE 12(b). A typical 4 × 4 CAM chip. (Courtesy of Fairchild Semiconductors.)

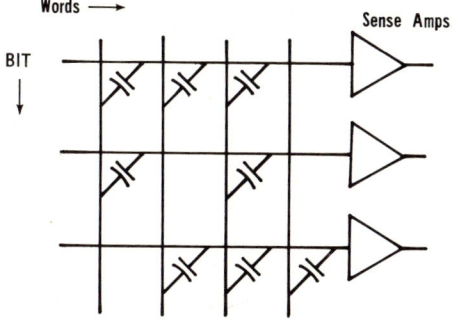

FIGURE 13(a). Capacitor ROM.

C. Ultraviolet Rays-Erasable-Programmable ROM (EPROM)

For this device, data can be written into, and mass-erased by ultraviolet rays. It is therefore reprogrammable and most desirable for applications of low quantity productions. Devices such as 1702, 2708 and 2716 are now available on the market with reasonable cost. 1702 has 256 × 8 b on one chip which requires +5 and −9 V power supplies with 650 nsec access time. 2708 has 1024 × 8 b per chip which requires +5, 12, and −5 V power supplies with 450 nsec access time. 2716 and 2732 have a capacity

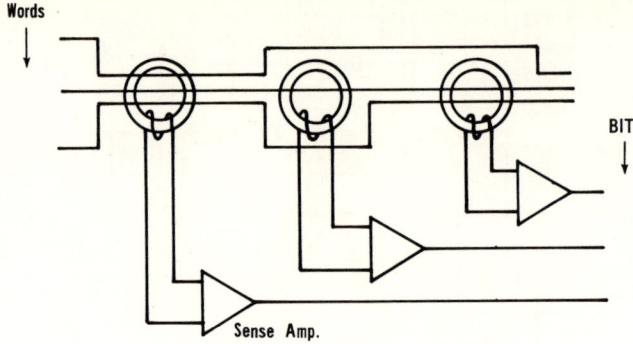

FIGURE 13(b).   Core ROM.

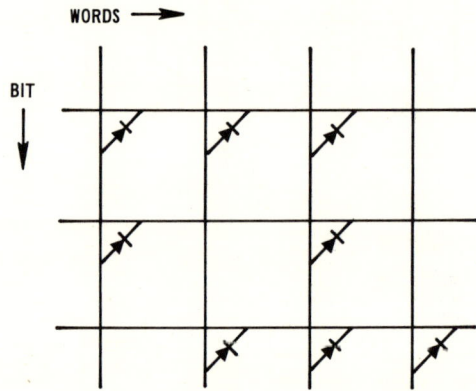

FIGURE 13(c).   Semiconductor ROM.

of 2048 × 8 and 4096 × 8 b per chip, respectively, which requires only +5 V power supply with an access time of 450. They all require 10 to 30 min of time to mass erase the data by ultraviolet rays. This type of ROM is ideal for applications in development/design of systems where short or medium term read-only-memories are essential. For long term memory application, the mask or fusible types ROM would be more desirable.

D. Electrically Alterable Read-Only-Memory (EAROM)

Note that all the different kinds of ROMs described so far are OFF-line in erase/write processes and ON-line in read process. However, it is most desirable that a ROM is electrically alterable or erasable. Since 1978 this type of ROM has been commercially available. Nippon Electric Corporation (NEC), NCR, and General Instrument Corporation among others have made these devices available to the designer. Although it appears that EAROM has all the desirable features, such as nonvolatile and nondestructive read-out in comparison with core and semiconductor RAM, it still requires a rather long time to write data in comparison to normal processor cycle time. This is why at the present time this device is still considered as a Read-Only-Memory device.

E. Read-Only-Memory System

Although the ROM is normally used in read-process, its system organization is similar to RAM. Again, there are three groups of input/output lines in the system, namely,

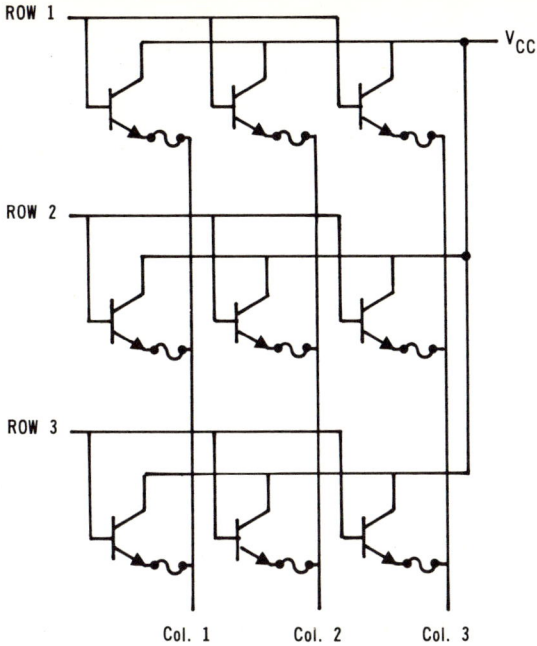

FIGURE 14. Field-programmable ROM.

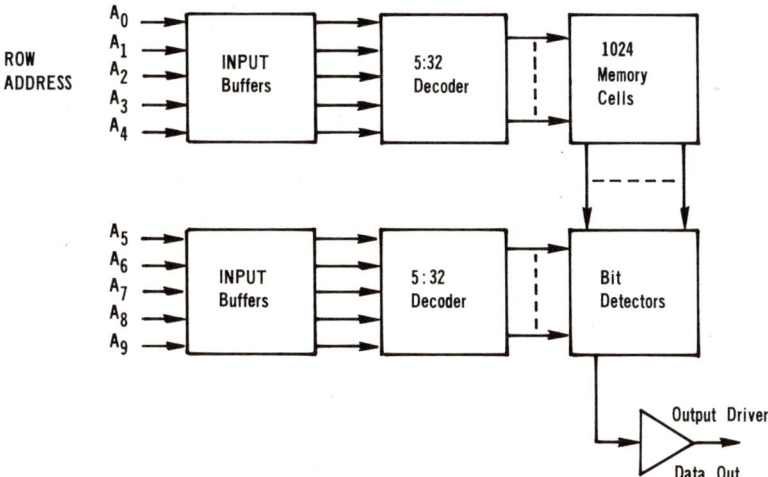

FIGURE 15. 1024 × 1 ROM system block diagram.

control, address, and data. For address lines, it is usually subdivided into ROW and COLUMN address lines. Not so clear-cut as the 3-D core memory system, in semiconductor ROM system, the column address lines, after decoded, are used to control the sense or output circuitry. Figure 15 illustrates a typical system organization of a 1024 × 1 ROM.

VII. MAGNETIC-BUBBLE MEMORIES

The concept of magnetic-bubble memories has been experimental in the laboratory over a decade. To date, there are already several major semiconductor firms providing

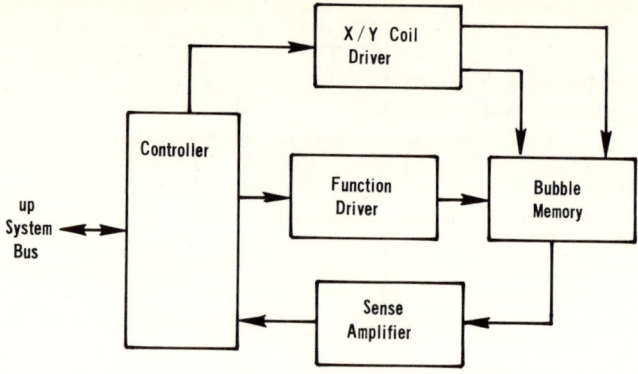

FIGURE 16. Bubble memory system block diagram.

samples and products of 256-K and 1-Mb magnetic-bubble memory chips to the system designers. Due to its extremely high packing density and nonvolatility, it is projected that bubble memories will play a major role as memory devices in system design in the 1980s. While the coil-access technology becomes a practical technique, the current-access technology is proposed by Bell Laboratories scientists, which will put the bubble-memories into the memory families which require only a +5 V power supply. Due to its physical properties, bubble memories do require special supporting circuitry for which the LSI technologies have been employed. Therefore, the bubble memories would be sold in a subsystem basis instead of in just a single chip at the present time. In the system designer's point of view, one may consider that a bubble-memory chip contains a group of magnetic bubble shift registers complemented with LSI supporting circuitry which carries out the functions of timing, addressing, and read/write data. As a typical example, Figure 16 shows a block diagram of a magnetic-bubble-memory subsystem. It is foreseeable in the eighties; major progress in bubble memory technologies can be expected. Bubble memories would eventually replace some application areas where disk memory devices are being employed.

REFERENCES

1. **Gilligan, T. J. and Persons, P. B.**, High-Speed Ferrite 2-1/2D Memory, Fall Jt. Computer Conf., 1965, Institute of Electrical and Electronics Engineers, New York, 1965.
2. **Gilligan, T. J. and Persons, P. B.**, Comparison of core memory system organization, *Comput. Des.*, May 1966.
3. **Vadsz, L. L., Chua, H. T., and Grove, A. S.**, Semiconductor random access memories, *IEEE Spectrum*, May 1971.
4. Special issue on semiconductor memories and optoelectronics, *IEEE J. Solid State Circuits*, SC-5 (5), 1970.
5. **Moore, G. E.**, Semiconductor RAMS, *Computer*, March/April 1971.
6. **Rudolph, J. A., et al.**, With associative memory, speed limits no barrier, *Electronics*, June 22, 1970.
7. **Bartlett, J., et al.**, Associative memory chips: fast, versatile and here, *Electronics*, August 17, 1970.
8. **Riley, W. B.**, Ed., *Electronic Computer Memory Technology*, McGraw-Hill, New York, 1971.
9. **Hodges, David A.**, Ed., *Semiconductor Memories*, IEEE Press, New York, 1972.
10. **Altman, L.**, Charge-coupled devices move in on memories and analog signal processing, *Electronics*, August 8, 1974.

11. Designer's Guide to Semiconductor Memories, Series I, *EDN,* August 5, 1975; II, *EDN,* August 20, 1975; III, *EDN,* September 5, 1975; IV, *EDN,* September 20, 1975; V, *EDN,* October 5, 1975; VI, *EDN,* October 20, 1975; VII, *EDN,* November 5, 1975; VIII, *EDN,* November 20, 1975.
12. **Bobeck, A. H., Bonyhard, P. I., and Geusic, J. E.,** Magnetic bubbles — an emerging new memory technology, *Proc. IEEE,* August 1975.
13. **Cohen, M. S. and Chang, H.,** The frontiers of magnetic bubble technology, *Proc. IEEE,* August 1975.
14. **Coe, J. E. and Oldham, W. G.,** Enter the 16,384-bit RAM, *Electronics,* Feb. 19, 1976.
15. **Sander, W. B., Shepherd, W. H., and Schinelle, R. D.,** Dynamic I^{2}L random-access memory competes with MOS designs, *Electronics,* August 19, 1976.
16. **Greene, R., Perlegos, G., Salsbury, P. J., and Morgan, W. L.,** The biggest erasable PROM yet puts 16,384 bits on a chip, *Electronics,* March 3, 1977.
17. **Proebsting, R.,** Dynamic MOS RAMs: an economic solution for many system designs, *EDN,* June 20, 1977.
18. **Mohan Rao, G. R. and Hewkin, J.,** 64-K dynamic RAM needs only one 5-volt supply to outstrip 16-k pats, *Electronics,* September 28, 1978.
19. **Waller, L.,** Has bubble memory's day arrived? *Electronics,* March 29, 1979.
20. **Bryson, D., Clover, D., and Lee, Dave,** Megabit bubble-memory chip gets support from LSI family, *Electronics,* April 26, 1979.
21. **Capece, R. P.,** The race heats up in fast static RAMs, *Electronics,* April 26, 1979.
22. **Bisset, S., Bristow, S., and Chen, T. T.,** Bubble memories demand unique test methods, *Electronics,* May 10, 1979.
23. **Wallace, C.,** Electrically erasable memory behaves like a fast, nonvolatile RMA, *Electronics,* May 10, 1979.
24. **Welch, T. A.,** Analysis of memory hierarchies for sequential data access, *Computer,* May 1979.
25. **Halsema, A. I.,** Bubble memories — a short tutorial, *Byte,* June 1979.
26. **Juliessen, J. E.,** Where bubble memory will find a niche, *Mini-Micro Systems,* July 1979.
27. **George, P. K. and Reyling, Jr., G.,** Bubble memories come to the boil, *Electronics,* Aug. 2, 1979.
28. **Lee, Dave and Spiegel, P.,** Ease bubble-system design with a few basic guidelines, *EDN,* Aug. 5, 1979.
29. **Greene, R. and Louie, F.,** E-PROM doubles bit density without adding a pin, *Electronics,* Aug. 16, 1979.

Chapter 7

ARITHMETIC LOGIC UNIT (ALU)

Just like the memory unit, arithmetic logic unit (ALU) is another important subsystem in digital system designs. In many applications such as numerical controls, process controls, instrumentations, etc., one would like to carry out binary arithmetic/logic operations within the system. Thanks to the LSI technology, different types of ALU chips are now available on the market, from which a proper type can be selected for a specific application. Generally, most of the ALUs available, however, are ADDER-based; subtraction and multiplication operations, etc., would need additional hardwares to implement them according to some algorithms. In this chapter, we shall review some basic element designs as well as some more popular algorithms. Examples and some typical LSI chips will be described.

I. BINARY ADDITION

A. Basic Element
Half-Adder

A half-adder can be defined by the following truth table:

A (Augend)	B (Addend)	S (Sum)	C (Carry)
0	0	0	0
0	1	1	0
1	0	1	0
1	1	0	1

and its switching functions derived from the truth table for Sum and Carry are $S = A\overline{B} + B\overline{A} = A \oplus B$, and $C = AB$. Figure 1(a) shows the logic diagram and symbol of a half-adder.

Full-Adder

A full-adder can be defined by the following truth table:

A_n	B_n	C_{n-1}	S_n	C_n
0	0	0	0	0
0	0	1	1	0
0	1	0	1	0
0	1	1	0	1
1	0	0	1	0
1	0	1	0	1
1	1	0	0	1
1	1	1	1	1

and its switching functions derived from the truth tables are $C_n = A'_nB_nC_{n-1} + A_nB_nC'_{n-1} + A_nB'_nC_{n-1} + A_nB_nC_{n-1} = A_nB_n + (A_n \oplus B_n)C_{n-1}$, and $S_n = A'_nB'_nC_{n-1} + A'_nB_nC'_{n-1} + (A'_nB_n + A_nB'_n)C'_{n-1} = (A'_nB'_n + A_nB_n)C_{n-1} + (A'_nB_n + A_nB'_n)C'_{n-1} = (A_n \oplus B_n)'C_{n-1} + (A_n \oplus B_n)C'_{n-1} = A_n \oplus B_n \oplus C_{n-1}$, where, the subscription n and n − 1 denote the nth and (n − 1)th binary bit of a binary number, respectively. Figure 1(b) shows the logic diagram and symbol of a full-adder.

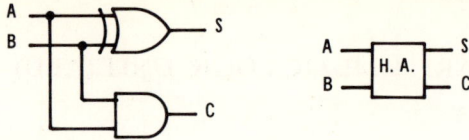

FIGURE 1(a). Logic diagram and symbol of a half-adder.

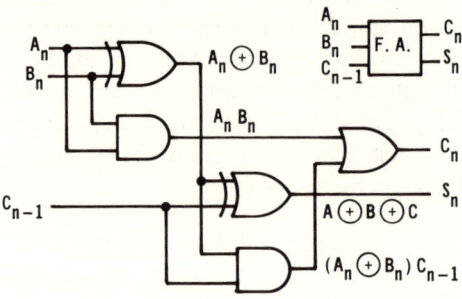

FIGURE 1(b). Logic diagram and symbol of a full-adder.

B. Multiple-Bit Addition
Carry-Ripple-Through Parallel Adder

The heart of an ALU is a multiple-bit adder which is a collection of full-adders shown in Figure 1(b). Figure 2(a) shows the diagram of a ripple-through parallel adder. Note that the outputs of the nth or the most significant bit F.A. unit shown, is a function of the $(n-1)$th, $(n-2)$th, ..., 1st F.A. units. Therefore, the final valid output values would not settle after $n\tau$ sec in the worst case, where τ is the propagation delay of each F.A. unit. This adder is thus called carry-ripple-through parallel adder. Due to its simplicity in organization, this type of adder is being used in many systems. For a fast processor, however, the $n\tau$ delay of time may not be desirable. Therefore, another type of adder known as carry-look-ahead parallel adder is recommended.

Carry-Look-Ahead Parallel Adder

To avoid the time delay caused by the carry-ripple-through, one can design a logic network of which the nth carry-bit is independent of the carry-bit of any other F.A. unit. The following mathematical derivation will clarify the concept.

From Section I(A), we have $C_n = A_n B_n + (A_n \oplus B_n) C_{n-1}$. Let $G_n \underline{\Delta} A_n B_n$ and $P_n \underline{\Delta} A_n \oplus B_n$, then $C_n = G_n + P_n C_{n-1} = G_n + P_n(G_{n-1} + P_{n-1}C_{n-2}) = G_n + P_n G_{n-1} + P_n P_{n-1} C_{n-2} = G_n + P_n G_{n-1} + P_n P_{n-1}(G_{n-2} + P_{n-2}C_{n-3}) \ldots\ldots\ldots = G_n + P_n G_{n-1} + \ldots + P_n P_{n-1} \ldots P_2 G_1 + P_n P_{n-1} \ldots P_1 C_0$.

In view of the last line of derivation, it reveals that C_n can be so designed that it is a function of G_n, P_n, and C_0 which are available at the first level of addition. Figure 2(b) shows the logic diagram of a 3-b-carry-look-ahead parallel adder, which is direct implementation of the switching function derived. Here we have $n = 3$.

n-Bit Serial Adder

In many applications, a serial adder is more appropriate than a parallel adder if operation speed is not a major concern. Figure 3 shows a logic diagram of a typical serial adder. The augend and addend are first loaded into a parallel-in/serial-out shift register, respectively. Then the clock pulse will right shift the registers and their serially

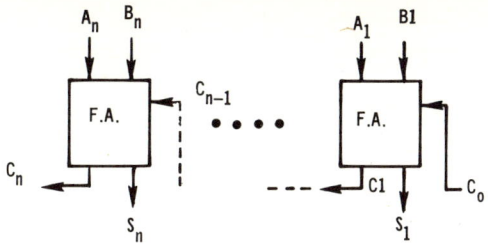

FIGURE 2(a). Ripple-through multibit parallel adder.

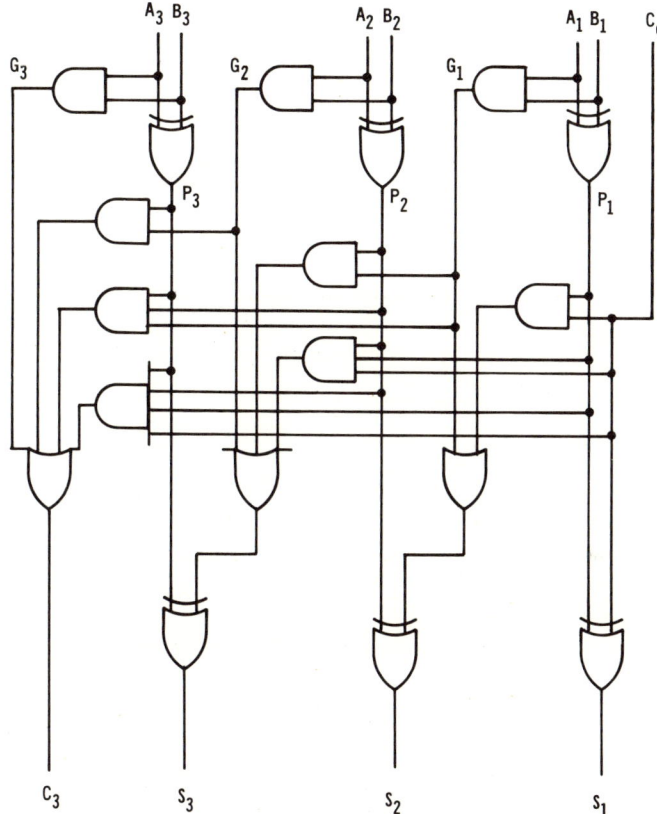

FIGURE 2(b). 3-Bit-carry-look-ahead parallel adder.

shifted outputs are fed to the full-adder as shown. The sum of the two numbers is then stored in the (n + 1)-bit serial-in/parallel-out register. The D flip-flop is used to delay the carry-bit for one clock period and merged into the carry-in of the full-adder.

Accumulator

An accumulator is a device which is able to perform multiple-number addition serially and retain the value of the total summation. Obviously it would normally contain adder and registers. Figure 4 shows a typical block of an accumulator. The D flip-flops are used as a storage register for the sum of the augend or the incoming data $A_n,...,A_1$ and the old result; and the number in the storage register is then fed back to become the addend of the next addition.

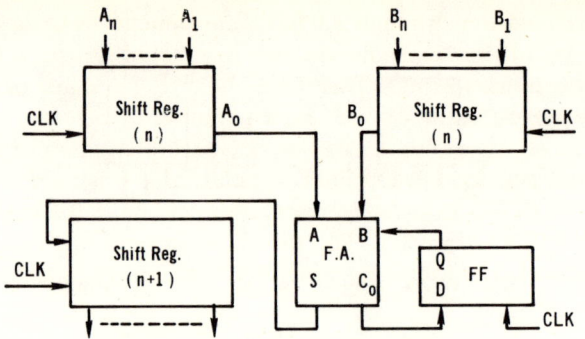

FIGURE 3. n-Bit binary serial adder.

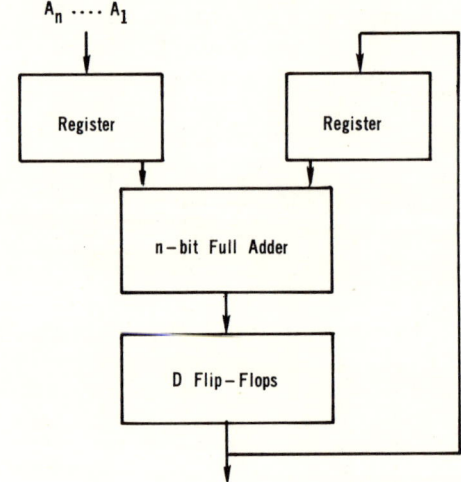

FIGURE 4. Block diagram of an accumulator.

II. SUBTRACTION BY COMPLEMENTARY ARITHMETIC

Although one can easily design a subtractor by following the same logical steps for designing an adder, it is normally and more convenient to do subtraction using the same hardware which is used for addition. In other words, an ALU normally contains the basic full-adder elements for add-operation and it is convenient then if the same elements can also be used to implement subtraction. To do this, a complementary arithmetic is normally utilized.

A. Background
Signed-Binary-Coded Integer

Conventionally an n-bit binary coded integer, $x \triangleq A_{n-1}, A_{n-2}, \ldots, A_0$; is interpreted as a positive integer with a value of $x = A_{n-1}2^{n-1} + A_{n-2} + \ldots + A_0 2^0$.

For a binary-coded signed-magnitude integer representation, however, the most significant bit, A_{n-1}, is reserved as a sign-bit. That is, when $A_{n-1} = 1$, the number is negative, otherwise, positive. For example, in signed-magnitude representation, $1111 = -7$, and $0111 = +7$. As another example, $0000 = 0$ and $1000 = -0$.

Therefore, the range of a 4-b binary integer is ±7 in which there exist positive and negative zero representation although ± 0 has no arithmetic meaning. Binary representation of a number is obviously the natural one in a digital system. It is, however, very inconvenient for the human beings. Therefore, the digital designer or digital computer user would often use different ways of representation of a binary integer. Octal or hexadecimal are most commonly used. For octal representation, the binary bits are grouped into threes and for hexadecimal, a group of 4 b is used. For example, 8-b binary integer: $(10100110)_2$ in octal representation will be, $(10,100,110)_2 \rightarrow (246)_8$; in hexadecimal, $(1010,0110)_2 \rightarrow (A6)_{16}$. The subscripts indicate the radix or base of the number representation. The following table shows the conversion of the three systems.

Binary				Octal		Hexadecimal
0	0	0	0	0	0	0
0	0	0	1	0	1	1
0	0	1	0	0	2	2
0	-	1	1	0	3	3
0	1	0	0	0	4	4
0	1	0	1	0	5	5
0	1	1	0	0	6	6
0	1	1	1	0	7	7
1	0	0	0	1	0	8
1	0	0	1	1	1	9
1	0	1	0	1	2	A
1	0	1	1	1	3	B
1	1	0	0	1	4	C
1	1	0	1	1	5	D
1	1	1	0	1	6	E
1	1	1	1	1	7	F

Complement Representations

Twos Complement Representation

The twos complement representation of a n-bit signed-magnitude binary integer is symbolically and mathematically defined as follows.

Let $Y_2 \triangleq$ n-bit signed-magnitude binary integer = $y_{n-1}, y_{n-2}, \ldots, y_0$ with $y_i = 1,0$; i = 0, ..., n − 1.

y_{n-1} = sign-bit, $1 \triangleq$ negative, $0 \triangleq$ positive, then symbolically,

$$Y_2(\overline{2}) \triangleq -y_{n-1}2^{n-1} + y_{n-2}2^{n-2} + \ldots + y_0 2^0 \quad (1)$$

Other Complement Representations

Similarly, we can define other complement representation symbolically and mathematically as follows.

$$Y_2(\overline{1}) \triangleq \text{ones complement of } Y_2$$

$$Y_2(\overline{1}) \triangleq Y(\overline{2}) + 1 \quad (2)$$

$$= -y_{n-1}2^{n-1} + y_{n-2}2^{n-2} + \ldots + y_0 2^0 + 1$$

$$Y_8(\overline{8}) \triangleq \text{eights complement of } Y_8$$

$$Y_8(\overline{8}) \triangleq -y_{n-1}8^{n-1} + y_{n-2}8^{n-2} + \ldots + y_0 8^0 \quad (3)$$

with $y_i = 0,1,2, \ldots, 7$; i = 0, ..., n − 1

$Y_8(\bar{7}) \triangleq$ sevens complement of Y_8

$$Y_8(\bar{7}) \triangleq Y_8(\bar{8}) + 1 \tag{4}$$

$$= -y_{n-1}8^{n-1} + y_{n-2}8^{n-2} + \ldots + y_0 8^0 + 1$$

$Y_{10}(\overline{10}) \triangleq$ tens complement of Y_{10}

$$Y_{10}(\overline{10}) \triangleq -y_{n-1}10^{n-1} + y_{n-2}10^{n-2} + \ldots + y_0 10^0 \tag{5}$$

with $y_i = 0, 1, \ldots, 9$; $i = 0, \ldots, n-1$

$Y_{10}(\bar{9}) \triangleq$ nines complement of Y_{10}

$$Y_{10}(\bar{9}) \triangleq Y_{10}(\overline{10}) + 1 \tag{6}$$

$$= -y_{n-1}10^{n-1} + y_{n-2}10^{n-2} + \ldots + y_0 10^0 + 1$$

Logical Complement Conversion Procedures

The mathematical definition of a complement number appears complicated. The complement version of a number however, can be obtained by a simple logical conversion procedure. The following examples can clarify the conversion process.

Let $|Y_2| \triangleq$ the magnitude of the signed-binary-coded number, or, $|Y_2| = y_{n-2}, y_{n-3}, \ldots, y_2, y_0$; then the logical conversion procedure for obtaining ones complement of Y_2 is simply the complement of each individual magnitude-bit, i.e.,

$$\text{L.C.}[Y_2(\bar{1})] = y_{n-1}, \bar{y}_{n-2}, \bar{y}_{n-3}, \ldots, \bar{y}_0 \tag{7}$$

where L.C. means "Logical Conversion of."

The logical conversion procedure for twos complement is adding one to the magnitude of the ones complement and ignoring the carry if it occurs, i.e.,

$$|Y_2(\bar{2})| = |Y_2(\bar{1})| + 1 \tag{8}$$

Note that in logical conversion, the sign-bit is always left alone or untouched.

Example 1: Numerical Example for Ones and Twos Complement
Logical Conversion Procedure

Let a binary coded signed-magnitude number, $Y_2 = 10\ 101\ 001$, then, its magnitude, $|Y_2| = 0,1,0,1,0,0,1$. Thus by ones complement logical operation,

$$Y_2(\bar{1}) = 11\ 010\ 110 \tag{9}$$

From Equation 8, we obtain the twos complement of Y_2 through the binary addition, i.e.,

$$
\begin{array}{r}
11\ 010\ 110 \longleftarrow Y_2(\bar{1}) \\
+)\quad\quad\quad\quad 1 \\
\hline
11\ 010\ 111 \longleftarrow Y_2(\bar{2})
\end{array}
$$

To represent this result in signed-decimal equivalent, we have $Y_2(\bar{2}) = -|Y_2(\bar{2})| = -|(1 \cdot 2^6 + 0 + 1 \cdot 2^4 + 0 + 1 \cdot 2^2 + 1 \cdot 2^1 + 1 \cdot 2^0)| = -|(87)| = -87$.

Similarly, the signed-decimal equivalent of ones complement in Equation 9 is then $Y_2(\bar{1}) = -|Y_2(\bar{1})| = -86$.

Conversion by Mathematical Definition

By Equation 1, we can convert the given number $Y_2 = 10\ 101\ 001$ into, $Y_2(\bar{2}) = -1 \cdot 2^7 + 0 + 1 \cdot 2^5 + 0 + 1 \cdot 2^3 + 0 + 0 + 1 \cdot 2^0 = -128 + 32 + 8 + 1 = -87$. By Equation 2, $Y_2(\bar{1}) = Y_2(\bar{2}) + 1 = -87 + 1 = -86$.

Example 2: Proof of the Equivalent of the Two Conversion Procedures

From Equation 7, L.C. $[Y_2(\bar{1})] = y_{n-1}, \bar{y}_{n-2}, \bar{y}_{n-3}, \ldots, \bar{y}_0$. But the bit-complement of a binary variable can be expressed by the following equation: $\bar{y}_i = y_i - 1$ thus

$$\text{L.C.}[Y_2(\bar{1})] = y_{n-1}, (y_{n-2} - 1), (y_{n-3} - 1),$$

$$\ldots, (y_0 - 1)$$

$$= -y_{n-1} 2^{n-1} + (y_{n-2} - 1) 2^{n-2} +$$

$$\ldots + (y_0 - 1) 2^0$$

$$= -y_{n-1} 2^{n-1} + y_{n-2} 2^{n-2} +$$

$$\ldots + y_0 2^0 - (2^{n-2} + \ldots + 2^0) \quad (10)$$

But $2^{n-1} = (2^{n-2} + 2^{n-3} + \ldots + 2^0) + 1$ or $-(2^{n-2} + 2^{n-3} + \ldots + 2^0) = -2^{n-1} + 1$. Substituting into Equation 10

$$\text{L.C.}[Y_2(\bar{1})] = -y_{n-1} 2^{n-1} + y_{n-2} 2^{n-2} +$$

$$y_0 2^0 - 2^{n-1} + 1 \quad (11)$$

Note that the most significant bit of the magnitude is 2^{n-2}, and the sign-bit which is 2^{n-1}, is to be left alone or untouched, the influence of (-2^{n-1}) term which is a carry, and thus logically disappears. We now can see the equivalent between Equation 11 and Equation 2.

Example 3: Double Complement Property

To illustrate the double complement property, let us complement the results obtained in Example 1 by logical conversion procedure as follows:

One complement of the ones complemented:

$$Y_2(\bar{1}) = 11\ 010\ 110$$

$$(\bar{1}) \text{ Operation}$$

$$Y_2 \longleftarrow 10\ 101\ 001$$

Twos complement of the twos complemented:

$$Y_2(\bar{2}) = 11\ 010\ 111$$

$$10\ 101\ 000 \quad (\bar{1})\ \text{Operation}$$

$$+)\qquad\qquad 1$$

$$Y_2 \longleftarrow 10\ 101\ 001$$

Therefore we can conclude that a complement of a complemented number will yield the original number.

Example 4: Logical Complement Procedure for Numbers of Nonbinary Representation

The logical complement procedure, although it has no apparent mathematical meaning, has been widely used due to its simplicity in manual operation and hardware implementation. Examples of complementing numbers of radix other than binary are now in order. Let $Y_8 = y_{n-1}, y_{n-2}, \ldots, y_0$. With $y_i = 0, \ldots, 7$, $i = 0, 1, \ldots, n-2$, $i \neq n-1$, then $Y_8(\bar{7}) = y_{n-1}, (7 - y_{n-2}), (7 - y_{n-3}), \ldots (7 - y_0)$. For example,

$$\text{If}\quad y_8 = -634$$
$$(\bar{7})\ \text{Operation}$$
$$\text{then}\quad Y_8(\bar{7}) = -143$$

And, $|Y_8(\bar{8})| = |Y_8(\bar{7})| + 1$, we have $Y_8(\bar{8}) = -144$. Similarly,

$$\text{Let}\quad Y_{10} = -987$$
$$(\bar{9})\ \text{Operation}$$
$$\text{then}\quad Y_{10}(\bar{9}) = -012$$

And, $|Y_{10}(\overline{10})| = |Y_{10}(\bar{9})| + 1$, we have $Y_{10}(\overline{10}) = -013$.

Example 5: Complement of a Positive Number

Let $Y_2 = y_{n-1}, y_{n-2}, y_{n-3}, \ldots, y_0$; then $Y_{n-1} = 0$, if Y_2 is a positive number. Thus, for $Y_2 > 0$, we have $Y_2 = 0, y_{n-2}, y_{n-3}, \ldots, y_0$. But by the mathematical definition, Equation 1, $Y_2(\bar{2}) = -0 \cdot 2^{n-1} + y_{n-2} 2^{n-2} + \ldots + y_2 2^0 =$ magnitude of Y_2.

Therefore, it is evident that the complement of a positive number is the positive number itself.

Example 6: Different Bases

A numerical example for complement number representations of different radixes or bases is shown in the following table.

Binary-coded signed-magnitude				Decimal equivalent	Ones complement				Twos complement				Sevens complement	Eights complement
0	0	0	0	0	0	0	0	0	0	0	0	0	0	0
0	0	1	1	1	0	0	0	1	0	0	0	1	1	1
0	0	1	0	1	0	0	1	0	0	0	1	0	2	2
0	0	1	1	3	0	0	1	1	0	0	1	1	3	3
0	1	0	0	4	0	1	0	0	0	1	0	0	4	4
0	1	0	1	5	0	1	0	1	0	1	0	1	5	5
0	1	1	0	6	0	1	1	0	0	1	1	0	6	6
0	1	1	1	7	0	1	1	1	0	1	1	1	7	7
1	0	0	0	−0	1	1	1	1	0	0	0	0	17	10
1	0	0	1	−1	1	1	1	0	1	1	1	1	16	17
1	0	1	0	−2	1	1	0	1	1	1	1	0	15	16
1	0	1	1	−3	1	1	0	0	1	1	0	1	14	15
1	1	0	0	−4	1	0	1	1	1	1	0	0	13	14
1	1	0	1	−5	1	0	1	0	1	0	1	1	12	13
1	1	1	0	−5	1	0	0	1	1	0	1	0	11	12
1	1	1	1	−7	1	0	0	0	1	0	0	1	10	11

B. Subtraction by Addition of Complements

Procedure for Subtraction by Addition of Twos Complements

Step 1: Convert the negative number or numbers into their twos complements.
Step 2: Apply binary addition operation to the two numbers including the sign-bit, and ignore carry, if it occurs.
Step 3: If the sign-bit of the sum is 1, the result is a negative number in its twos complement representation. To obtain its signed-magnitude representations, it should be twos-complemented again.

If the sign-bit of the sum is 0, the result is a positive number, no further operation is necessary.

Example 1: Determine $(18)_{10} - (14)_{10} = ?$

```
                          Binary-coded signed-
                                magnitude
         Decimal
            18         [0]ᵃ  1  0  0  1  0
           −14         [1]ᵃ  0  1  1  1  0
```

ᵃ Enclosed-bit denotes sign-bit.

Step 1: Twos complement of −14:

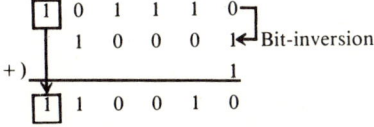

Step 2:

```
    [0]  1   0   0   1   0
    [1]  1   0   0   1   0
+) ─────────────────────────
    [0]  0   0   1   0   0
```

Step 3:   Sum = 4

Example 2: $(6)_{10}, -(14)_{10} = ?$

Step 1:

```
(6)  → [0]  0   0   1   1   0
(-14)→ [1]  1   0   0   1   0
```

Step 2:

```
    [0]  0   0   1   1   0
    [1]  1   0   0   1   0
+) ─────────────────────────
     1   1   1   0   0   0   ← Negative number in twos
                                complement
```

Step 3:

```
    [1]  1   1   0   0   0 ─┐
         0   0   1   1   1 ←┘ Inversion
+)  ─────────────────────1
    [1]      1   0   0   0   ← Signed-magnitude = −8
```

Example 3: $(-17)_{10} - (14)_{10} = ?$

Step 1:

```
-17 → [1]  1   0   0   0   1 ─┐
           0   1   1   1   0 ←┘ Inversion
+)  ───────────────────────1
      [1]  0   1   1   1   1
-14 → [1]  1   0   0   1   0
```

Step 2:

```
    [1]  0   1   1   1   1
    [1]  1   0   0   1   0
+) ─────────────────────────
(1) [1]  0   0   0   0   1
 ↑
 └─ Carry-bit should be ignored
```

Step 3:

$$\begin{array}{cccccc} \boxed{1} & 0 & 0 & 0 & 0 & 1 \\ & 1 & 1 & 1 & 1 & 0 \end{array} \Bigg] \text{Inversion}$$
$$+ \quad 1$$
$$\boxed{1} \; 1 \; 1 \; 1 \; 1 \; 1 \quad \leftarrow \text{Signed-magnitude} = -31$$

Overflow Error of Twos Complement Addition
Example 1: $(18)_{10} + (16)_{10} = ?$

Step 1:

$$18 \rightarrow \boxed{0} \; 1 \; 0 \; 0 \; 1 \; 0$$
$$16 \rightarrow \boxed{0} \; 1 \; 0 \; 0 \; 0 \; 0$$

Step 2:

$$\boxed{0} \; 1 \; 0 \; 0 \; 1 \; 0$$
$$+) \; \boxed{0} \; 1 \; 0 \; 0 \; 0 \; 0$$
$$\boxed{1} \; 0 \; 0 \; 0 \; 1 \; 0$$

Step 3:

$$\begin{array}{cccccc} \boxed{1} & 0 & 0 & 0 & 1 & 0 \\ & 1 & 1 & 1 & 0 & 1 \end{array} \Bigg] \text{Inversion}$$
$$1$$
$$\boxed{1} \; 1 \; 1 \; 1 \; 1 \; 0$$

SUM = −30 ← Error!!

Example 2: $(-18)_{10} - (16)_{10} = ?$

Step 1:

$$\begin{array}{cccccc} -18 \rightarrow \boxed{1} & 1 & 0 & 0 & 1 & 0 \\ & 0 & 1 & 1 & 0 & 1 \end{array} \Bigg] \text{Inversion}$$
$$+ \quad 1$$
$$\boxed{1} \; 0 \; 1 \; 1 \; 1 \; 0$$

$$\begin{array}{cccccc} -16 \rightarrow \boxed{1} & 1 & 0 & 0 & 0 & 0 \\ & 0 & 1 & 1 & 1 & 1 \end{array} \Bigg] \text{Inversion}$$
$$+ \quad 1$$
$$\boxed{1} \; 1 \; 0 \; 0 \; 0 \; 0$$

Step 2:

$$\boxed{1} \; 0 \; 1 \; 1 \; 1 \; 0$$
$$+) \; \boxed{1} \; 1 \; 0 \; 0 \; 0 \; 0$$
$$(1) \; \boxed{0} \; 1 \; 1 \; 1 \; 1 \; 0$$

Step 3:   SUM = 30 ← Error!!

From these two examples, we found that the errors are due to the fact that both of the true answers fall outside of the range of ±31 which is the maximum magnitude for a 6-b ALU; therefore, this kind of error is called *overflow error*. It is thus important that an overflow error detection circuit should be designed.

Let x = sign-bit of the augend, y = sign-bit of the addend, z = sign-bit of the resulting sum, and then a truth table can be constructed, i.e.,

x	y	z	Overflow
0	0	0	0
0	0	1	1[a]
0	1	0	0
0	1	1	0
1	0	0	0
1	0	1	0
1	1	0	1[b]
1	1	1	0

[a] Addition of two positive numbers, results in a negative sum.
[b] Addition of two negative numbers, results in a positive sum.

According to the truth table, the switching function of the overflow, $F = x'y'z + xyz'$. A detection circuit can be designed accordingly to caution the error results.

Shifting of Twos Complement

Right shift (RS) — FROM Equation 1, we have $Y_2(\bar{2}) = -y_{n-1}2^{n-1} + y_{n-2}2^{n-2} + \ldots + y_0 2^0$. To shift the binary bits of $Y_2(2)$ to the right by one position, we have y_{n-1} moved to the position of 2^{n-2}, and y_{n-2} to that of 2^{n-3}. However, the sign of the number should not be changed. Therefore, in shifting a negative number in twos complement form, the sign-bit which is always equal to 1, should be copied to its right position. For example, if it is desirable to right-shift the number, $\boxed{1}$ 0 1 1 0 1 three times, we will have: $\boxed{1}$ 0 1 1 0 1 $\xrightarrow{RS}$ and $\boxed{1}$ 1 0 1 1 0 $\xrightarrow{RS}$ $\boxed{1}$ 1 1 0 1 1 $\xrightarrow{RS}$ and $\boxed{1}$ 1 1 1 0 1.

Left shift (LS) — SINCE for left shift, y_{n-2} is being shifted to the 2^{n-1} position, therefore, one can shift the number to the left only if $y_{n-2} = y_{n-1}$, etc. For example, we can only shift the following number by two times at the most. $\boxed{1}$ 1 1 0 1 1 $\xrightarrow{LS}$ $\boxed{1}$ 1 0 1 1 0 $\xrightarrow{LS}$ $\boxed{1}$ 0 1 1 0 0.

Procedure for Subtraction by Addition of Ones Complements

Step 1: Convert the negative number or numbers into their ones complements.
Step 2: Apply binary addition operation to the two numbers including sign-bits. If carry occurs, add 1 to the resulting sum.
Step 3: A. If the sign-bit of the result is 1, apply ones complement operation again to obtain signed-magnitude representation of the result.
B. If the sign-bit of the result is 0, the result is a positive number, no further operation is necessary.

Example 1: $(34)_{10} - (23)_{10} = ?$

Step 1:

```
  34 →[0] 1  0  0  0  1  0
 -23 →[1] 0  1  0  1  1  1⏋
       [1] 1  0  1  0  0  0←Inversion
```

Step 2:

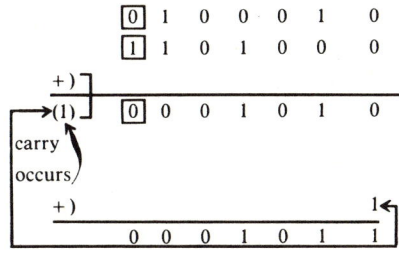

Step 3:   Sum = 11.

Example 2: $(17)_{10} - (23)_{10} = ?$

Step 1:

```
  17 →[0] 0  1  0  0  0  1
 -23 →[1] 1  0  1  0  0  0
```

Step 2:

```
      [0] 0  1  0  0  0  1
      [1] 1  0  1  0  0  0
  +) ─────────────────────
      [1] 1  1  1  0  0  1
```

Step 3:

```
      [1] 1  1  1  0  0  1⏋
      [1] 0  0  0  1  1  0←Inversion
```

Sum = −6.

Example 3: $(-22)_{10} - (8)_{8} = ?$

Step 1:

```
 -22 →[1] 0  1  0  1  1  0⏋
      [1] 1  0  1  0  0  1←Inversion
  -8 →[1] 0  0  1  0  0  0⏋
      [1] 1  1  0  1  1  1←Inversion
```

Step 2:

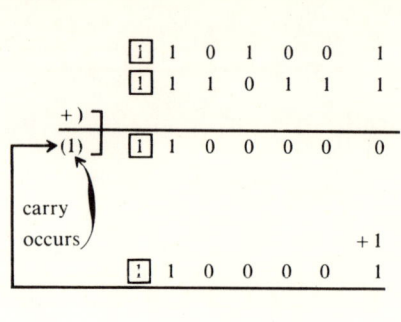

Step 3:

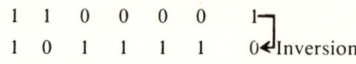

Sum = −30

Subtraction by Addition of Eights Complements

By following the same logic, subtracting two numbers of radix other than one or two can be carried out by addition of its complements. For example, to determine $(715)_8 - (234)_8 = ?$, we can proceed as follows:

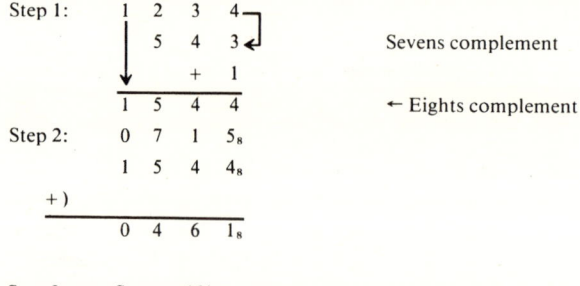

III. A TYPICAL ALU CHIP

Figure 5 shows a typical ALU chip (54/74181) which has been commercially available for a number of years. It is a 4-b ALU. By controlling the M-pin the chip will operate at logic/arithmetic mode. There are four functional control pins, s_0, s_1, s_2, and s_3 which can provide 16 functions for each type of the input data (high or low true). It can also be used as a comparator by referring to the logic levels at C_n and $C_n + {}_4$. Note that both ripple-carry C_n, $C_n + {}_4$, and look-ahead-carry $\overline{P}$, $\overline{G}$ pins are provided. To implement a 16-b with look-ahead-carry operation, one may use four ALU chips and a look-ahead-carry generator chip (54/74182) which also is available on the market.

IV. BINARY MULTIPLICATION

A. Multiplication by Iterative Addition

This method is quite straightforward. It can easily be illustrated by a numerical

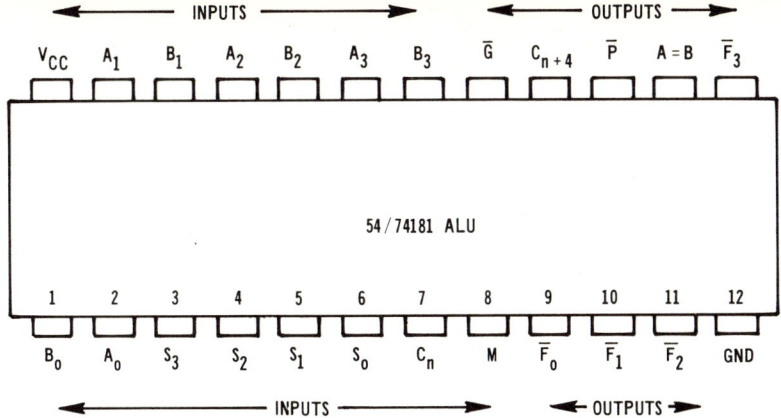

FIGURE 5. A typical ALU chip 54/74181.

example. For example, one can implement 23 × 3 by adding 23 three times. This is exactly how it is done with binary numbers. Figure 6 shows how this method can be implemented with digital hardware. First, the multiplicand and multiplier are loaded, respectively, to the register and down counter. As the multiply command set the flip-flop, the system clock enables the ADD operation in the accumulator as well as the count down operation in the down counter, which determines the number of times that the addition of multiplicand is required. As the down counter reaches zero, the operation is finished.

B. Multiplication by Add and Shift

The algorithm is the same as we would do manually as shown in the following example.

```
               1 0 1 1 0 1
          X)   1 1 0 0 1
          ─────────────────
               1 0 1 1 0 1
             0 0 0 0 0 0
           0 0 0 0 0 0
         1 0 1 1 0 1
    X) 1 0 1 1 0 1
    ─────────────────────────
       1 0 0 0 1 1 0 0 1 0 1
```

Figure 7 shows the exact hardware implementation of the algorithm. Since the diagram shown is self explanatory, detail description is omitted here. Basically, the multiplier controls the action. That is, if [1] occurs, the multiplicand is added to the accumulator and then shifted, otherwise just shifted and with no addition operation.

Multiplication of Twos Complement Binary Number

Booth's algorithm — Let X $\underline{\Delta}$ multiplicand = $(x_{n-1}, x_{n-2}, ..., x_o)$ Y $\underline{\Delta}$ multiplier = $(y_{n-1}, y_{n-2}, ..., y_o)$.

Then

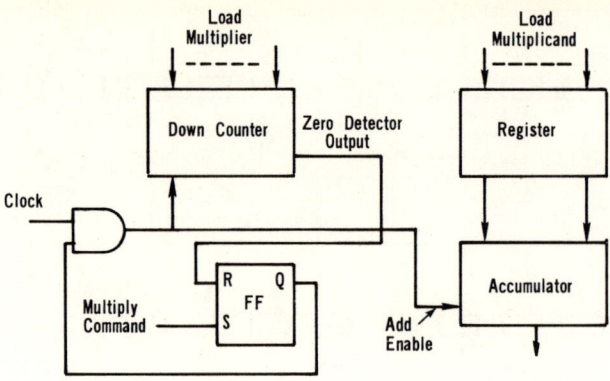

FIGURE 6. Multiplication by iterative addition.

$$Y(\bar{2}) = -y_{n-1}2^{n-1} + y_{n-2}2^{n-2} + \ldots + y_0 2^0 + y_{-1} 2^0$$

where $y_{-1} = 0$, is inserted for mathematical derivation. Now, let us add more dummy terms to it, we have:

$Y(\bar{2})$

$$= -y_{n-1}2^{n-1} + y_{n-2}2^{n-2} + (y_{n-2}2^{n-2} - y_{n-2}2^{n-2}) + y_{n-3}2^{n-3}$$

$$+ (y_{n-3}2^{n-3} - y_{n-3}2^{n-3}) + \ldots + y_1 2 + (y_1 2 - y_1 2) + y_0 2^0 + (y_0 2^0 - y_0 2^0) + y_{-1} 2^0$$

$$= -y_{n-1}2^{n-1} + 2y_{n-2}2^{n-2} - y_{n-2}2^{n-2} + 2y_{n-3}2^{n-3} - y_{n-3}2^{n-3} + \ldots +$$

$$2y_0 2^0 - y_0 2^0 + y_{-1} 2^0$$

$$= 2^{n-1}(y_{n-2} - y_{n-1}) + 2^{n-2}(y_{n-3} - y_{n-2}) + \ldots + 2(y_0 - y_1) + 2^0(y_{-1} - y_0)$$

$X \cdot Y$

$$= 2^{n-1}(y_{n-2} - y_{n-1})X + 2^{n-2}(y_{n-3} - y_{n-2})X + \ldots + 2(y_0 - y_1)X + 2^0(y_{-1} - y_0)X$$

$$= p^{n-1} + \ldots + p^i + \ldots + p^0$$

where $p^j \triangleq j^{th}$ partial product of $X \cdot Y = 2^j(y_{j-1} - y_j) X$.

$2^j(y_{j-1} - y_j) X$ = the product of $(y_{j-1} - y_j) X$ being shifted to the left by j places.

However, there are only four possible combinations of a pair of y_js, i.e.,

y_j	y_{j-1}	p_j	Remark
0	0	$2^j(0)X$	Zero being shifted to the left j places
0	1	$2^j(1)X$	X being shifted to the left j places
1	0	$2^j(-1)X$	$-X$ being shifted to the left j places
1	1	$2^j(0)X$	Zero being shifted to the left j places

In view of the following manual operation:

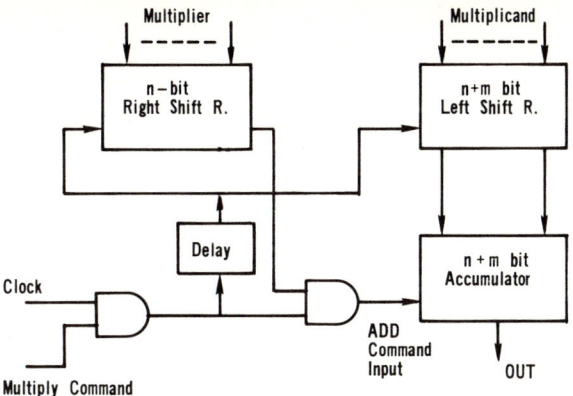

FIGURE 7. Multiplication by shifting.

$$x) \; 2^{n-1}(y_{n-2} - y_{n-1}) + 2^{n-2}(y_{n-3} - y_{n-2}) + \cdots + 2^0(y_{-1} - y_0)$$

$$+ 2^0(y_{-1} - y_0)$$
$$+ 2(y_0 - y_1)$$

$$+ 2^{n-1}(y_{n-2} - y_{n-1})X$$

$$X \cdot Y = \sum_{j=0}^{n-1} 2^j(y_{j-1} - y_j)X$$

it reveals that $2(y_0 - y_1)X$ being shifted to the left one place, is equivalent to shifting $2^0(y_{-1} - y_0)X$ to the right one place, and the last term should not be shifted, however. Hence the Booth's algorithm for binary multiplication can be formulated as follows:

Multiplier bits		Operation
y_j	y_j-1	
0	0	Shift partial product right 1 b
0	1	Add X then shift right 1 b
1	0	Subtract X then shift right 1 b
1	1	Shift partial product right 1 b

Example 1: $(-9) \times 11 = ?$

	Decimal	Signed magnitude	+X in twos compl.	−X in twos compl.
X	−9	[−1] 0 0 1	[1] 0 1 1 1	[0] 1 0 0 1
Y	+11	1 0 1 1	—	—

j	y_j	y_{j-1}	Operation	ACCUMULATOR (A)
				[0]00000000
0	1	0	(A) −X → (A)	[0]10010000
			shift right	[0]01001000
1	1	1	(A) + 0 → (A)	[0]01001000
			shift right	[0]00100100
2	0	1	(A) + X → (A)	[1]10010100
			shift right	[1]11001010
3	1	0	(A) −X → (A)	[0]01011010
			shift right	[0]00101101
4	0	1	(A) + X → (A)	[1]10011101

Inversion

[1]01100010
+ 1
[1]01100011

Thus XY = [1 + 2 + 32 + 64] = −99.

Example 2: *(−5) × (−7) = ?*

$$X = -5 = -(0\ 1\ 0\ 1)_2$$

$$-X = 5 = 0\ 0\ 1\ 0\ 1$$

$$X(\overline{2}) = 1\ 0\ 1\ 1\ 1$$

$$Y = -7 = -(0\ 1\ 1\ 1)$$

$$Y(\overline{2}) = 1\ 1\ 0\ 0\ 1$$

j	y_j	y_{j-1}	Operation	ACCUMULATOR (A)
				[0]00000000
0	1	0	(A) −X → (A)	[0]01010000
			shift right	[0]00101000
1	0	1	(A) + X → (A)	[1]11011000
			shift right	[1]11101100
2	0	0	(A) to → (A)	[1]11101100
			shift right	[1]11110110
3	1	0	(A) −X → (A)	[0]01000110
			shift right	[0]00100011
4	1	1	(A) to → (A)	[0]00100011

Thus X · Y = 1 + 2 + 32 = 35.

Booth's algorithm vs. conventional — The conventional multiplication of two twos complement numbers can be derived as follows:
Since

$$X(\bar{2}) = -2^{n-1}x_{n-1} + \sum_{k=0}^{n-2} 2^k x_R = 2^{n-1}x_{n-1} + |X|$$

$$Y(\bar{2}) = -2^{m-1}y_{m-1} + \sum_{j=0}^{m-2} 2^j y_j = -2^{m-1}y_{m-1} + |Y|$$

then

$$X(\bar{2}) \cdot Y(\bar{2}) = (-2^{n-1}x_{n-1} + |X|)(-2^{m-1}y_{m-1} + |Y|)$$

$$= |X||Y| - 2^{m-1}y_{m-1}|X| - 2^{n-1}x_{n-1}|Y| + 2^{m+n-2}x_{n-1}y_{m-1}$$

In view of the last equation, it reveals that if x_{n-1}, y_{m-1} are zero, that is, if X and Y are both positive numbers, the last three terms will disappear and the answer will be correct. If not, the addition of the last three terms will require more calculation. In Booth's algorithm, however, the correction is not necessary.

Features of Booth's Algorithm are

1. Serial-parallel operation
2. Faster than add and shift multiplication
3. No correction is needed for twos complement numbers
4. It can easily be implemented with ALU chip to be applied on signal processing and digital filters, etc.

V. BINARY RATE MULTIPLIER

To describe its principle of operation, a 3-b-binary rate multiplier is shown in Figure 8(a) where m_1, m_2, m_3, is the 3-b multiplier. The output frequency F_{out} is the function of the input frequency F_{in} and the multiplier, m_1, m_2, m_3. From the logic diagram shown in Figure 8(a), we have, $F_o = (m_3Q_1 + m_2Q_1Q_2 + m_1 Q_1Q_2Q_3) \cdot$ pulse. The principle of operation for this network can easily be described by referring to the timing diagram shown in Figure 8(b). Note that the number of pulses at the output within the eight-pulse frame of F_{in} which is determined by the 3-b counter, is equal to the setting value of m_1, m_2, m_3, i.e.,

$$F_o = \frac{(m_3 m_2 m_1)_2}{8} F_{in}$$

In general, if the multiplier is an n-bit binary number and the counter is an n-bit counter, then we have,

$$F_o = \frac{(m_n, \ldots, m_1)}{2^n} F_{in}$$

or the output frequency count is always a fraction of the input frequency within a frame of 2^n. It is important to point out the nonuniform distribution of the pulse-position. For example, as shown in Figure 8(b), when $m_3m_2m_1 = 101$, there are three pulses clustered around the center of the eight-pulse frame. This nonuniform pulse position property may not be desirable in some applications.

176 Handbook of Digital System Design for Scientists and Engineers

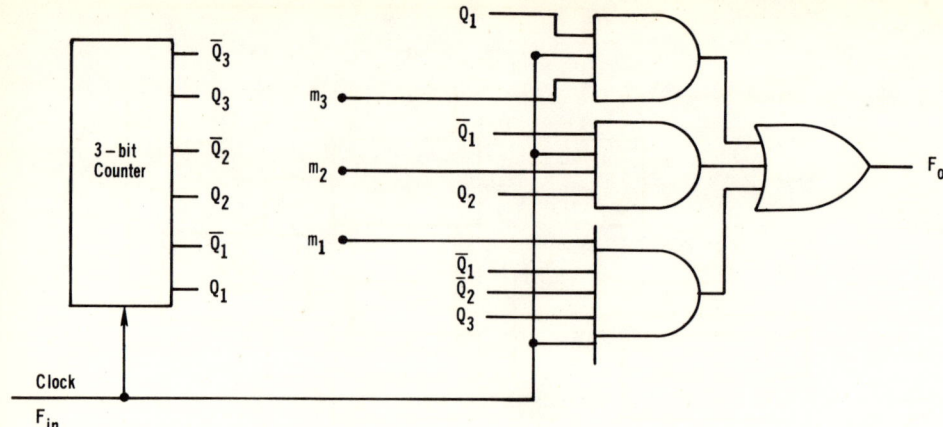

FIGURE 8(a). Binary rate multiplier.

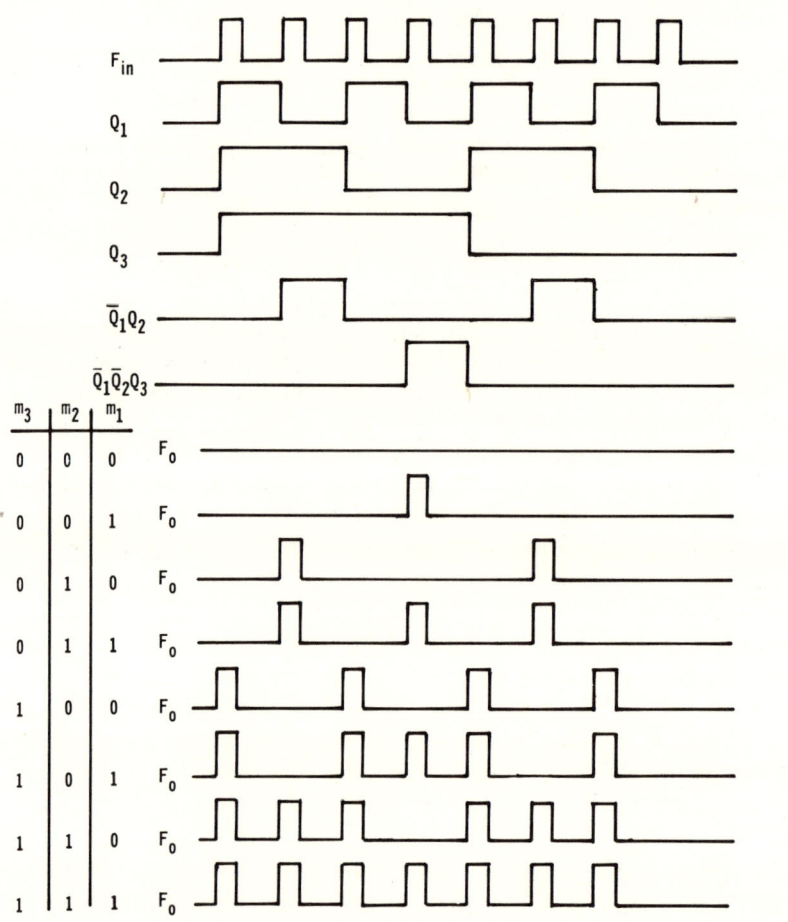

FIGURE 8(b). Timing diagram of the binary rate multiplier.

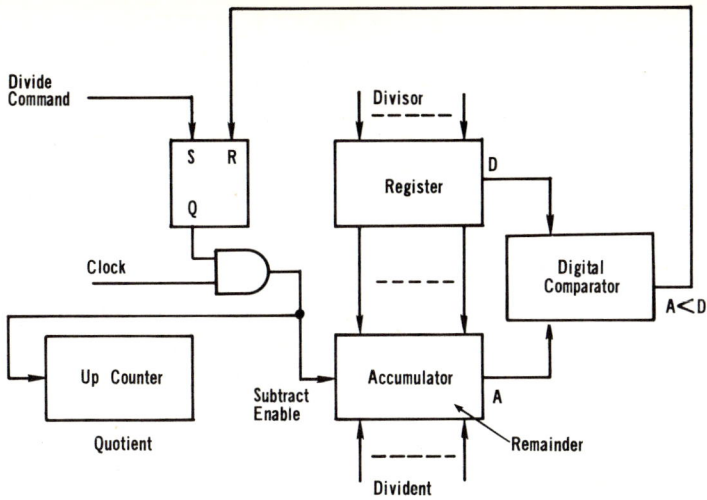

FIGURE 9.   Logic diagram of a binary divider.

VI. BINARY DIVISION

Figure 9 shows the hardware implementation of the algorithm known as the division by iterative subtraction. It is similar to manual long hand division. A simple numerical example will clarify the algorithm. Let us divide 11 by 2 as follows.

```
2 | 11      | 1
    2
    9       | 1
    2
    7       | 1        Total count of ones is the
    2                  Quotient. Here, Quotient = 5
    5       | 1        Remainder = 1
    2
    3       | 1
    2
    1    ← Remainder
```

Note that the up-counter is used to count the total number of subtractions. The dividend has been subtracted by the divisor before the accumulator has a value less than that of divisor. The digital comparator detects this criterion. When this occurs, subtract-operation is ended by resetting the flip-flop. At this point, we have the quotient in the up-counter and the remainder in the accumulator.

REFERENCES

1. **Chu, Yaohan,** *Digital Computer Design Fundamentals,* McGraw-Hill, New York, 1962.
2. **Rabiner, L. R. and Gold, B.,** *Theory and Application of Digital Signal Processing,* Prentice-Hall, Englewood Cliffs, N.J., 1975.
3. **Mitchell, J. N., Jr.,** Computer multiplication and division using binary logarithms, *IRE Trans. Electron. Comput.,* Aug. 1962.
4. **Brubaker, T. A. and Becker, J. C.,** Multiplication using logarithms implemented with ROM, *IEEE Trans. Comput.,* Aug. 1975.
5. **Peatman, J. B.,** *The Design of Digital System,* McGraw-Hill, New York, 1972.
6. **Moshos, G. J.,** Error analysis of binary rate multiplier, *NASA Tech. Note,* NASA TND-3124, Lewis Research Center, December, 1965.
7. **Parasuraman, B.,** Hardware multiplication techniques for microprocessor systems, *Comput. Des.,* April, 1977.
8. **Waser, S.,** High-speed monolithic multipliers for real-time digital signal processing, *Computer,* October 1978.
9. TTL Data Book, National Semiconductor, Santa Clara, Calif., 1976.
10. TTL Data Book, Texas Instruments, Dallas, 1977.
11. Data Manual, Signetics Corp., Sunnyvale, Calif., 1976.
12. **Mick, J. R.,** Digital Signal Processing Handbook, Advanced Micro Devices, Sunnyvale, Calif., 1976.

Chapter 8

ANALOG-DIGITAL-ANALOG CONVERSION

I. INTRODUCTION

A. General Consideration of Analog and Digital System Design

As we all know, digital circuits in general are considerably more reliable and less sensitive to variations of circuit components; but in the real world, output signals generated by the transducers are mostly in analog or continuous forms. To take full advantage of digital technology, one would first need to convert the analog signals into digital and convert them back to analog form afterwards. Depending on the degree of sophistication required in applications, selection of an appropriate A/D/A conversion technique or device for a system becomes a nontrivial problem to the system designer. Furthermore, consideration of trade-off between analog and digital techniques in system design becomes an important step to a competent designer. In this chapter, the techniques of A/D/A and major components used to implement the techniques will be described. First, an understanding of the features of analog and digital techniques is in order. The following table which compares the two systems in a broad sense would clarify some of the confusing issues.

Analog systems	Digital systems
Directly applicable to the real world	Requires A/D/A Converters
Suitable for fast and real time applications	Comparatively slower than its counterpart of analog devices
Simple and low cost for parallel processing	More expensive for simple systems as well as parallel processing
Requires frequent and generally complicated system calibration	More reliable and requires least frequent calibration
Sensitive to temperature and component aging effects	Least sensitive to temperature and component aging effects
Limitation on accuracy; it becomes very expensive if high accuracy is required	Can be designed for high accuracy with reasonable cost
Difficult to devise an analog random access memory	Easy to provide random access memory device with low cost

B. A Typical System Organization of a Digital Data Acquisition System

Figure 1 shows a typical data acquisition system where A/D/A technologies are used. While the functions of most of the blocks are quite evident, functions of the analog guard filter, sample/hold, and sampling clock require clarification. Since an analog signal is continuously varying, a technique to "freeze" the amplitude of a single for a period just long enough to perform A/D conversion is necessary. The block of sample/hold is simply for this purpose. The next question would then be at what frequency should the signal be frozen. In other words, what should be the sampling rate of the system. This, of course, depends on the nature of the analog signal to be processed. If the frequency spectrum of the signal is known, one can use a sampling frequency at least greater than twice the highest frequency component of the signal fre-

180 Handbook of Digital System Design for Scientists and Engineers

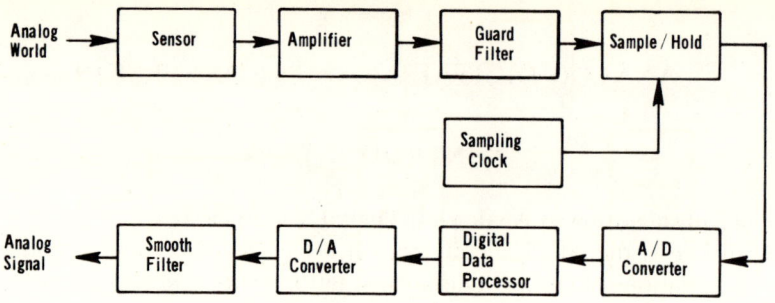

FIGURE 1. Block diagram of a typical data acquisition system.

quency spectrum. It is not unusual to have the sampling frequency chosen at five times the maximum frequency component of the signal, if the digitized data yielded by the A/D converter is not more or faster than that which the digital data processor can manage. It is important that a system designer should be aware of the fact that a so-called aliasing problem will occur if this guideline is not observed. As a result, the recovered signal after D/A conversion will be distorted and produce errors. The guard filter is used to assure that the incoming signal frequency spectrum is within specification, and that the frequency components higher than the expected maximum frequency are considered as noise, will be eliminated or attenuated to a negligible level.

II. DIGITAL-TO-ANALOG CONVERSION

A. Parallel Conversion

The following equation describes the conversion: $v(analog) = [w_1x_1 + w_2x_2 + \ldots + w_nx_n]v_r$, where, $w_1, w_2, \ldots, w_n$ = weight, $x_1, x_2, \ldots, x_n$ = binary number or digital bits, and v_r = reference voltage.

The block diagram shown in Figure 2(a) realizes this method, where D/A decoder is usually a resistor network. The following example will clarify the concept.

Example — In reference to Figure 2(b), a 4-b binary number, x_4, x_3, x_2, x_1 can be converted into analog signal v_0. Here, x_4 is the sign-bit and x_1 is the least significant bit. Then, since the operational amplifier is in shunt-to-shunt feedback configuration, we have,

$$v_0 = (w_1x_1 + w_2x_2 + w_3x_3)x_4 V_r$$

$$= \left(\frac{R_F}{R_1}x_1 + \frac{R_F}{R_2}x_2 + \frac{R_F}{R_3}x_3\right) x_4 V_r$$

Let $R_1 = 2^3 R_F$, $R_2 = 2^2 R_F$, $R_3 = 2 R_F$, $V_R = 8$ V, we have

$$v_0 = \left(\frac{1}{2^3}x_1 + \frac{1}{2^2}x_2 + \frac{1}{2}x_3\right) x_4 \cdot 8$$

$$= x_4 (2^2 x_3 + 2^1 x_2 + 2^0 x_1)$$

B. Serial Conversion

The mathematical expression,

$$\cdot v_0(t + T) = \frac{1}{2}\left\{v_0(t) + x(t + T)V_r\right\}$$

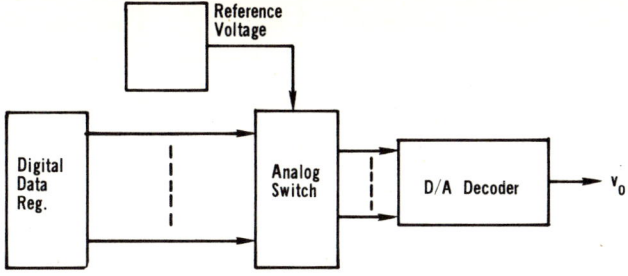

FIGURE 2(a). Block diagram of a parallel D/A converter.

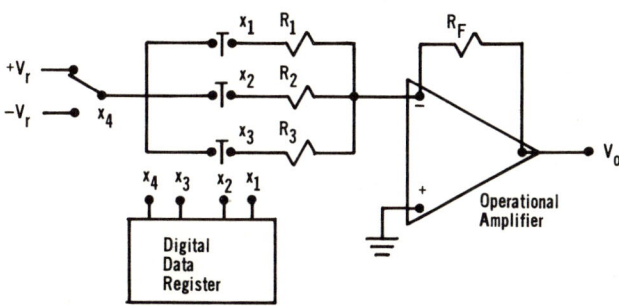

FIGURE 2(b). A simple 4-b D/A converter.

where $v_o(t)$: the output voltage at t, $x(t + T)$: digital input information, [1] or [0], at $(t + T)$, where T = constant and V_r: reference voltage. To illustrate, let the input be a 3-b signal, i.e., x_1, x_2, x_3.

$$\text{Let } v_o(t) = v_o(0) = 0$$

$$\text{then } v_o(0 + T) = \frac{1}{2}\left\{v_o(0) + x_1 V_r\right\}$$

$$v_o(0 + 2T) = \frac{1}{2}\left\{v_o(T) + x_2 V_r\right\}$$

$$= \frac{1}{2}\left\{\frac{1}{2}x_1 V_r + x_2 V_r\right\}$$

$$= \left(\frac{1}{2}\right)^2 x_1 V_r + \frac{1}{2}x_2 V_r$$

$$v_o(0 + 3T) = \frac{1}{2}\left\{v_o(2T) + x_3 V_r\right\}$$

$$= \left[\left(\frac{1}{2}\right)^3 x_1 + \left(\frac{1}{2}\right)^2 x_2 + \frac{1}{2}x_3\right] V_r$$

In general for a n-bit digital signal

$$v_o(0 + nT) = V_r\left\{2^{-1} x_n + 2^{-2} x_{n-1} + \ldots + 2^{-n} x_1\right\}$$

Figure 3(a) shows the system, and Figure 3(b) shows the block diagram of this type of converter.

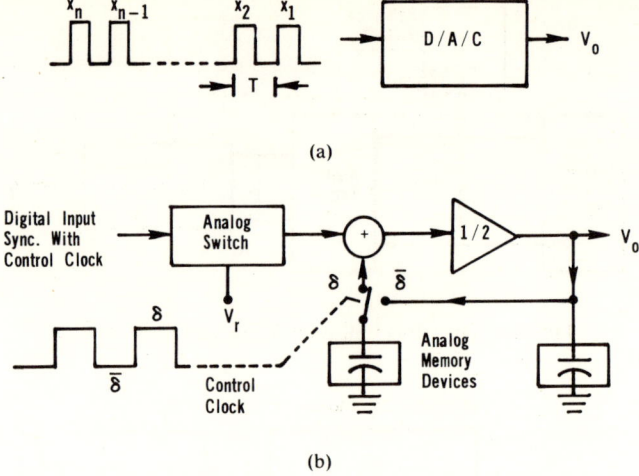

FIGURE 3. Serial D/A conversion.

III. ANALOG-DIGITAL CONVERSION

A. Parallel Conversion

As shown in Figure 4, this technique has extremely high conversion speed. It, however, requires n comparators although the conversion speed is not a function of the width of the digital output. By setting the values of the reference voltage a precision and nonlinear conversion can be achieved. This is thus most likely to be used in some special applications.

B. Counter Ramp Converter

As shown in Figure 5, the analog signal, after sample and hold, is input at the positive terminal of the comparator. The counter is originally set to zero, therefore the D/A converter output which inputs at the negative terminal of the comparator, is zero. The output comparator is HIGH, the clock signal is gated into the up-counter. As a result, there is a ramp signal at the output of the D/A converter. When this output slightly exceeds the analog signal, the comparator output switches to LOW; clock signal is blocked from the counter. Thus the value retained in the counter is the digital equivalent of the analog. This technique is accurate and reliable, since the circuit components used are mostly digital devices. However, the conversion speed is obviously restricted by the magnitude of the analog signal and counter speed. When the input is in full scale, it will require a full count of the counter. For example, if an 8-b A/D or counter is used here, and the clock is 1 MHz, it will require 2^8, or 256 μsec to reach full count. It is obvious that during this RAMP period, no new analog data can be accepted. In other words, this system has a blind period of 256 μsec, which the designer should take into consideration. Normally, there is a digital buffer register at the output. Therefore, as soon as the clock-gate is disabled, the data in the up-counter will be transferred to the buffer. The up-counter will then be reset to zero as soon as a new data in sample/hold is ready for conversion. In the system point of view, the converter may be considered as a black box which expects a control pulse, called "start-to-convert" command from the system and generates an end-of-conversion pulse when data is ready in the buffer register. Most of the A/D converters commercially available now have these provisions for system control purposes.

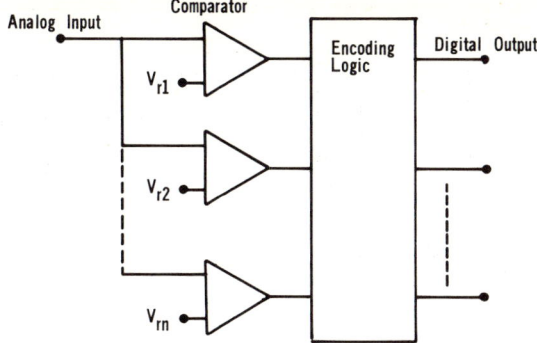

FIGURE 4. Parallel A/D converter.

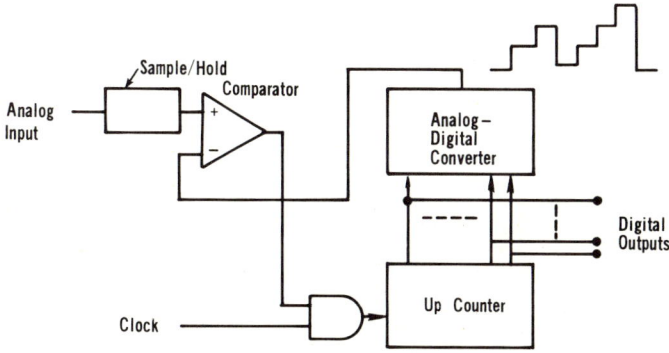

FIGURE 5. Counter ramp A/D converter.

C. Successive Approximation Converter

Figure 6(a) shows the logic diagram of a successive approximation converter. This type of converter is currently the most popular one on the market. In principle, it is similar to counter ramp converters except that it has a shorter blind period. The ring-counter in Figure 6(a) is basically a (N + 1)-b shift register whose output has one and only 1 b at a time is logically HIGH. The relative timing diagram of the two-phase-clock, ϕ_1 and ϕ_2 as well as the partial ring-counter output is shown in Figure 6(b). Initially, all flip-flops and ring-counters are set to zero, the output of the D/A converter is thus also zero. the incoming analog signal would cause the output of the comparator z to go LOW. The clock ϕ_1 would set N-bit of the ring-counter HIGH and thus the most significant bit (MSB) of the D/A is set HIGH. As a result, Y would yield an analog signal with a magnitude of one half of the full scale, or one half of the maximum expected analog signal. At this point, if X < Y, then Z goes HIGH, ϕ_2 would reset the MSB. The next ϕ_1 would shift the ring-counter and (N − 1)-b goes HIGH; consequently, the next MSB would be set HIGH. Now, if X > Y, Z goes LOW which blocks off ϕ_2. As the ring-counter moves downward, this second MSB-bit would remain HIGH. The process continues until the output of the D/A converter matches the value of the unknown analog signal at X. Basically the algorithm is analogous to a systematic guessing game. The first question is whether the input is greater or less than one half of the full scale. If the answer is "greater than," the next question is whether the input is greater or less than, (½ + ¼) full scale, etc. Just like any other A/D converter, this one, of course, also has the ability to accept start-to-convert command and generate the end-of-conversion pulse for system control.

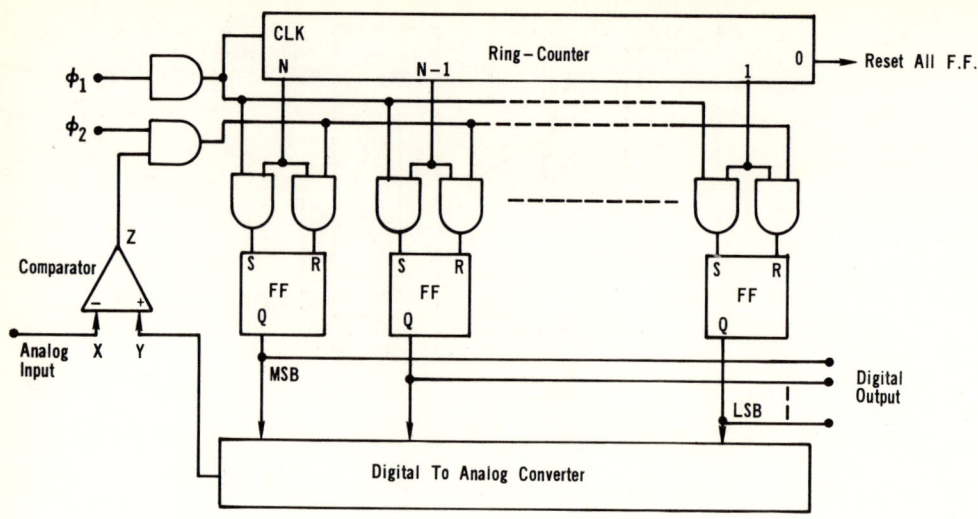

FIGURE 6(a). Successive approximation A/D converter.

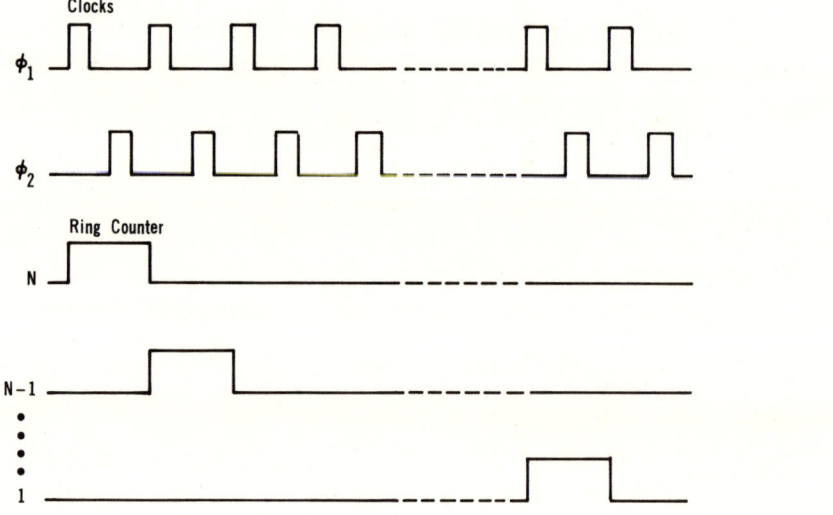

FIGURE 6(b). Timing diagram of the successive approximation A/D converter.

D. Dual-Slope Integrator Converter

The circuit diagram of this type converter is shown in Figure 7(a). The output waveform of the integrator, e_o is shown in Figure 7(b). If the reference voltage V_R is positive, the analog input voltage e_{in} has to be negative. The conversion starts when $\bar{s}$-switch is closed and $\bar{s}$-switch is open. The e_o begins to rise and clock-gate enabled. Binary counter counts upward until all output-bits are one except Nth bit, the carry or control bit. At this point, one more clock pulse would set the control-bit to one and all of the other bits to zero. The control-bit would then close s-switch and open $\bar{s}$-switch. The V_r with opposite polarity becomes the input and e_o decreases until it reaches zero or slightly negative. As a result, the clock-gate is disabled and the reading of the binary counter is thus the digital value equivalent to the analog input signal. The mathematical

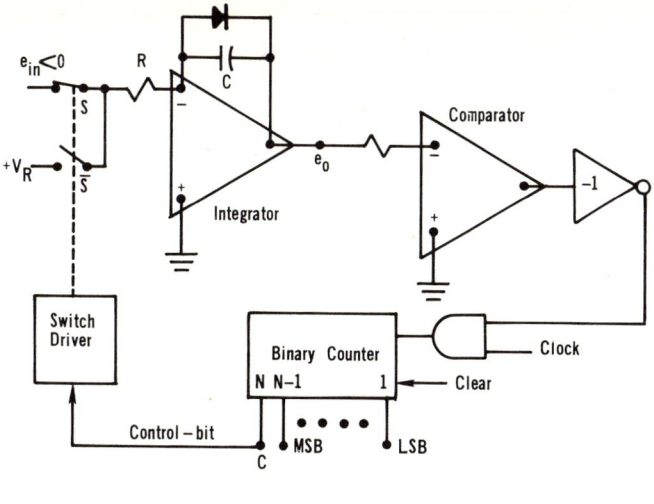

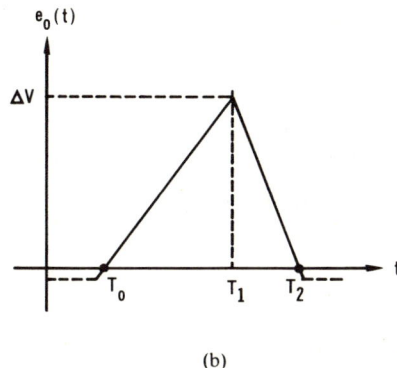

FIGURE 7. Dual-slope integrator A/D converter.

derivation follows. In reference to Figure 7(b), consider the binary counter is set to zero. The s-switch is thus closed. Assume that the input $e_{in} < 0$, is applied at T_0, then the integrator output,

$$e_o(t) = -\frac{1}{RC} \int_{T_o}^{t} e_{in} dt$$

Recall that e_{in} is the output voltage of the sample/hold unit, which can be considered as a constant, thus for $T_o \leq t \leq T_1$,

$$e_o(t) = \frac{1}{RC} \int_{T_o}^{t} e_{in} dt = \frac{1}{RC} e_{in}(t - T_o) \tag{1}$$

Where T_1 is the time at which the output 1, ..., (n − 1)-b of the counter are all [1] except the carry-bit. Assume at this point,

$$e_o(t = T_1) = \frac{1}{RC} e_{in}(T_1 - T_o) = \Delta V \tag{2}$$

Now, an extra clock pulse would set the carry-bit to one and the others to zero, which in turn closes s-switch and opens s-switch, thus

For $T_1 < t \leq T_2$

$$e_o(t) = -\frac{1}{RC} \int_{T_1}^{t} V_R dt + \Delta V \quad (3)$$

$$= -\frac{1}{RC} V_R (t - T_1^+) + \Delta V$$

Substituting Equation 2 into Equation 3,

$$e_o(t) = \frac{1}{RC} V_R (t - T_1^+) + \frac{1}{RC} e_{in}(T_1 - T_o) \quad (4)$$

But at $t = T_2$, as shown in Figure 7(b), we have $e_o(t = T_2) = 0$ which would disable the clock-gate and the counter stop counting. Equation 4 becomes,

$$-\frac{1}{RC} V_R (T_2 - T_1^+) + \frac{1}{RC} e_{in}(t_1 - T_o) = 0 \text{ or}$$

$$e_{in} = V_R \frac{(T_2 - T_1^+)}{(T_1 - T_o)}$$

However, $T_1 - T_o$ is equal to $(2^N - 1)$ clock periods, if the counter is an N-bit counter including the carry-bit. Thus

$$e_{in} = K (T_2 - T_1^+) \quad (5)$$

$$K = \frac{V_R}{T_1 - T_o}$$

But $(T_2 - T_1)$ is the counter value at $t = T_2$. Therefore, the digital equivalent of e_{in} is proportional to the value in the counter at $t = T_2$.

Examine Equation 5; it is interesting to point out that this method of conversion is not a function of the circuit components R and C, nor the accuracy of the clock rate. It is, however, sensitive to voltage drift of the operational amplifier used as an analog integrator.

IV. KEY ELEMENTS COMMONLY USED IN ANALOG-DIGITAL-ANALOG CONVERSION

A. Reference Voltage

This element is important in a sense that all analog values yielded are depending on it. Normally, a zener diode is used as the basic device. A more sophisticated one would need temperature compensation. For low voltage reference which can be a fraction of 1 V, one may use forward biased diode as the voltage reference source.

B. Analog Comparator

As described in Section IV of Chapter 2, a comparator is a special purpose operational amplifier. The key parameters to be considered are: slewing-rate, input/output impedances, settling time, and open-loop gain. The open-loop gain, A, should be greater than the logical swing voltage divided by the equivalent voltage of one least significant bit. The slewing rate in volts per second specifies the response speed of the amplifier.

C. Analog Switch

The desirable characteristics of an analog switch to be used for A/D/A conversion, are

1. Switch-ON resistance = zero
2. Switch-OFF resistance = ∞
3. Switches which do not bounce
4. Zero settling time
5. High switching speed
6. Bilateral

Electromechanical Switch

Electromechanical switches such as mercury-wet relays would satisfy the requirements. The major drawback of the electromechanical switches is their switching speed. It would not respond to high driving frequency. Of course, when the physical size of the analog switch is important, the electromechanical switch may not be desirable either.

Semiconductor Analog Switch

With the exception of nonlinearity and nonzero switch-on resistance, the CMOS analog switches can closely meet the desirable characteristics. The nonzero switch-on resistance characteristic of the CMOS, however, may not be a serious problem in many cases, if input/ output impedance of other associated devices are known; but the nonlinearity property needs careful consideration. The thorough understanding of how a CMOS switch functions in general is thus important. Figure 8(a) shows a CMOS analog switch controlled by a CMOS inverter. Nominal voltages for $V_{DD} = +15$, and $V_{SS} = -15$. When C = +15 V, Q_1 is OFF and Q_2 is ON, thus $\bar{C} = -15$ V. As a result, Q_3 and Q_4 are both conducting. The equivalent channel resistances for Q_3 and Q_4 are functions of the voltage across the channel V_{SD}. The dotted lines in Figure 8(b), respectively, show channel resistance of each device. Since both Q_3 and Q_4 are conducting, the solid line which is the resultant parallel channel resistance of Q_3 and Q_4 shows the equivalent switch-on resistance R_{ON} of the analog switch. Typically, R_{ON} varies between 45 to 55 Ω for $-10 < V_{DS} < +10$ V. If $R_L \gg R_{ON}$, then the variation of the parallel channel resistance may become insignificant for processing the analog signal. When C = -15 V, Q_1 conducts and Q_2 is OFF, then $\bar{C} = +15$ V. This will turn off both Q_3 and Q_4, and the analog switch becomes a high impedance device; the e_{in} is then practically disconnected from R_L. The signal voltage, e_{in}, however, should not exceed the range of $V_{DD} \rightarrow V_{SS}$. Figure 8(c) shows another point of view on how a CMOS analog switch function by means of the Binary-State-Analysis (BSA) concept introduced earlier in this book. As shown, the composite $V_{DS} - I_D$ piece-wise-linear curves of the switch ON/OFF mode can be used to plot the response of the device to the input signal e_{in}. The e_{in} shown is a triangular waveform, which is also the voltage across the switch and the load resistor R_L. As the magnitude of e_{in} changes, the load-line locus sweeps through the V-I space as shown. Accordingly, the i_D through the switch can be determined graphically as shown. The load-line loci, i_D and V_{DC} should satisfy the following loop equation: $e_{in} = V_{DS} + i_D R_L$. For example, when $e_{in} = +$max, we may determine graphically the corresponding $i_D = I_1$, $V_{DS} = V_1$; similarly when $e_{in} = -$max, we have $i_D = I_2$, and $V_{DS} = V_2$. Note that as R_L increases, V_{DS} decreases. Recall that, ideally,

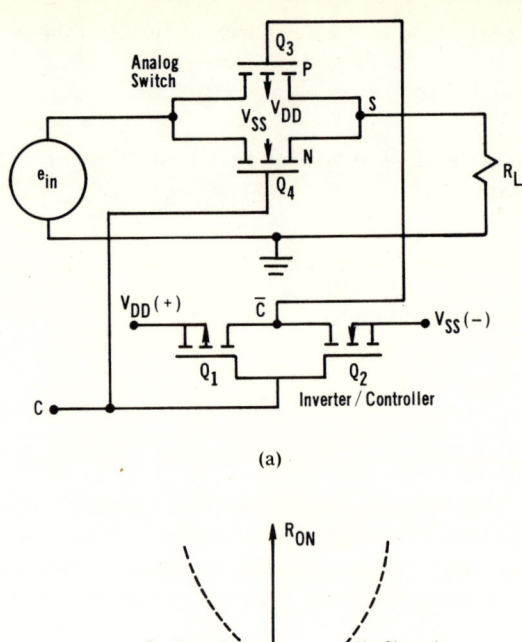

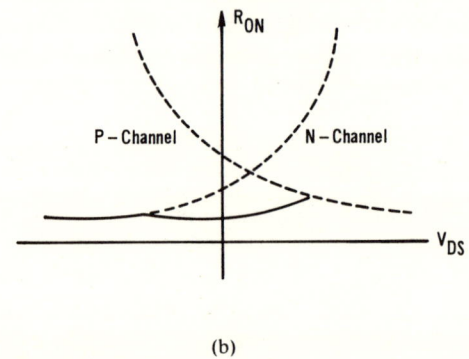

FIGURE 8. CMOS analog switch.

R_{ON} should be negligible with respect to R_L, so that the voltage drop across the switch, F_{DS} is zero when the switch is ON.

D. D/A Decoder Resistor Networks
Weight-Resistor D/A Decoder

As shown in Figure 9, we have

$$V_o = -\left(\frac{R_f}{R_o}x_o + \ldots + \frac{R_f}{R_j}x_j + \ldots + \frac{R_f}{R_{n-2}}e_{n-2}\right)$$

$$= x_{n-1}V_R\left(\frac{R_f}{R_o}x_o + \ldots + \frac{R_f}{R_j}x_j + \ldots + \frac{R_f}{R_{n-2}}x_{n-2}\right)$$

where, digital inputs = $x_{n-1}, x_{n-2}, \ldots, x_j, \ldots x_o$. $x_j = 1; 0$. $j = n - 2, \ldots, 0$. Sign-bit, x_{n-1}: logic [0] = 1; logic [1] = -1.

For $R_j = 1/2^j R_f$, we have

$$V_o = \pm V_R\left(2^o x_o + \ldots + 2^j x_j + \ldots + 2^{n-2} x_{n-2}\right) \quad (6)$$

Mathematically, Equation 6 appears no problem. In practice, however, when n is large,

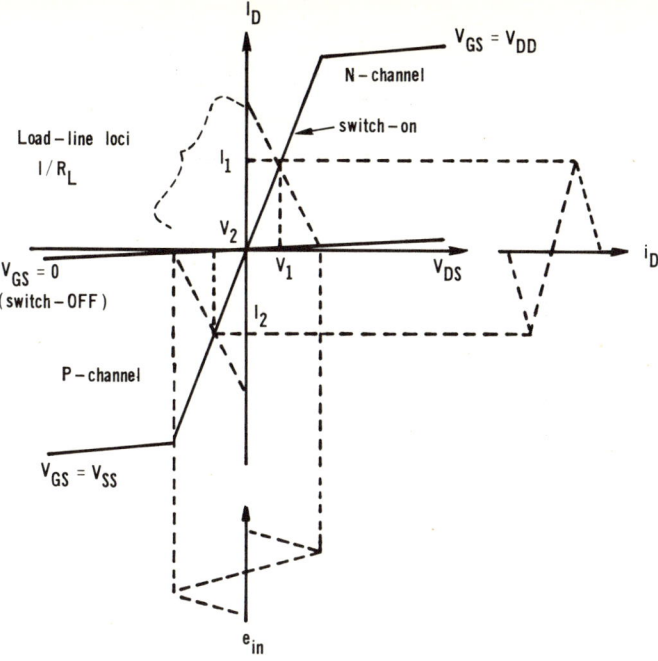

FIGURE 8(c). Graphical analysis of analog switch.

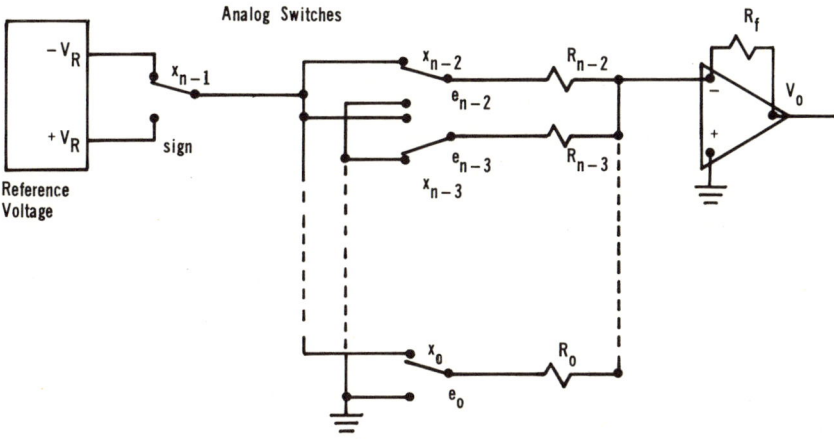

FIGURE 9. Using weight-resistor D/A decoder.

say 12, it may cause problems. To illustrate, let R_{on} of the analog switch = 50 Ω. In order to make the variation of R_{on} be insignificant to this system, one may select R_{n-2} = 200 × R_{ON} = 10 K. For n = 12, $R_f = 2^{n-2}R_{n-2} = 2^{10} \times 10$ K. This is obviously not practical. Therefore, this decoder is more appropriate to be used for a system with digital word of low number of bit.

Resistor-Ladder D/A Decoder

The D/A converter using resistor-ladder as its decoder is shown in Figure 10(a). The outstanding feature of this network is that the impedance of each of the three branches

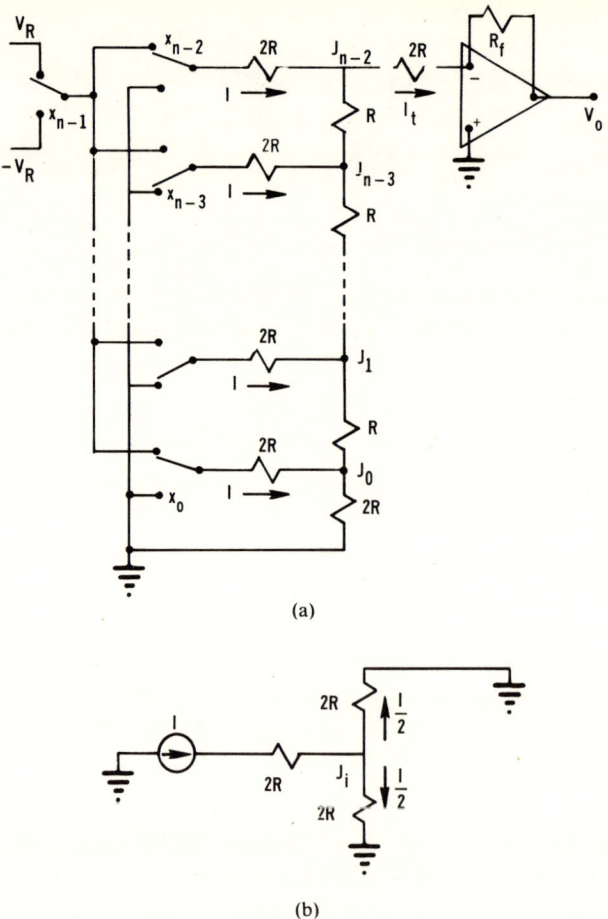

FIGURE 10. Resistor ladder D/A decoder.

at any junctions, J_i, $i = 0, ..., n - 2$, is always equal to 2R, and the whole network uses only R and 2R as circuit components which in practice is the most desirable feature, regardless if the network is to be implemented by discrete components or integrated circuit technology. The principle of operation follows. Note that the network has the following properties:

1. The total resistance of each branch is 2R.
2. The current due to each bit is the same, i.e., $I = V_R/3R$.
3. The incoming current at any junction J_i is always split into two equal outgoing currents $(1/2) I$ as shown in Figure 10(b).

Based on these properties, one can apply the superposition theory to obtain the input current of the operational amplifier.

$$I_t = I \left(\frac{1}{2^{n-2}} x_0 + \frac{1}{2^{n-3}} x_1 + \ldots + \frac{1}{2} x_{n-2} \right)$$

$$= \frac{V_R}{3R} \left(\frac{1}{2^{n-2}} x_0 + \frac{1}{2^{n-3}} x_1 + \ldots + \frac{1}{2} x_{n-2} \right)$$

But $V_o = -R_f I_t$, thus,

$$V_o = \frac{V_R}{3R} \cdot R_f \left(\frac{1}{2^{n-2}} x_o + \frac{1}{2^{n-3}} x_1 + \ldots + \frac{1}{2} x_{n-2} \right)$$

Let $R_f = 3R$, then

$$V_o = -V_R \left(\frac{1}{2^{n-2}} x_o + \frac{1}{2^{n-3}} x_1 + \ldots + \frac{1}{2} x_{n-2} \right)$$

E. Sample and Hold Element

Circuit

Figure 11 is a typical sample and hold circuit on a chip, commercially available on the market. Figure 11(a) shows a circuit of noninvert configuration with an overall gain,

$$G = 1 + \frac{R_2}{R_1}$$

and Figure 11(b), a circuit of invert configuration with an overall gain,

$$G = -\frac{R_2}{R_1}$$

Operational amplifier A_1 is used to charge the capacitor; while operational amplifier A_2 is a buffer amplifier which has the capacitor at its input for A/D conversion process.

Definitions

There are two important terminologies regarding the sample and hold circuit which should be defined here.

Aperture time — The delay time between the HOLD command and the instant at which the analog switch is really OPEN. A typical aperture time is 50 nsec.

Acquisition time — The time between HOLD command and the instant at which the output voltage e_o settles to ±1% of its final value. A typical acquisition time is 5 μsec.

V. SPECIAL PURPOSE D/A CONVERSION CIRCUITS

There are two special purpose D/A Conversion Circuits worthy of being described.

A. Inverse D/A Converter

Figure 12(a) shows an inverse D/A converter. That is, the output voltage is inversely proportional to the binary-coded input number. Note that the operational amplifier at the output functions as a high input impedance buffer amplifier with unity gain. The equivalent resistance shunting the amplifier input is the total parallel resistance of the D/A decoder network, i.e.,

$$R_t = \frac{1}{\sum_{j=0}^{n-1} \frac{1}{R_j} x_j}$$

Thus,

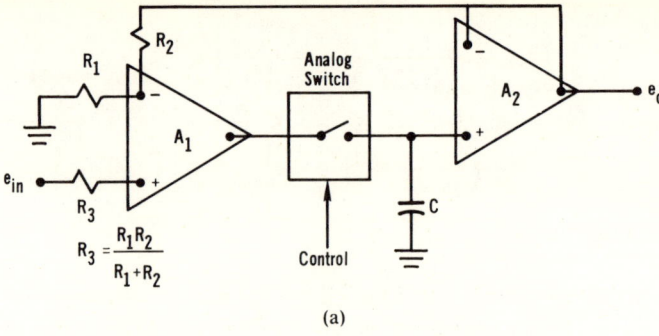

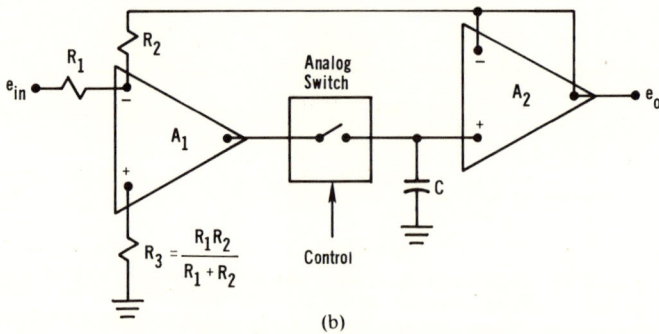

FIGURE 11. Sample and hold circuits.

$$e_o = IR_t = I \frac{1}{\sum_{j=0}^{n-1} \frac{1}{R_j} x_j}$$

where, I = a constant current; $x_j = 1, 0$; $j = 0, ..., n - 1$.

B. Multiply D/A Converter

The circuit shown in Figure 12(b) is similar to the parallel D/A converter described in Section II.A and Section IV.D except that V_r in this circuit is no longer a constant. That is, the reference voltage V_r has been changed to V_x which becomes a variable carrying information. Therefore, in this configuration,

$$V_o = V_x \left(2^0 x_0 + ... + 2^j x_j + ... + 2^{n-2} x_{n-2} \right)$$

VI. CODES EMPLOYED BY A/D CONVERTERS

As we know, the function of an A/D converter is to assign a digital code in one and zero to represent an analog value. There are many kinds of coding systems being used in many fields. In this section we shall describe the three most popular ones by examples of 4-b words to represent an analog voltage which ranges from 0 to 1024 mV. In general, one least-sigificant-bit (LSB) of an n-bit word is equivalent to,

$$\frac{\text{Total Voltage Range}}{2^n}$$

For n = 4, total voltage range = 1024 mV and 1 LSB = 64 mV.

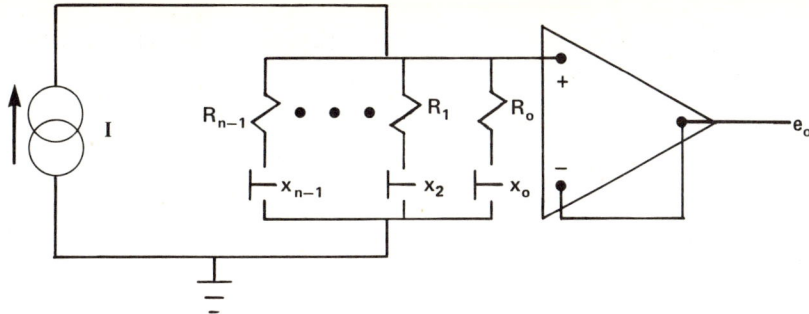

(a) Inverse D/A Converter

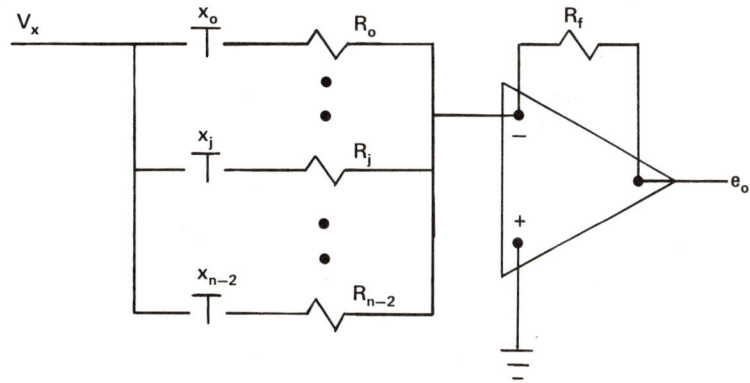

(b) Multiply D/A Converter

FIGURE 12.   Special purpose D/A conversion circuits.

A. Binary Codes

Tables 1 and 2 show, respectively, a unit polar binary code and bipolar offset binary code. The latter is called offset code, because the analog equivalent voltage is so offset by −512 mV that the zero voltage is now in the middle of the binary code.

B. Complement Codes

Ones complement and twos complement codes for bipolar analog voltage are shown, respectively, in Table 3 and Table 4. In Table 3, the eight negative binary words are the ones complement of the eight positive ones with the most significant bit as sign bit. They are symmetric and have separate codes for +0 and −0 as shown. In twos complement code shown in Table 4, however, there is only one 0 V equivalent, i.e., 0000 = −32 to +32. The negative code is the twos complement of its correspondent positive ones, but it lost the symmetry with respect to zero in contrast to ones complement code. Again, its most significant bit is the sign-bit.

C. Gray Code

Table 5 shows gray code equivalent of the same analog voltage range. The outstanding feature of gray code is that there is always only 1-b difference between adjacent digital words. This is most desirable for coding a rotating shaft. The rule for gray code is that starting from LSB column, there are 2, 4, 8, 16, etc., ones alternating with 2, 4, 8, 16, etc., zeros downward except both ends of the column split half of zeros of the 2, 4, 8, 16, etc., when there is enough space to fit in.

Table 1
UNIT POLAR BINARY CODE

Voltage (mV)	Binary code
0 to 64	0 0 0 0
64 to 128	0 0 0 1
128 to 192	0 0 1 0
192 to 256	0 0 1 1
256 to 320	0 1 0 0
320 to 384	0 1 0 1
384 to 448	0 1 1 0
448 to 512	0 1 1 1
512 to 576	1 0 0 0
576 to 640	1 0 0 1
640 to 704	1 0 1 0
704 to 768	1 0 1 1
768 to 832	1 1 0 0
832 to 896	1 1 0 1
896 to 960	1 1 1 0
960 to 1024	1 1 1 1

Table 2
BIPOLAR OFFSET BINARY CODE

Voltage (mV)	Offset binary code
−512 to −448	0 0 0 0
−448 to −384	0 0 0 1
−384 to −320	0 0 1 0
−320 to −256	0 0 1 1
−256 to −192	0 1 0 0
−192 to −128	0 1 0 1
−128 to −64	0 1 1 0
−64 to 0	0 1 1 1
0 to 64	1 0 0 0
64 to 128	1 0 0 1
128 to 192	1 0 1 0
192 to 256	1 0 1 1
256 to 320	1 1 0 0
320 to 384	1 1 0 1
384 to 448	1 1 1 0
448 to 512	1 1 1 1

Table 3
ONES COMPLEMENT CODE

Voltage (mV)	Ones complement	Signed-magnitude equivalent
−512 to −448	1 0 0 0	−7
−448 to −384	1 0 0 1	−6
−384 to −320	1 0 1 0	−5
−320 to −256	1 0 1 1	−4
−256 to −192	1 1 0 0	−3
−192 to −128	1 1 0 1	−2
−128 to −64	1 1 1 0	−1
−64 to 0	1 1 1 1	−0
0 to 64	0 0 0 0	+0
64 to 128	0 0 0 1	+1
128 to 192	0 0 1 0	+2
192 to 256	0 0 1 1	+3
256 to 320	0 1 0 0	+4
320 to 384	0 1 0 1	+5
384 to 448	0 1 1 0	+6
448 to 512	0 1 1 1	+7

VII. MAJOR ERRORS IN CONVERSION

A. Quantization Error

Figure 13 illustrates the quantization error of a 3-b binary code for an analog voltage range 0 to 1024 mV. Note that each binary word represents a range of 128 mV. For example, 001 represents the analog value ranging from 128 to 256. That is, any analog

Table 4 TWOS COMPLEMENT CODE			Table 5 GRAY CODE	
Voltage (mV)	Twos complement	Signed-magnitude equivalent	Voltage (mV)	Gray code
			−512 to −448	0 0 0 0
−480 to −544	1 0 0 0	−8	−448 to −384	0 0 0 1
−416 to −480	1 0 0 1	−7	−384 to −320	0 0 1 1
−352 to −416	1 0 1 0	−6	−320 to −256	0 0 1 0
−288 to −352	1 0 1 1	−5		
			−256 to −192	0 1 1 0
−224 to −288	1 1 0 0	−4	−192 to −128	0 1 1 1
−160 to −224	1 1 0 1	−3	−128 to −64	0 1 0 1
−96 to −160	1 1 1 0	−2	−64 to 0	0 1 0 0
−32 to −96	1 1 1 1	−1		
			0 to 64	1 1 0 0
−32 to +32	0 0 0 0	0	64 to 128	1 1 0 1
322 to 96	0 0 0 1	+1	128 to 192	1 1 1 1
96 to 160	0 0 1 0	+2	192 to 256	1 1 1 0
160 to 224	0 0 1 1	+3		
			256 to 320	1 0 1 0
224 to 288	0 1 0 0	+4	320 to 384	1 0 1 1
288 to 352	0 1 0 1	+5	384 to 448	1 0 0 1
352 to 416	0 1 1 0	+6	448 to 512	1 0 0 0
416 to 480	0 1 1 1	+7		

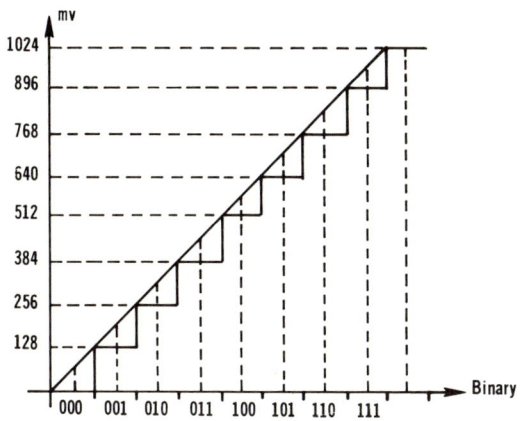

FIGURE 13.   Quantization error.

voltage within the range of 128 to 256 mV will be assigned 001. When converting 001 back, only one value, say 192 mV will be assigned. Therefore, logically, each binary word inherently has a ±1/2 LSB error.

B. Analog Component Error

This error is due to the fact that in reality there is no perfect analog component. Their properties or values change as environmental conditions vary. Therefore, analog errors inherently exist in systems using analog components. Fortunately, most A/D/ C/ manufacturers restrict the analog errors of their products within a ±1/2 LSB equivalent, so that the total error will be less than ±1 LSB.

C. Aperture Error

Figure 14 shows the effect of aperture in accuracy of data obtained. Note that the

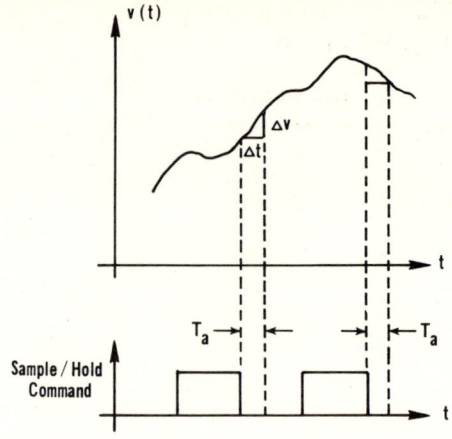

FIGURE 14. Aperture error.

uncertainty of switching time during the transition of sample and hold would cause error if the analog signal changed rapidly within this aperture time period. Thus the aperture error can be estimated by the following equation.

$$\text{Aperture Error} = \frac{\Delta v}{\Delta t} \times T_a$$

where $\Delta v/\Delta t$ is the time slope of the signal as shown and T_a is the aperture time. Normally, the worst case estimation is employed. That is the maximum $\Delta v/\Delta t$ of the analog signal is used. For example, if v(t) is the analog signal, then $(\Delta v/\Delta t)_{max}$ is determined by equation

$$\frac{dv(t)}{dt} = 0$$

VIII. KEY FACTORS FOR SYSTEM DESIGN CONSIDERATION USING A/D/C

Consider that in designing the system shown in Figure 1, a designer should first understand the characteristics of the analog signal, that is, the voltage range, tolerable total error, and the resolution required. Based on this information, the number of bit of the A/D/C can be determined. For example, if the range is 1 V and the resolution is 1 mV, then at least a 10-b A/D/C is required, since

$$1 \text{ LSB} = \frac{1 \text{ V}}{2^{10}} = \frac{1}{1024} = 0.977 \text{ mV}$$

Now, if the maximum frequency component of the signal is 100 Hz, its frequency spectrum ranges DC to 100 Hz, then a low-pass filter cut-off at 100 Hz with sharp attenuation beyond 100 Hz should be used as the guard or anti-aliasing filter. At this point, the sampling frequency of 200, preferably more, should be selected. If 1000 Hz sampling frequency is chosen, then the maximum hold time would be 500 μsec. This figure restricts the maximum conversion time specification of the A/D/C. Finally, one can total the tolerable error of the system. Based on this information, the specifications of aperture time for sample/hold circuit, and that of total quantization and analog errors of A/D/C can be determined.

REFERENCES

1. **Naylor, J. R.**, Digital and analog signal applications of operational amplifier I — multiplexer and converters, *IEEE Spectrum*, May 1971.
2. **Naylor, J. R.**, Digital and analog signal applications of operational amplifier II — sample/hold modules, peak detectors and comparators, *IEEE Spectrum*, June 1971.
3. **Hoeschele, D. F., Jr.**, *A/D and D/A Conversion Techniques*, John Wiley & Sons, New York, 1968.
4. **Susskind, A. K.**, *Notes on A/D Conversion Techniques*, MIT Press, Cambridge, 1956.
5. **Schmid, H.**, A practical guide to D/A conversion, *Electron. Des.*, October 24, 1968.
6. **Schmid, H.**, A practical guide to A/D conversion, I, *Electron. Des.*, December 5, 1968; II, *Electron. Des.*, December 19, 1968; and III, *Electron. Des.*, January 4, 1969.
7. **Lin, Wen C.**, A precision A/D converter for nonlinear conversion, *Control Eng.*, April 1970.
8. **Graeme, J. G., Tobey, G. E., and Huelsman, L. P.**, *Operational Amplifiers, Design and Application*, McGraw-Hill, New York, 1971, chap. 9.
9. **Gordon, B. M.**, Noise-effects on A/D conversion accuracy. I, *Comput. Des.*, March 1974.
10. **Risch, D.**, Design D/A/D interfaces for your computer, *EDN*, April 5, 1974.
11. **Fogarty, J. D.**, Know your A/D converter's dynamic range, *Electron. Des.*, February 1, 1975.
12. Guide to Analog CMOS Switches and Multiplexers, Analog Devices, Norwood, Massachusetts, 1974.
13. Analog-Digital Conversion Notes, Analog Devices, Norwood, Massachusetss, 1977.
14. **Thibodeaux, E.**, Getting the most out of C-MOS devices for analog switching jobs, *Electronics*, Dec. 25, 1975.

Chapter 9

NOISE IN DIGITAL SYSTEMS

I. INTRODUCTION

Any experienced circuit or system designer one way or the other goes through the nightmare of noise problems in equipment design. It has been a severe problem to experimenters for centuries and yet has never been appreciated in classrooms. Fact after fact shows manufacturers cannot deliver their products because of their apparently unsolvable noise problems. Although there is no theory or sure way to cope with noise problems, careful planning plus following a set of guidelines or rules may avoid learning things in the hard or expensive ways. Generally speaking, the noise problems appearing in analog systems are quite different from digitals. Electronic systems containing electromechanical devices would behave differently in comparison with pure electronic systems. Unfortunately, a digital system designer would most likely be challenged to design systems involving all of the three different types of circuitries or devices. In this chapter, an attempt will be made to provide the system designer with analytical background and a set of guidelines to cope with the real problems in the real world.

II. EXTERNAL OR RADIATION NOISE

There are cases showing the sources of noise causing malfunction of equipment are external in origin. In this respect, the analog circuits and regenerative modules within the system (such as operational amplifiers, comparators, one shot, Schmitt trigger, flip-flops) are more sensitive to them, especially if high gain or high input impedance amplifiers are employed. These noise sources would either be high voltage or high current in nature. The noise induced within the system would either be through the electrostatic induction from the high voltage or fast-rise spike sources or through the electromagnetic induction from high current sources. For example, lightning and high voltage switching devices would belong to the former, while welding and starting or stopping heavy duty motors would belong to the latter. Figure 1(a) shows the concepts of electrostatic and electromagnetic induction. Proper shielding of the sensitive devices would mostly solve these kinds of problems, Figure 1(b) suggests some methods to achieve the shielding effect. For electromagnetic noise, some kind of high μ ferromagnetic material should be used for shielding purposes. These materials are expensive but fortunately high current sources are not as popular as the high voltage noise sources. For electrostatic noise, some metal such as aluminum shielding would be satisfactory in most cases. The coaxial shielded cable is the most effective among the three shown. Twisted pair-wire is better than the parallel grounded wire. As for ribbon cable, if possible, one may tie every other one of the conductors to ground. For more critical, low-signal analog circuits, one should consider enclosing the whole section with an aluminum box which should be tied to the system ground with a heavy conductor. If necessary, differential amplifiers should be used in conjunction with two-conductor coaxial cables. Note that the diagram shown has only one end of the shielded wire connected to ground, and the other end is floating. This configuration is recommended only for low frequency and low level analog signals. If both ends are grounded, the shielded wire and the ground-plane would form a loop which would pick up noise and be amplified as a signal. For digital signals where no amplifier is being used, however, grounding both ends is recommended. High frequency fast rise-time clock in a digital

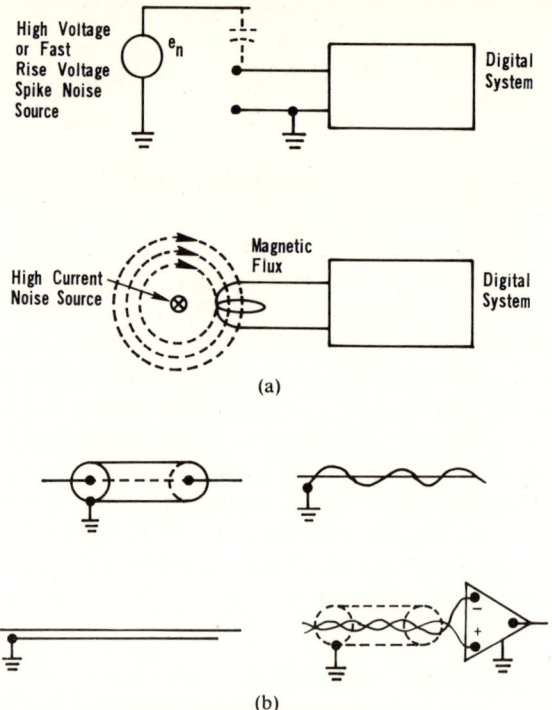

FIGURE 1. Shielding techniques.

system usually causes cross-talk. In this case, it is desirable to shield the clock-line using one of the methods shown in Figure 1(b). This would confine the clock from causing cross-talk.

III. INTERNAL NOISE

A. Decoupling Technique

Often in both analog and digital circuits, noise may generate from one circuit to the other when both are sharing the same power supplies as shown in Figure 2(a). R_o is a lump resistance which may include the internal resistance of the power supply and the line resistance (impedance). While Circuit 1 is in operation, it may draw a current pulse from the power supply. This pulse may be transmitted along the line down to Circuit 2 and become a noise to it. This noise is a common problem to both analog and digital circuits. Therefore, it is recommended that a decoupling capacitor C_d always be employed to each circuit board, and for amplifiers or one shot, Schmitt trigger, flip-flop chips, a decoupling capacitor should be connected directly across its power supplies and ground pins. An example will clarify the problem. As shown in Figure 2(b), let the noise pulse produced by Circuit 1 be an amplitude of $-E_N$ and have a pulse width τ. R is the Thevenin resistance of power supply and Circuit 1. A decoupling capacitor C_d is connected directly across Circuit 2.

Then $V_2(t) = V_2(\infty) + [V_2(O^+) - V_2(\infty)]e^{-t/RC_d}$, and $V_2(\tau) = V_{cc} - E_N) + [V_{cc} - (V_{cc} - E_N)]e^{-\tau/RC_d}$.

If, $RC_d \gg \tau$; $e^{-\tau/RC_d} \simeq 1$; $V_2(\tau) = V_{cc}$.

That is, Circuit 2 will not see the noise, if RC_d is much greater than the pulse-width τ. It is, however, not practical to control R, therefore, an electrolytic capacitor of tens or even hundreds of microfarads is generally used for this purpose. It is important

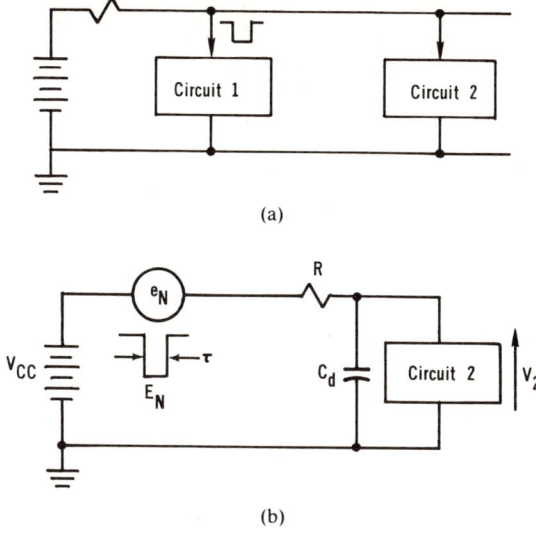

FIGURE 2. Noise in power supply line.

to point out that in very high frequency or switching speed circuitry, a small disc capacitor of 0.1 μF is connected in parallel with a good size electrolytic capacitor. The reason for addition of the small capacitor is that the electrolytic capacitor has inherent inductance which would not fill out radio-frequency, thus the small capacitor used here is sometimes called RF capacitor.

B. Grounding Technique

Experience has shown us that a well-planned grounding system should eliminate a lot of noise troubles in system design, especially for a system which contains a mixture of analog, digital, and electromechanical subsystems. As shown in Figure 3(a), there are two important points that should be emphasized, i.e.,(1) each subsystem should have its own power supplies, and (2) there should be two grounding systems, namely chassis ground and AC power ground. Normally, if each subsystem has its own chassis, then the common of the power supplies should have one and only one point connected to its individual chassis. The three chassis should then be connected together with a heavy high conductance conductor and finally electrically tied to the AC ground. Optical coupling devices are strongly recommended for signal coupling between electromechanical and other subsystems. If separate power supplies for analog and digital circuitries are not economically feasible, one should at least use separate voltage regulator chips which are now commercially available at reasonable cost. Of course, a good decoupling network is necessary before the regulators. Another problem as illustrated in Figure 3(b) is that there are too many devices sharing the same return-line. In this case, $i_t = i_s + i_1 + \ldots + i_j + \ldots + i_n$. If z_0 is the impedance of the return wire segment as shown, the input signal of D_1 would be, $e_s - i_t z_0$ instead of pure e_s, the signal voltage. Figure 3(c) shows the center-point configuration for dealing with this kind of problems. This same recommendation should be practiced on the power distribution layout for the back-plane wiring.

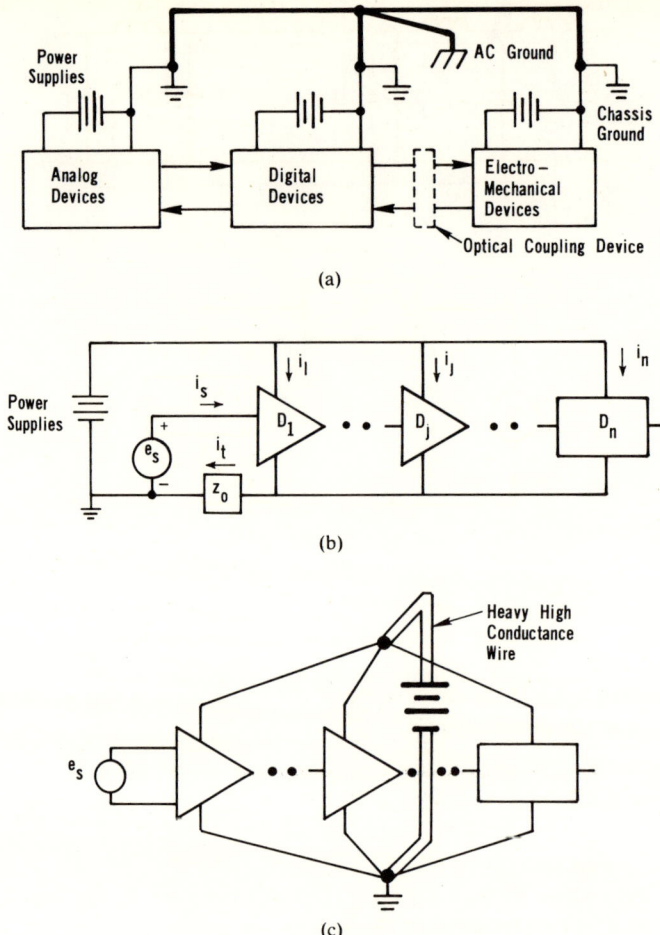

FIGURE 3. Grounding technique.

IV. TRANSMISSION-LINE REFLECTION

A. Introduction

A transmission line need not be wires of miles in length. Generally speaking, if the propagation delay of a segment of a wire is greater than the transition time of a pulse to be sent, it can be considered a transmission line, where line-reflection may result in ringing, overshooting, and reducing noise margin. A brief review of transmission line theory follows.

Figure 4(a) shows the equivalent circuit of a segment of transmission line where, L: inductance in henry per unit length of wire, and C: capacitance in farad per unit length of wire. Both L and C are functions of width, thickness, and spacing of conductors, and dielectric constant of the insulation materials. Time delay per unit length = $\sqrt{LC}$ = τ. Characteristic impedance of a transmission line is defined as

$$z_o = \sqrt{\frac{L}{C}}$$

Figure 4(b) shows the equivalent circuit of a transmission line with sending voltage V_s,

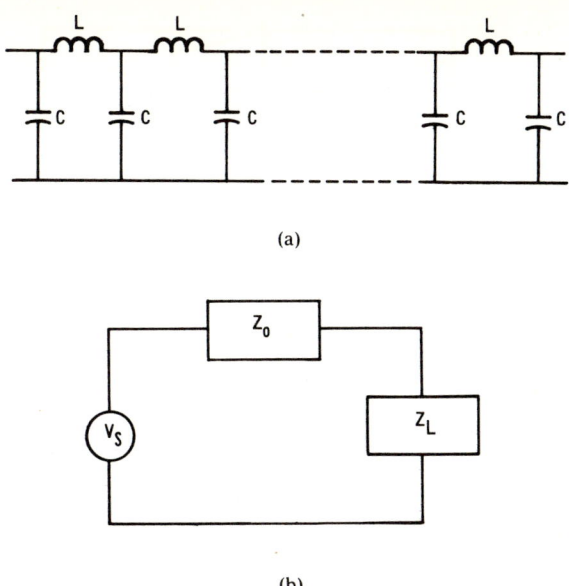

(a)

(b)

FIGURE 4. Equivalent circuits of a transmission line.

and a load impedance z_L. If the delay time of the transmission line is 1 T, the reflection voltage for t = T at the receiving end,

$$V_r(t = T) = \frac{z_L - z_o}{z_L - z_o} V_s = \rho V_s \tag{1}$$

where

$$\rho \triangleq \text{reflection coefficient} \triangleq \frac{z_L - z_o}{z_L + z_o} \tag{2}$$

B. Transmission Lines Terminated by Linear Devices

As a numerical example, consider a transmission line driven by a 1 V source which has an output impedance R_s = 25 Ω. As shown in Figure 5(a), we have $z_L = \infty$. As the switch S is closed at t = 0; we are interested in the time waveforms of the driving voltage, $V_D(t)$ and receiving voltage, $V_R(t)$ at the driving and receiving ends of the transmission line which has a delay time of T. Basically we are studying the transient response of the transmission line to a step-function. Figure 5(b) shows a graph describing the reflection process of the transmission line. The lower line represents the time axis of $V_D(t)$, and the upper line, $V_R(t)$. The triangle-shaped line with arrows depicts the traveling process of the step function. That is, at t = 0, $V_D(0)$ starts to travel down the line and reaches the end of the line when t = T. Therefore, $V_R(0 < t < T)$ is the receiving end voltage before the traveling wave arrives at the end of the line, therefore, we have $V_R(0 < t < T) = 0$. At t = 2 T the reflection wave would arrive at the driving end and reflect again. The wave is then reflected back and forth as the arrow-line shown. It is clear that the magnitudes of $V_D(t)$ and $V_R(t)$ will change, respectively, whenever t = 2 T, 4 T, ..., and t = T, 3 T, 5 T, ..., etc. Eventually, $V_D(\infty) = V_R(\infty)$ = 1 V. To plot $V_D(t)$ and $V_R(t)$, we have to determine the values at these critical times. It is important to point out that all the voltages at the critical times are composed of three components, namely, (1) original voltage V_1, (2) incoming voltage, V_2 and (3) reflected voltage V_3. Calculation of the voltages follows.

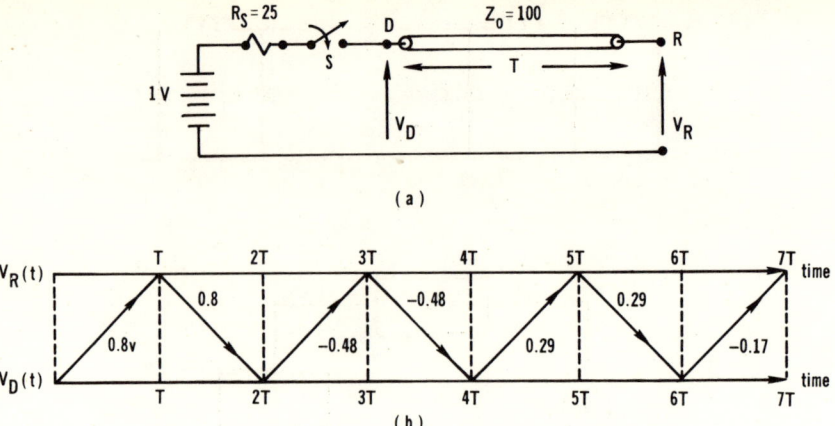

FIGURE 5. A. transmission line terminated by linear device.

Reflection coefficients:

$$\rho_D(\text{driving end}) = \frac{R_s - z_o}{R_s + z_o} = \frac{25 - 100}{25 + 100} = -0.6$$

$$\rho_R(\text{receiving end}) = \frac{z_L - z_o}{z_L + z_o} = \frac{\infty - 100}{\infty + 100} = 1$$

Voltage Calculations

Since, at point D, the load impedance is z_0, regardless of how the transmission line is terminated at the receiving end, $V_D(0^+) = V_1 + V_2 + V_3$ can be determined as follows.

For $t = 0^+$

$$V_1 = \frac{z_O}{R_s + z_O} \; 1v = \frac{100}{25 + 100} \times 1 = 0.8$$

$$V_2 = 0$$

$$V_3 = 0$$

$$V_D(0^+) = 0.8 \text{ V}$$

For $t = T$

$$V_1 = 0$$

$$V_2 = 0.8$$

$$V_3 = \rho_R \, 0.8 = 1 \times 0.8 = 0.8$$

$$V_R(T) = 0 + 0.8 + 0.8 = 1.6$$

For t = 2 T

$V_1 = V_D(0^+) = 0.8$

$V_2 = 0.8$

$V_3 = \rho_D \times 0.8 = -0.6 \times 0.8 = -0.48$

$V_D(2\,T) = 0.8 + 0.8 - 0.48 = 1.12$

For t = 3 T

$V_1 = V_R(T) = 1.6$

$V_2 = -0.48$

$V_3 = -0.48$

$V_R(3\,T) = 1.6 - 0.48 - 0.48 = 0.64$

For t = 4 T

$V_1 = V_R(3\,T) = 0.64$

$V_2 = 0.29$

$V_3 = 0.29$

$V_R(5\,T) = 0.64 \text{ to } 0.29 + 0.29 = 1.22$

For t = 5 T

$V_1 = V_D(2T) = 1.12$

$V_2 = -0.48$

$V_3 = (-0.6) \times (-0.48) = 0.288$

$V_D(4T) = 1.12 - 0.48 + 0.288 = 0.928$

For t = 6 T

$V_1 = V_D(4\,T) = 0.93$

$V_2 = 0.29$

$V_3 = (-0.6) \times 0.29 = -0.14$

$V_D(6\,T) = 0.93 + 0.29 - 0.17 = 1.05$

For t = 7 T

$V_1 = V_R(5T) = 1.22$

$V_2 = -0.17$

$V_3 = -0.17$

$V_R(7\,T) = 1.22 - 0.17 - 0.17 = 0.88$

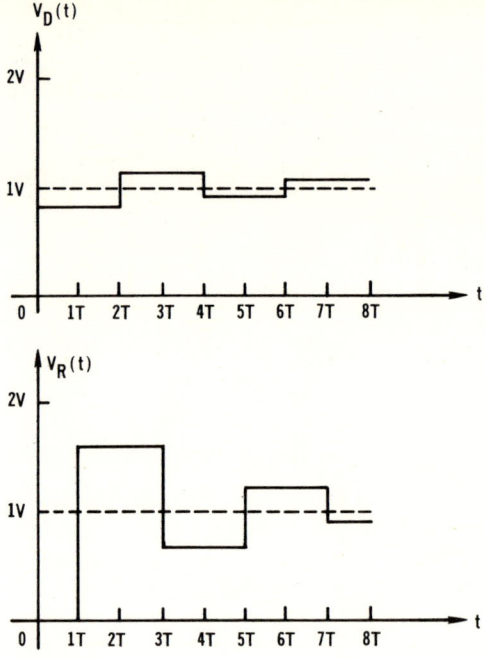

FIGURE 6. Waveforms of driving and receiving voltages.

Figure 6 shows the plots of $V_D(t)$ and $V_R(t)$ according to the calculated results obtained.

C. Transmission Lines Terminated by Nonlinear Devices

As described in preceding sections, the analysis of transmission lines terminated by linear devices appears to be straightforward. In digital systems, however, practically all transmission lines are terminated by nonlinear devices such as TTL gates, etc. In this section we shall illustrate how the piece-wise linear graphical techniques can be utilized for analysis of TTL-terminated transmission lines. Consider a transmission line which is driven and terminated by TTL gates shown in Figure 7(a). It is evident that when the driver is in logic [1] or HIGH state, the input of the receiving gate is also high; therefore, practically no current is flowing in the line. When the driver is in LOW state, it will SINK current from receiver. For this reason, we shall conveniently assign that the current flow out of the receiver is positive and into the driver, negative. First, let us determine the input V-I curve of the receiver. As shown in Figure 7(a), one can assume that a variable DC voltage source is used, and it starts at +5 V and decreases until reaching the safety margin at some negative voltage. The corresponding I_i and V_i readings can be used to construct the input V-I curve. For example, by means of the Binary-State-Analysis (BSA), the base-emitter diode junction will not conduct until $V_i \leqslant 2.1$ V. The current increases and reaches

$$\frac{5 - 0.7}{2.8 \text{ K}} = 1.5 \text{ mA}$$

when $V_i = 0$. As V_i continuously decreases, the current increases until $V_i \leqslant -0.7$. At this point, the clamping diode conducts, therefore, $I_i = I + I'$ and increases rapidly. The $V_i - I_i$ curve is shown in Figure 7(b). Following the same logic, the output V-I curves, respectively, for Logic [1] and [0] are shown in Figures 8(a) and (b). In logic

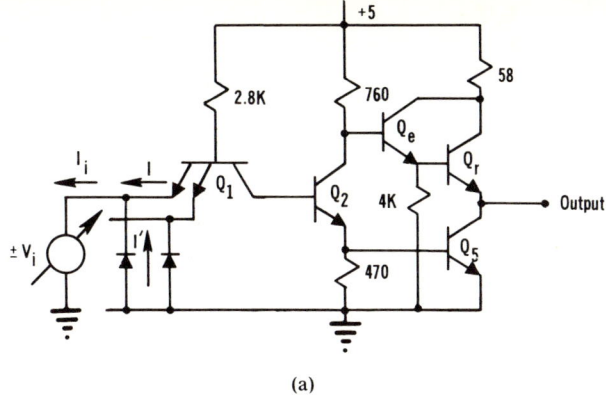

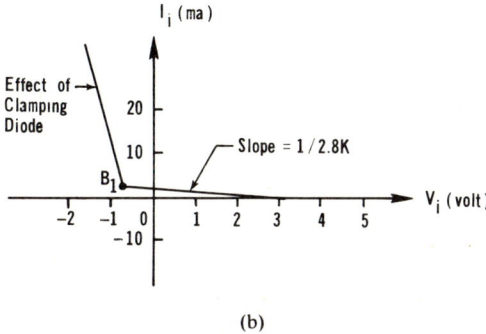

FIGURE 7. TTL input V-I curve.

[1] case, Q_5 is OFF, Q_3 and Q_4 are ON. The Thévenin output voltage is approximately equal to $5 - 1.4 = 3.6$ V, where 1.4 V drop is contributed by the B-E junctions of Q_3 and Q_4. Therefore, the current I changes to negative direction or flows outwardly when $V \leq 3.6$ V. When

$$V = 0, \quad I = \frac{5 - 0.9}{58} = 70 \text{ mA}$$

where the 0.9 V drop is due to the C-E junction of Q_3 (0.2) and B-E junction of Q_4 (0.7). As the source voltage, V, decreases to -0.7 and more, the B-C junction of Q_5 begins to be forward biases, current will flow from ground through the 470 Ω resistor. The output V-I curve for logic [1] is shown in Figure 8(a), where B_1 is the break point as Q_5 conducts. The output of the driver in logic [0] case is shown in Figure 8(b). Here, Q_3 and Q_4 are OFF. A constant currant I_1 flows into the base of Q_5 and thus Q_5 conducts. For $V \geq 0$, the V-I curve will be simply the $I_c - V_{cE}$ curve at $I_B = I_1$, which is one of the output characteristic curve families with which we are familiar. When $V \leq 0.7$, both B-E and B-C junctions begin to conduct and current increases more rapidly. The break point is B_1 as shown in Figure 8(b).

The composite V-I curves of two outputs and one input of a transmission line terminated by this TTL gate, are shown in Figure 9(a). Figure 9(b) shows the circuit configuration where the transmission line has a characteristic impedance of 100 Ω. Figure 9(a) shows the two steady-state points P_1 and P_0, where P_0 is the intersection of input curve and output LOW curve, while P_1 is the intersection of input curve and output HIGH curve. Referring to Figure 9(b), consider the driver is in HIGH state

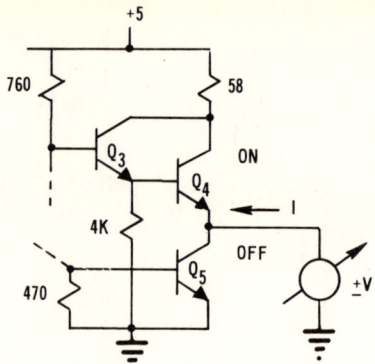

Output Circuit of the Driver

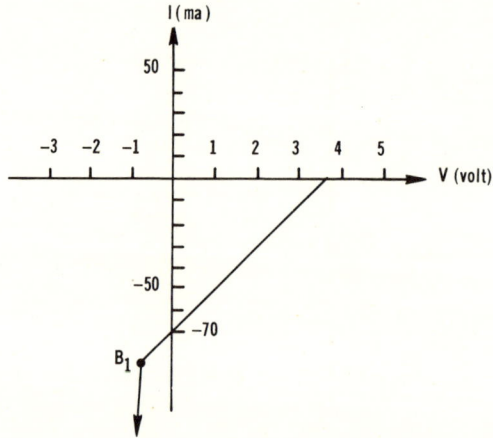

FIGURE 8(a). TTL output V-I curve for logic [1] (High).

and it is to be switched to LOW. We know that the initial point will be P_1 and the final point will be P_o as shown in Figure 9(a). The question is what is the operational path in the composite V-I curves during the transition of logic HIGH to logic LOW. From Figure 9(b), we have

$$V_R(t) - I(t)Z_o = V_D(t) \qquad (3)$$

For switching from HIGH to LOW, at $t = t_o$, $V_R(t)$ can not change instantly; thus we could consider $V_R(t)$ at this moment as a constant. By differentiating Equation 3, we have

$$0 - \frac{dI(t)}{dt} Z_o = \frac{dV_D(t)}{dt}$$

$$\frac{dI(t)}{dV_D(t)} = -\frac{1}{Z_o}$$

That is, the direction of the operational path is $-1/Z_o$. When $t = t_o + T$, the traveling pulse arrives at the receiving end and $V_R(t)$ changes suddenly while $V_D(t)$ can be considered a constant; thus by differentiating Equation 3 again, we have

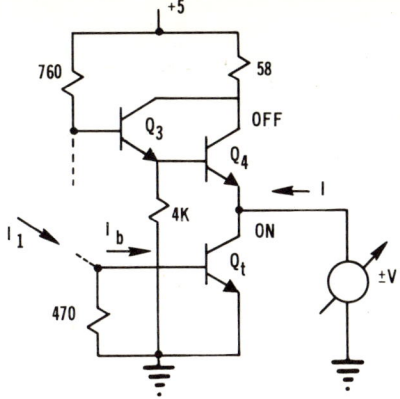

Output Circuit of the Driver

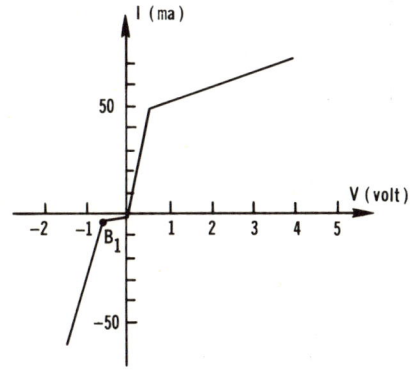

FIGURE 8(b). TTL output V-I curve for logic 0 (Low).

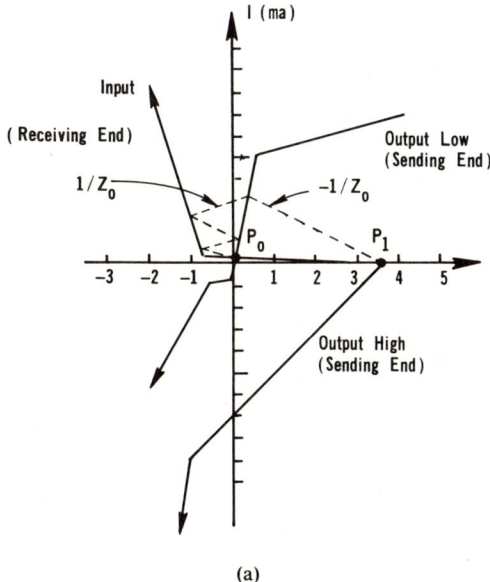

(a)

FIGURE 9(a)(b). Transmission lines terminated by TTLs.

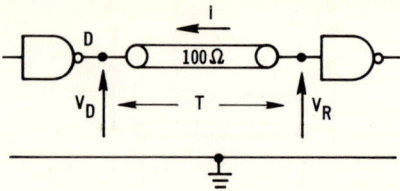

FIGURE 9(b).

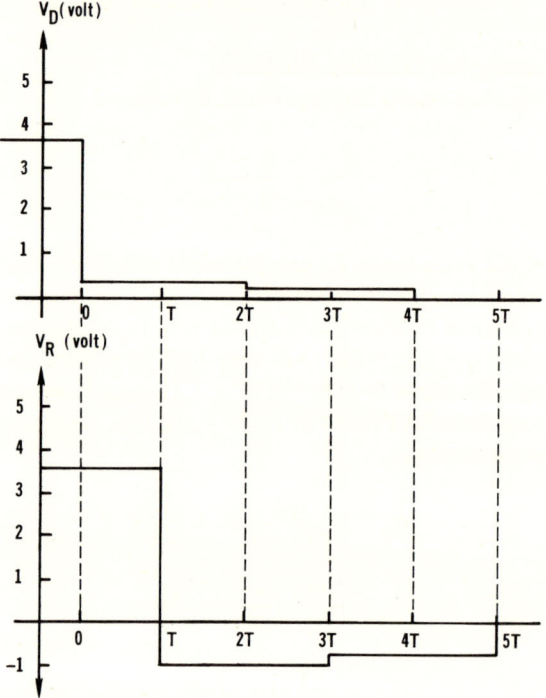

FIGURE 9(c). V_D and V_R waveforms of TTL terminated transmission line.

$$\frac{dV_R(t)}{dt} - \frac{I(t)}{dt} Z_o = 0$$

$$\frac{dI(t)}{dV_R(t)} = \frac{1}{Z_o}$$

which has a positive slope. Since we now know the direction of the operational paths and the starting/ending points, as well as the V-I curves, we can conclude that the reflecting pulse will travel back and forth between input curve and output LOW curve as shown in dotted lines in Figure 9(a). From which, the voltage magnitudes of $V_D(t)$ and $V_R(t)$ at t_o, T, 2 T, 3 T, ... can be read from the graphs, and their waveforms are shown in Figure 9(c). By the same principle, one can analyze the response of the transmission line for switching from LOW to HIGH.

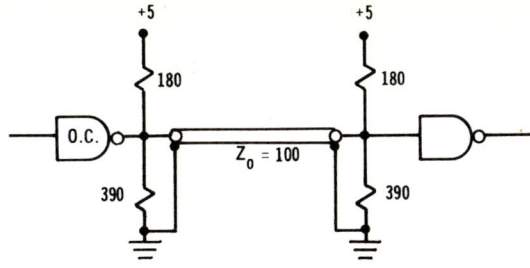

FIGURE 10. Matching transmission line impedance.

D. Impedance Matching for Transmission Lines

In view of the reflection coefficient equation, Equation 2

$$\rho = \frac{Z_L - Z_o}{Z_L + Z_o}$$

reveals that if the load impedance Z_L can be made nearly or exactly equal to Z_o, the reflection voltage can be reduced or eliminated. This concept has indeed been implemented in practice. Figure 10 shows a practical way to realize the idea. Note that the driver is an open collector gate which has very high impedance when in HIGH state. At the receiving end, the input impedance of a TTL gate is very high in HIGH state, and a couple or more thousand ohms in LOW state. The resistor dividers shown have Thévenin voltage and resistance,

$$V_{th} = \frac{390}{180 + 390} \times 5 = 3.42 \text{ V}$$

$$R_{th} = \frac{390 \times 180}{180 + 390} = 123 \text{ }\Omega$$

Therefore, except when the drive is in LOW state, the transmission line is terminated at both ends by 123 Ω which nearly matches the $Z_o = 100$. Experiment results show that the resistor networks indeed significantly reduce the reflection voltage of the transmission line.

V. DIFFERENTIAL LINE DRIVER AND RECEIVER

For long distance transmission line and noisy environment, a system designer should consider using the differential line driver and receiver as shown in Figure 11. These devices are now commercially available at a very reasonable cost. Note that the driver simultaneously generates a pair of signals with opposite phase and the receiver output is the difference of the input pair. Therefore, it has the desirable common mode rejection feature which would minimize the radiation type noise induces in the transmission line. The input impedance of the receiver is normally fairly high, thus a transmission line impedance matching network can be inserted at the receiving end to reduce the reflection voltage.

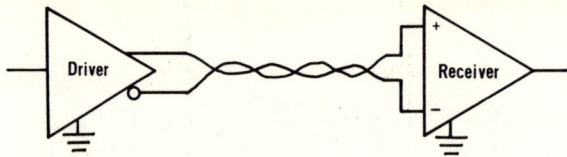

FIGURE 11. Differential transmission line driver and receiver.

REFERENCES

1. **Oh, Henry W.**, *Noise Reduction Techniques in Electronic Systems,* John Wiley & Sons, New York, 1976.
2. **Jones, J. P.**, *Causes and Curves of Noise in Digital Systems,* Computer Design Publishing, W. Concord, Mass., 1964.
3. **Heniford, B.**, Noise in 54/74 TTL Systems, TI Application Report CA-108, Texas Instruments, Dallas, 1980.
4. **Blood, W. R.**, MECL System Design Handbook, Motorola Semiconductor, Phoenix, 1900.
5. **Morris, R. L. and Miller, J. R.**, Eds., *Designing with TTL Integrated Circuits,* McGraw-Hill, New York, 1971.
6. **Morrison R.**, *Grounding and Shielding Techniques in Instrumentation,* John Wiley & Sons, New York, 1977.
7. **deFalco, J. A.**, Reflection and crosstalk in logic circuit interconnections, *IEEE Spectrum,* July 1970.
8. **Cushman, R. H.**, Make the most of noise — correlate it, I, *EDN,* March 1, 1971; II, *EDN,* April 15, 1971.
9. **Balph, T.**, Drive 50 ohm transmission lines with TTL and ECL gate outputs, *EDN,* Jan. 20, 1979.
10. **Matick, R. E.**, *Transmission Lines for Digital and Communication Network,* McGraw-Hill, New York, 1969.

Chapter 10

RANDOM LOGIC DESIGN USING PROGRAMMABLE LOGIC ELEMENTS

I. GENERAL BACKGROUND

Random logic is sometimes called control logic, which is a logic network that has no definite pattern in contrast to array or tree logic network. In any digital system design, one is always involved in designing some subsystem which controls the signal paths and schedules the time sequence of logical events. The design is usually customized and irregular in fabrication, therefore it is inherently a time consuming task and prone to fabrication errors. In this chapter, we shall describe the techniques of realizing random logic using regular or programmable logic elements which have been commercially available for a number of years. A brief review of logical design concepts follows.

The conventional procedure for the logic design of switching network of digital systems is shown in Figure 1. Figures 2(a) and (b) show, respectively, the general model of combinational and sequential logic networks. For combinational logic, the output at present time, $z \underline{\Delta} (z_1, z_2, ..., z_m)$, is and only is, a function of the input $X \underline{\Delta} (x_1, x_2, ..., x_n)$ at the present time; for sequential logic, however, the output is a function of the present and the past history of the input. The additional feedback or delay network in Figure 2(b) is responsible for the past history of the system. To clarify the concept, a simple design example is now in order. Following the procedure shown in Figure 1, one can show how a 1-b full-adder is designed. Figure 3 illustrates the design in three steps, (a), (b), and (c). However, the last step or the step of hardwired logic implementation connecting discrete logic elements, such as AND and EXCLUSIVE-OR gates, in a customized fashion can be replaced by programmed logic elements such as MULTIPLEXER, or programmable logic elements such as ROM, Programmable-Logic-Array (PLA), etc. Thanks to the LSI technology, a logic designer now has a choice of realizing his design by one of the five methods, namely hardwired or discrete logic, multiplexers, ROM, PLA, or microprocessor.

A. Design with Multiplexers

A typical 4:1 multiplex is shown in Figure 4. It was originally designed for multi-digital data. However, it is important to point out that a designer can use or borrow it to perform its unintended task, such as realization of some general switching functions of several variables.

Referring to Figure 4 again, note that the device is really a realization of the following switching function:

$$F = \overline{S}_1 \overline{S}_0 I_0 + \overline{S}_1 S_0 I_1 + S_1 \overline{S}_0 I_2 + S_1 S_0 I_3 \tag{1}$$

In view of the equation, it reveals that the device has all the four possible combinations of two switching variables, i.e., S_0, S_1. Thus it can potentially realize 2^{2^n} switching function, where n = 2, in this example. The following are some design examples.

Example 1

For realization of $F = x_1\overline{x}_2 + \overline{x}_1 x_2$, with reference to Equation 1, if one just simply tied I_0, I_3 to ground, or logic [0]; I_1, I_2 to V_{cc}, or logic [1]; and x_1 to S_0, X_2 to S_1, the device will realize the given switching function.

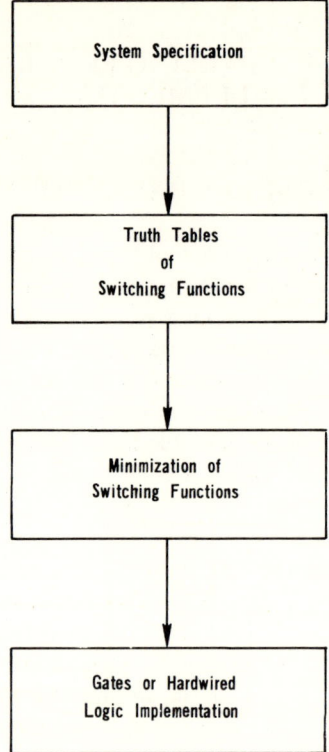

FIGURE 1. Logic design procedure.

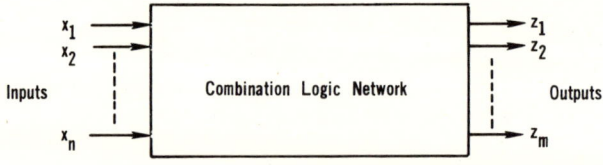

(a) Combination Logic Model

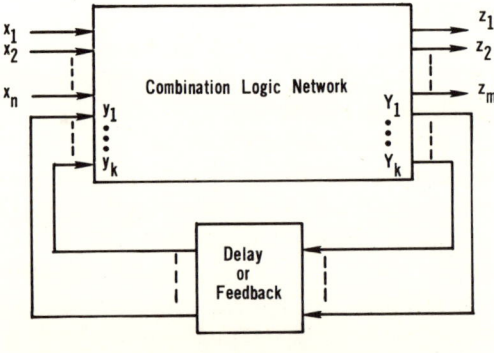

(b) Sequential Logic Model

FIGURE 2. System model of logic network.

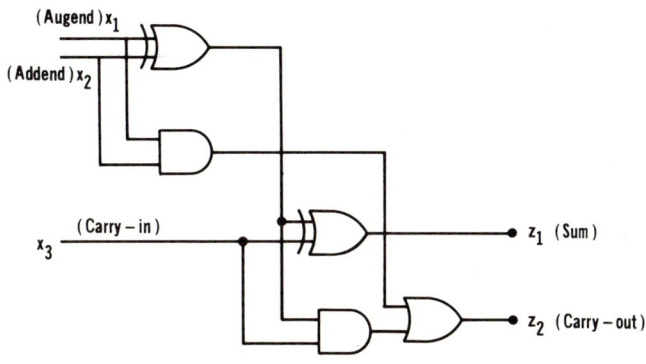

(a) Truth Table of a Full Adder

Before Minimization:

$z_1 = \bar{x}_1\bar{x}_2 x_3 + \bar{x}_1 x_2 \bar{x}_3 + x_1 \bar{x}_2 \bar{x}_3 + x_1 x_2 x_3$

$z_2 = \bar{x}_1 x_2 x_3 + x_1 \bar{x}_2 x_3 + x_1 x_2 \bar{x}_3 + x_1 x_2 x_3$

After Minimization:

$z_1 = x_1 \oplus x_2 \oplus x_3$

$z_2 = x_1 x_2 + (x_1 \oplus x_2) x_3$

(c) Hardwired Logic Implementation

FIGURE 3. Conventional combination logic design procedure.

Example 2

By following a simple procedure, one can use the same element to realize a switching function of three variables.

Problem

Realization of $F = x_1 \bar{x}_2 + \bar{x}_1 x_2 x_3$.

Solution

Step 1: Let us define a "template" in Karnaugh map format for this 4:1 multiplexer shown in Figure 5(a).

Step 2: Expand the switching function back to the expression before minimization, i.e., $F = x_1 \bar{x}_2 x_3 + x_1 \bar{x}_2 \bar{x}_3 + \bar{x}_1 x_2 x_3$. Then, construct its corresponding Karnaugh map as shown in Figure 5(b).

Step 3: Obtain Figure 5(c) by overlaying (b) on (a).

Step 4: Specify connection of multiplexer according to the shaded pattern, whose result is shown in Figure 5(d). To be more specific, for Figure 5(c), we proceed as follows.

  (1) Since the whole columns of I_0 and I_3 are unshaded, we tie I_0 and I_3 pins to ground.
  (2) Since the whole column of I_1 is shaded, we tie I_1 to V_{cc} or logic [1].
  (3) Since the lower section of the I_2 column is shaded, we tie I_2 pin to x_3. Note: if the upper section of I_2 column was shaded, then I_2 pin would be tied to $\bar{x}_3$.

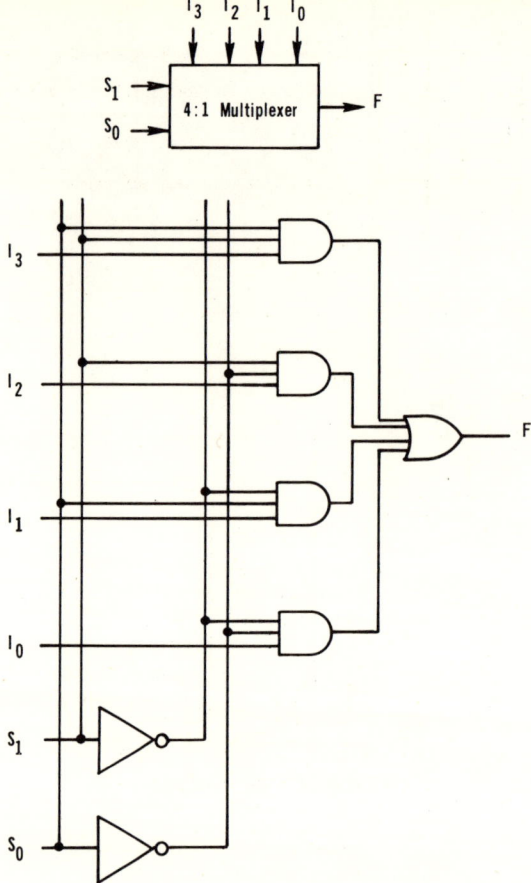

FIGURE 4. 4:1 Digital multiplexer.

(4) Finally, we tie x_1 to S_0, and x_2 to S_1.

Step 5: Verify the results of step 4, through Equation 1; we have $F = 0 + x_1\bar{x}_2 + \bar{x}_1 x_2 x_3$.

Example 3

By following the procedure proposed in Example 2, one can implement the full-adder shown in Figure 3 by the same multiplexer used in Examples 1 and 2 as follows.

Karnaugh map for SUM, $S = \bar{x}_1\bar{x}_2 x_3 + \bar{x}_1 x_2 \bar{x}_3 + x_1 x_2 x_3 + x_1\bar{x}_2\bar{x}_3$, is $x_2 x_1$

	00	0	11	10
x_3 0	0	1	0	1
1	1	0	1	0

Therefore we have,

$x_1 = S_0,$ $x_2 = S_1,$ $x_3 = I_0,$
$x_3 = I_1,$ $\bar{x}_3 = I_2,$ $x_3 = I_3.$

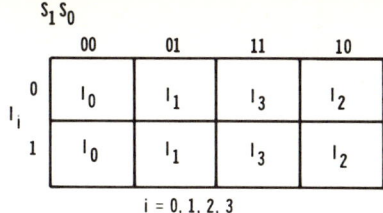

(a) Multiplexer Template

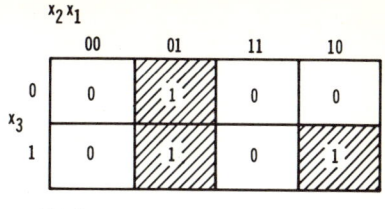

(b) Karnaugh Map of the Switching Function

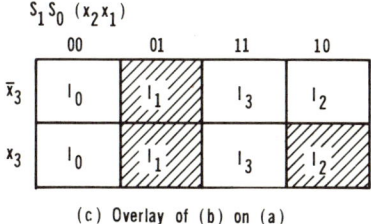

(c) Overlay of (b) on (a)

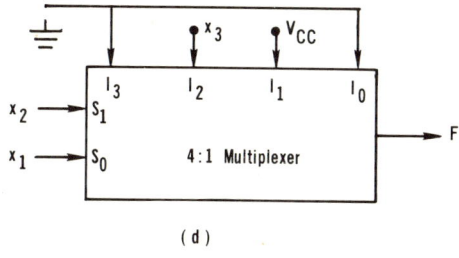

(d)

FIGURE 5. Procedure for using multiplexer.

Karnaugh map for Carry, $C = \overline{x}_1 x_2 x_3 + x_1 \overline{x}_2 x_3 + x_1 x_2 \overline{x}_3 + x_1 x_2 x_3$, is $x_2 x_1$

	00	01	11	10
x_3 0	0	0	1	0
1	0	1	1	1

We have,

$x_1 = S_0$, $x_2 = S_1$, $0 = I_0$,
$x_3 = I_1$, $x_3 = I_2$, $1 = I_3$.

The circuit diagram is shown in Figure 6(a).

At this point, one may argue, what is the advantage of using multiplexers in comparison with using gates to implement a full-adder? For a simple logic network, the hardwired logic implementation using gates may still be the best one. In this example we merely use the design of full-adder as a vehicle to illustrate the concept. However, if more complicated switching functions are to be implemented, one would gain reliability, simplify assembly operation in production lines, and reduce maintenance expense. Next example shows how a five variable switching function can be realized by a four-control, 16:1 multiplexer.

Example 4
Problem

Realization of a five-variable switching function, which has the following Karnaugh maps:

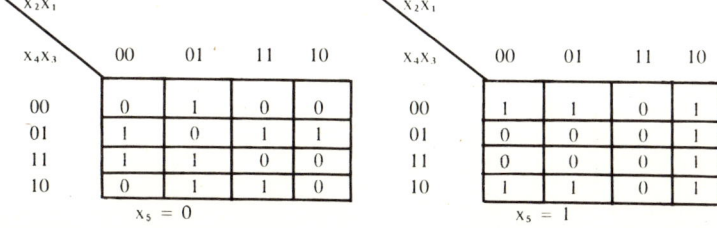

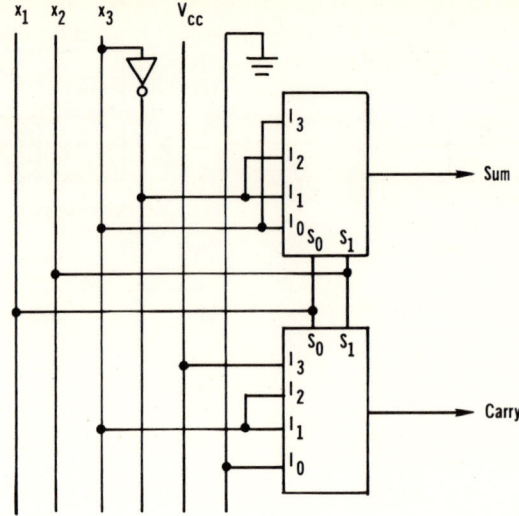

FIGURE 6(a). Implementation of a full-adder with multiplexers.

Solution

As the proposed procedure, the Karnaugh map template of the multiplexer of a four-control 16:1 can be constructed as follows.

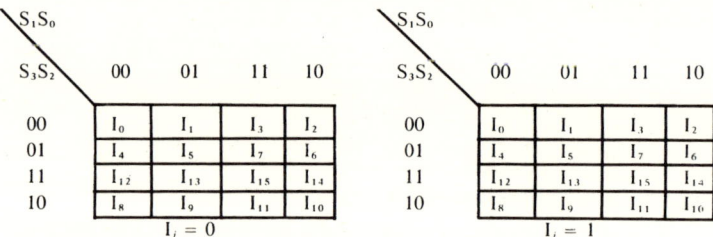

$i = 0, 1, 2, 3, 4, ..., 15$

Overlaying the Karnaugh map of switching function on the template, we have,

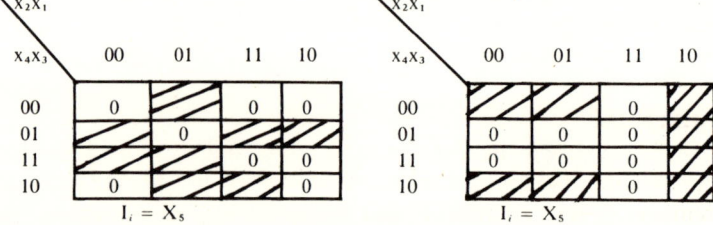

Accordingly, we can define the connection as follows.

$x_1 = S_0,$ $x_2 = S_1,$ $x_3 = S_2,$ $x_4 = S_3$
$I_0 = x_5,$ $I_1 = 1,$ $I_2 = x_5,$ $I_3 = 0$
$I_4 = \overline{X}_5,$ $I_5 = 0,$ $I_6 = 1,$ $I_7 = \overline{X}_5$
$I_8 = x_5,$ $I_9 = 1,$ $I_{10} = x_5,$ $I_{11} = \overline{X}_5$
$I_{12} = \overline{X}_5,$ $I_{13} = \overline{X}_5,$ $I_{14} = x_5,$ $I_{15} = 0$

This result is verified by the following verification table. Note that the W column is the corresponding outputs of the multiplexer which are totally in agreement with the values of F, the desired outputs of the switching function.

Switching function						MUX controls				MUX		
F	X_5	X_4	X_3	X_2	X_1	S_3	S_2	S_1	S_0	Inputs		Output (W)
0	0	0	0	0	0	0	0	0	0	I_0	X_5	0
1	0	0	0	0	1	0	0	0	1	I_1	1	1
0	0	0	0	1	0	0	0	1	0	I_2	X_5	0
0	0	0	0	1	1	0	0	1	1	I_3	0	0
1	0	0	1	0	0	0	1	0	0	I_4	$\overline{X}_5$	1
0	0	0	1	0	1	0	1	0	1	I_5	0	0
1	0	0	1	1	0	0	1	1	0	I_6	1	1
1	0	0	1	1	1	0	1	1	1	I_7	$\overline{X}_5$	1
0	0	1	0	0	0	1	0	0	0	I_8	X_5	0
1	0	1	0	0	1	1	0	0	1	I_9	1	1
0	0	1	0	1	0	1	0	1	0	I_{10}	X_5	0
1	0	1	0	1	1	1	0	1	1	I_{11}	$\overline{X}_5$	1
1	0	1	1	0	0	1	1	0	0	I_{12}	$\overline{X}_5$	1
1	0	1	1	0	1	1	1	-	1	I_{13}	$\overline{X}_5$	1
0	0	1	1	1	0	1	1	1	0	I_{14}	X_5	0
0	0	1	1	1	1	1	1	1	1	I_{15}	0	0
1	1	0	0	0	0	0	0	0	0	I_0	X_5	1
1	1	0	0	1	0	0	0	1	0	I_2	X_5	1
0	1	0	0	1	1	0	0	1	1	I_3	0	0
0	1	0	1	0	0	0	1	0	0	I_4	$\overline{X}_5$	0
0	1	0	1	0	1	0	1	0	1	I_5	0	0
1	1		1	1	0	0	1	1	0	I_6	1	1
0	1	0	1	1	1	0	1	1	1	I_7	$\overline{X}_5$	0
1	1	1	0	0	0	1	0	0	0	I_8	X_5	1
1	1	1	0	0	1	1	0	0	1	I_9	1	1
1	1	1	0	1	0	1	0	1	0	I_{10}	X_5	1
0	1	1	0	1	1	1	0	1	1	I_{11}	$\overline{X}_5$	0
0	1	1	1	0	0	1	1	0	0	I_{12}	$\overline{X}_5$	0
0	1	1	1	0	1	1	1	0	1	I_{13}	$\overline{X}_5$	0
1	1	1	1	1	0	1	1	1	0	I_{14}	X_5	1
0	1	1	1	1	1	1	1	1	1	I_{15}	0	0

B. Design with ROM

Using ROM to realize switching function is another way of implementing random logic with programmable logic element. As described in Chapter 6, ROM is a special type of memory. It has an address and a content for each memory word, but the content or the information is prestored in the mailbox or memory cells in the design phase. In the operation phase, the user has no way to change the content. All the user can do is to find what message has been stored in a mailbox specified by its unique address. The user cannot write a message in it. This is why this type of memory is called read-only-memory. As we know, there are many ways to realize a ROM. For demonstration purposes, we shall use a diode matrix to implement a full-adder and serial-adder. Figures 6(b) and 7 show the schematic diagram of these adders, respec-

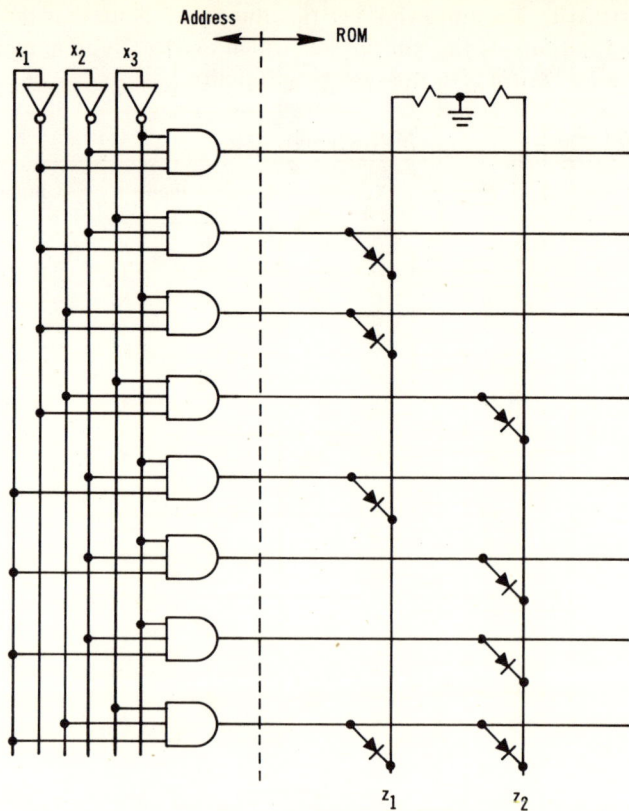

FIGURE 6(b). ROM realization of a full-adder.

tively. Notice that the diodes matrix are prewired. The user should not, need not, and in fact, cannot change the content of the ROM. In Figure 6(b), there are eight mailboxes addressed by 000, 001, ..., 111. Each mailbox has two memory-cells. For convenience, the schematic diagram shown in Figure 6(b) can be in table form as shown in Figure 8. Notice that the truth table in Figure 3(a) and ROM memory map in Figure 8 are identical; surprisingly, the ROM realizes the switching functions before minimization. Here, each address is equivalent to a midterm of the switching function and the combination of the first memory cell of each memory word (message) constitutes the switching function, z_1; similarly, that of the second cell, constitutes the switching function z_2. Thus we have (SUM) $Z_1 = \bar{x}_1\bar{x}_2x_3 + \bar{x}_1x_2\bar{x}_3 + x_1\bar{x}_2\bar{x}_3 + x_1x_2x_3$ and (CARRY) $Z_2 = \bar{x}_1x_2x_3 + x_1\bar{x}_2x_3 + x_1x_2\bar{x}_3 + x_1x_2x_3$. It is important to point out that, although the examples shown in Figures 6 and 7 appear to be hardwired, they can be fabricated on a chip. In fact, a chip can contain many circuits. For mass production and more complex circuitry, it is undoubtedly desirable to implement ROM with LSI technologies. Imagine the labor savings on wiring or soldering, to say nothing of fabrication time and system reliability.

Inspecting Figure 6(b), we observe that there are eight 2-b memory words, in which the first word at the address, 000, is not being used or no diode is connected at location 000. Recall that each word represents a minterm. We therefore have three switching variables, x_1, x_2, and x_3 or 2^3 minterms of which several are used. Consider a logic network with ten variables; the requirement is a ROM of 2^{10} or 1024 words. It is unlikely that the switching functions of ten variables to be realized would have 1024 minterms. Therefore, using ROM to realize the switching function has a larger number

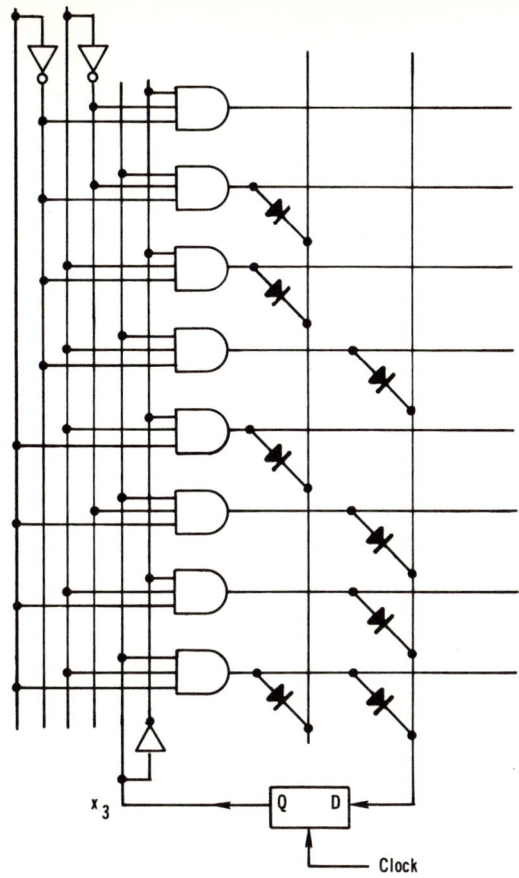

FIGURE 7. Serial adder realized by a ROM.

Address (Inputs)			Contents (Outputs)	
x_1	x_2	x_3	z_1	z_2
0	0	0	0	0
0	0	1	1	0
0	1	0	1	0
0	1	1	0	1
1	0	0	1	0
1	0	1	0	1
1	1	0	0	1
1	1	1	1	1

FIGURE 8. ROM memory map.

of switching variables is not efficient. PLA (Programmable-Logic-Array) would however, be able to avoid this problem. A brief description of PLA follows.

C. Design with PLA
Background

Figure 9(a) shows a conceptual P-MOS, n-input NAND gate, where $V_{DD} < V_{GG} < 0$, and logic [1] = 0V, logic [0] = V_{DD} which is negative for P-MOS. Figure 9(b) shows

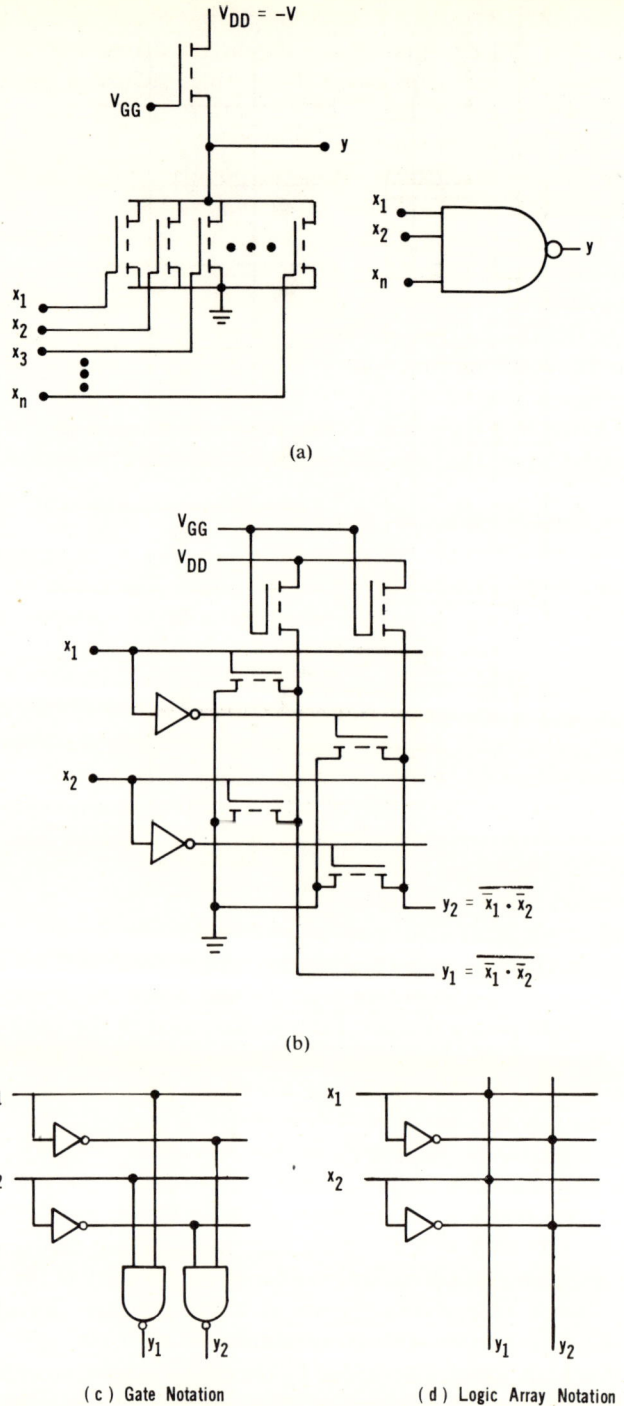

FIGURE 9. MOS NAND gate.

two two-input NAND gates connected in a ROM arrangement. Here, we have a ROM having two inputs x_1, x_2 and two outputs y_1 and y_2. It can be described in a gate notation as shown in Figure 9(c). Since we shall soon be dealing with logic array for convenience, the gate notation is simplified into DOT-notation. The equivalent DOT or more formally, LOGIC-ARRAY notation of the circuit shown in Figure 9(c) is shown in Figure 9(d). Figure 10(a) shows the circuit diagram of a typical two-level ROM configuration. It represents two two-variables switching functions y_1 and y_2. Figure 10(b) shows the logic diagram which directly describes the circuit shown in Figure 10(a). Figure 10(c) is the equivalent logic diagram of Figure 10(b). Here, we begin to see that the two-level ROM yields the popular sum of the product switching function. Figure 10(d) is the block diagram of the two-level ROM, and finally, Figure 10(e) shows the PLA notation of these two switching functions. We may now conclude that PLA is basically a two-level ROM device which can realize any switching function in AND-OR expression. The dots in the AND-matrix show the PRODUCT terms, while the dots in the OR-matrix, show the OR terms. As shown in Figure 10(e), one can easily write the switching functions by inspection, i.e., $y_1 = z_1 + z_2 = x_1 \cdot x_2 + \overline{x}_1 \cdot \overline{x}_2$ and $y_2 = z_1 = x_1 \cdot x_2$.

Just like the hardwired logic and ROM, PLA can be used to realize combination logic and sequential logic. The following examples illustrate the designs.

Example 1: PLA Realization of a Full-Adder

Figure 11 shows the schematic of a full-adder using PLA. Note that both switching function Z_1 (sum), and Z_2 (carry) have four OR-terms. Therefore, we see four dots on each line in OR-Matrix and that each AND-term has three variables, thus we can find three dots on each vertical line in AND-Matrix.

Example 2: Sequential Logic Realization

Figure 12 shows the block diagram of PLA which realize a 1-b serial-adder. Compare Figure 12 with Figure 11. They have the identical PLA, except that in Figure 12, a D flip-flop is inserted to form a feedback loop. It is interesting to point out that Figure 12 fits nicely to the sequential logic model shown in Figure 2(b).

Example 3: Random-Sequence Counter

In this example, we shall show how a random-sequence counter can be designed with PLA. Let us decide that we want a binary counter which will count in the sequence of 0-3-5-7-9-1-2-4-8-10-12-0 shown in Figure 13(a). Since the count exceeds 8 but is below 16, four flip-flops are needed. Let's assume that D flip-flops Q_0, Q_1, Q_2, and Q_3 are chosen. Based on the truth table, the Karnaugh maps are shown in Figure 13(b). Recall that the Q_0, Q_1 Q_2, Q_3 code of the Karnaugh maps is the present-state, and it corresponding entry is the next state. For example, if present state is, Q_3, Q_2, Q_1, Q_0 = 0000, then the next state, according to the truth table, is Q_3, Q_2, Q_1, Q_0 = 0011. Therefore, the next state for Q_0 is 1, which is the entry at 0000 of Q_0 — Karnaugh map, i.e., Q_0^{n+1} (at 0000) = 1. Similarly, we have Q_1^{n+1} (at 0000) = 1, Q_2^{n+1} (at 0000) = 0, Q_3^{n+1} (at 0000) = 0, etc. Note that there are four don't care entries. After the routine of grouping minimization, the input equation of D flip-flop for Q_0, Q_1, Q_2, and Q_3 is shown, respectively, below its Karnaugh map. Based on these input equations, the PLA realization is obtained and shown in Figure 13(c).

Just like ROM, PLA can be fabricated by the manufacturer according to designer's specification. Field programmable bipolar PLAs are now commercially available. Typically, 82S104/105 Field-Programmable-Logic-Array by Signetics has 15 inputs, 48 AND-gates, 28 OR-gates plus 14 flip-flops. Many switching functions can be realized with this device. In conclusion, a PLA network can be generalized in a block diagram as shown in Figure 13(d). One may use this model to design any switching functions, be it combination or sequential.

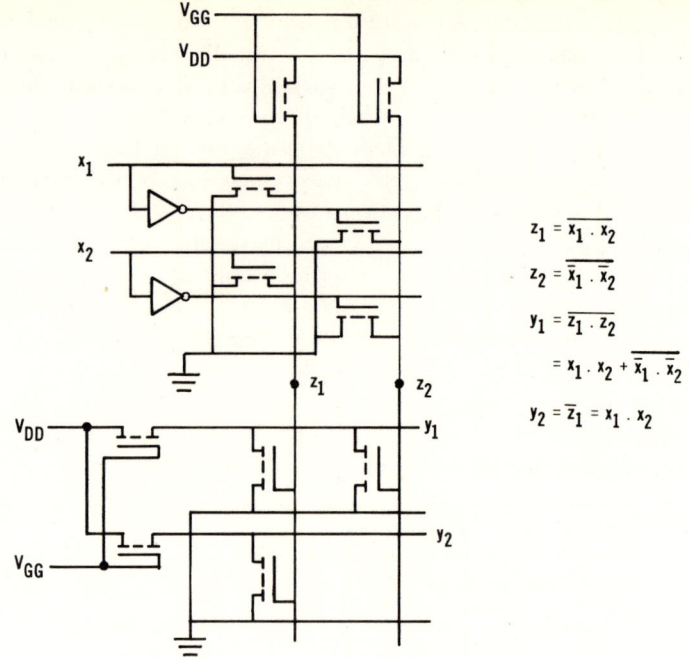

(a) Two-level ROM circuit configuration.

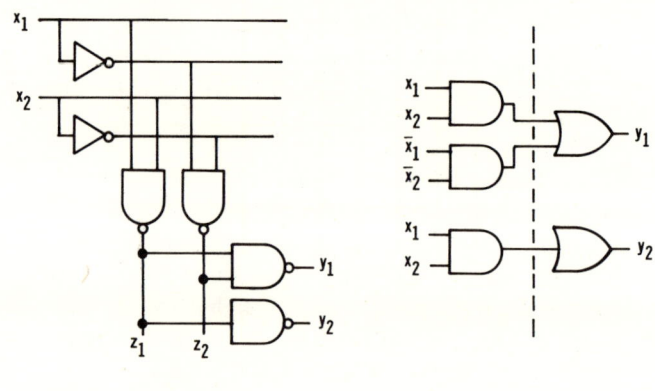

(b) Logic diagram.   (c) Equivalent logic diagram.

FIGURE 10. Derivation of PLA notation.

D. Microprogramming

In the preceding sections, we have demonstrated that one can use hardwired-logic, or multiplexer, or ROM, or PLA to realize switching functions. Except for hardwired-logic, all others are nonrandomly connected together in systematic patterns to implement random logic functions. That is, they are all organized on an address-content basis. Even the multiplexers, the control inputs S_0, S_1, ..., etc. are actually the address variables of the contents, I_0, I_1, ..., etc. Thus, for a 16:1 multiplexer, it is really a 16×1 memory chip. In other words, the multiplexer is a memory chip which has 16 words and each word is 1-b wide. For convenience in discussion, we shall call the multiplexer, ROM, and PLA as programmable logic elements in contrast with hardwired-logic.

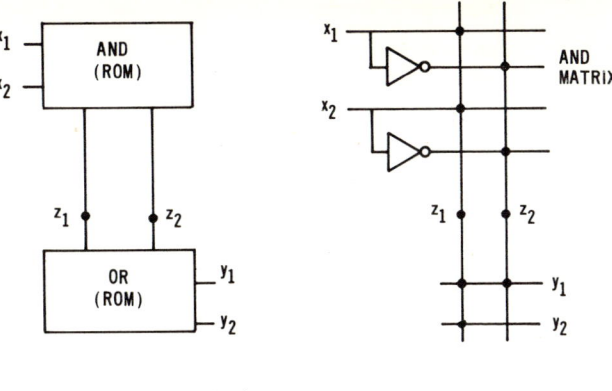

FIGURE 10(d). Block diagram.

FIGURE 10(e). PLA notation.

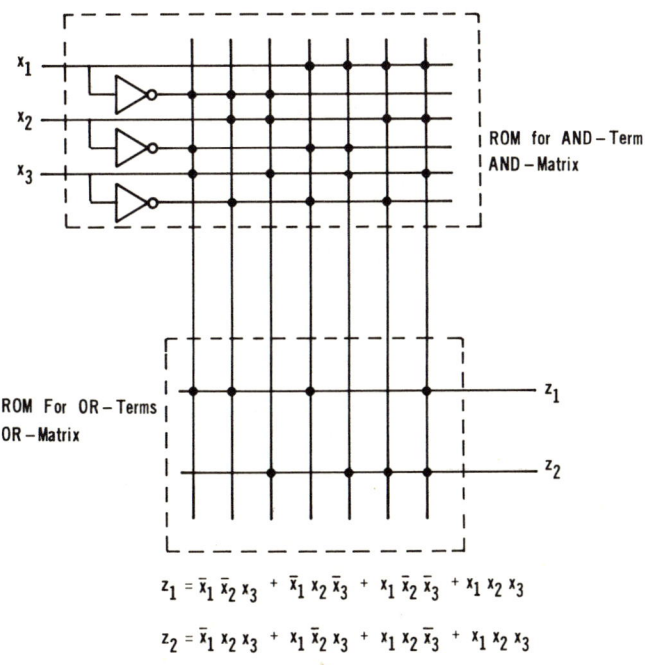

$$z_1 = \bar{x}_1 \bar{x}_2 x_3 + \bar{x}_1 x_2 \bar{x}_3 + x_1 \bar{x}_2 \bar{x}_3 + x_1 x_2 x_3$$

$$z_2 = \bar{x}_1 x_2 x_3 + x_1 \bar{x}_2 x_3 + x_1 x_2 \bar{x}_3 + x_1 x_2 x_3$$

FIGURE 11. PLA realization of a full-adder.

Figure 14 shows the block diagrams of a general sequential logic network. In comparison with Figure 2(b), this model has the provision of injecting conditional information into the feedback path. Note that the two major blocks shown are combination logic networks. Thus it is evident that any sequential network can be realized by the programmable logic elements. Sequential logic design techniques had been developed in the fifties. Many papers and textbooks on this subject have been published. The question is how one can integrate the new realization techniques with the old, by well developed design procedure or algorithm. As one may recall, the traditional sequential logic design procedure can be described in a flow diagram as shown in Figure 15. A general Mealy state diagram is shown in Figure 16. Where, X_0, Z_0, respectively, denote the binary input and output vector of a sequential network at Q_0 state. As the input changes from X_0 to X_1, the network jumps to next state Q_1 with output vector Z_1. The

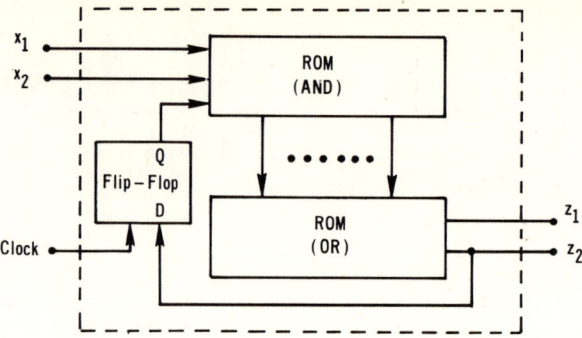

FIGURE 12. PLA realization of a serial adder.

Counting Sequence	Binary Equivalent			
	Q_3	Q_2	Q_1	Q_0
0	0	0	0	0
3	0	0	1	1
5	0	1	0	1
7	0	1	1	1
9	1	0	0	1
1	0	0	0	1
2	0	0	1	0
4	0	1	0	0
8	1	0	0	0
10	1	0	1	0
14	1	1	1	0
12	1	1	0	0

FIGURE 13(a). Truth table.

local arrow loop indicates the network is in its stable state, i.e., the input and output conditions remain unchanged. Examine Figure 15 again, when the internal state assignment is completed, the coding process follows. Here, n flip-flops are normally used to code the internal states. If the total number of the internal states is m, then n is the minimum integer which satisfies the following inequality, $2^n \geq m$. That is, if m = 4, then n = 2; if m = 5, or 6, or 7, or 8, then n = 3. Consider m = 4, then the internal states will be coded as, 00, 01, 10, and 11 for two flip-flops.

Once the number of flip-flops is determined, the excitation matrix and output matrix are defined and hardwired logic realization follows. In terms of the programmable logic elements, however, we need to specify the addresses and their corresponding contents. By careful examination of the two technologies, one can appreciate the equivalence of state codes and output matrix in hardwired-logic, vs. the addresses and contents in programmable logic. That is, we may consider that a specific address of the memory devices, be it ROM or PLA, is a specific state of the machine, while its content, loosely speaking, is the output of that state. Let us reexamine the general model of a sequential logic network shown in Figure 14. We may consider that $\{x_j s\}$ and $\{y_k s\}$ specify the addresses and $\{z_j s\}$ and $\{Y_k s\}$ constitute the contents. Conditional information can be used to modify $\{Ys\}$.

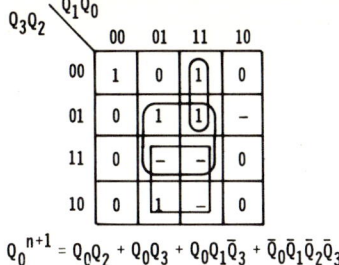

$Q_0^{n+1} = Q_0Q_2 + Q_0Q_3 + Q_0Q_1\bar{Q}_3 + \bar{Q}_0\bar{Q}_1\bar{Q}_2\bar{Q}_3$

$Q_1^{n+1} = \bar{Q}_1\bar{Q}_2\bar{Q}_3 + Q_0\bar{Q}_1\bar{Q}_3 + \bar{Q}_0\bar{Q}_2Q_3$

$Q_2^{n+1} = Q_1Q_3 + Q_1\bar{Q}_2 + Q_0\bar{Q}_1Q_2$

$Q_3^{n+1} = Q_1Q_2 + Q_1Q_3 + \bar{Q}_0Q_2\bar{Q}_3 + \bar{Q}_0\bar{Q}_2Q_3$

FIGURE 13(b). Karnaugh maps.

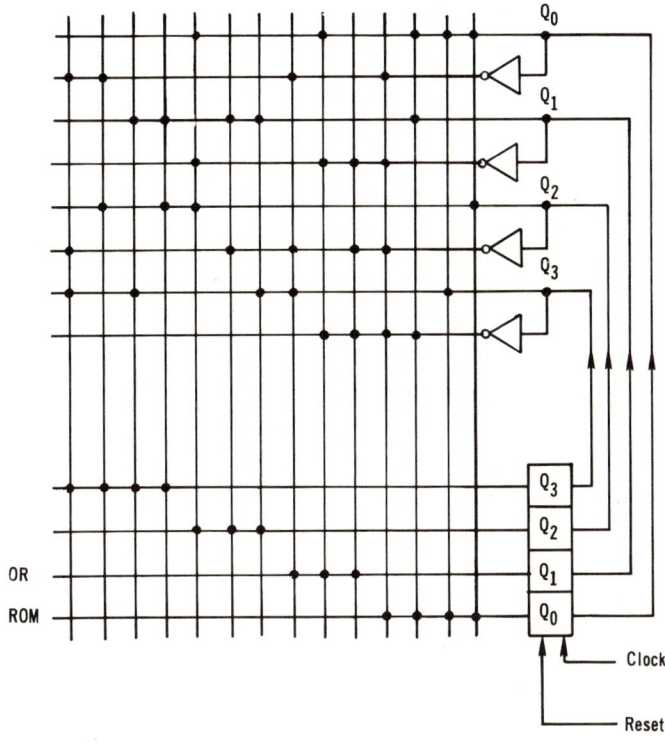

FIGURE 13(c). PLA realization of the random-sequence counter.

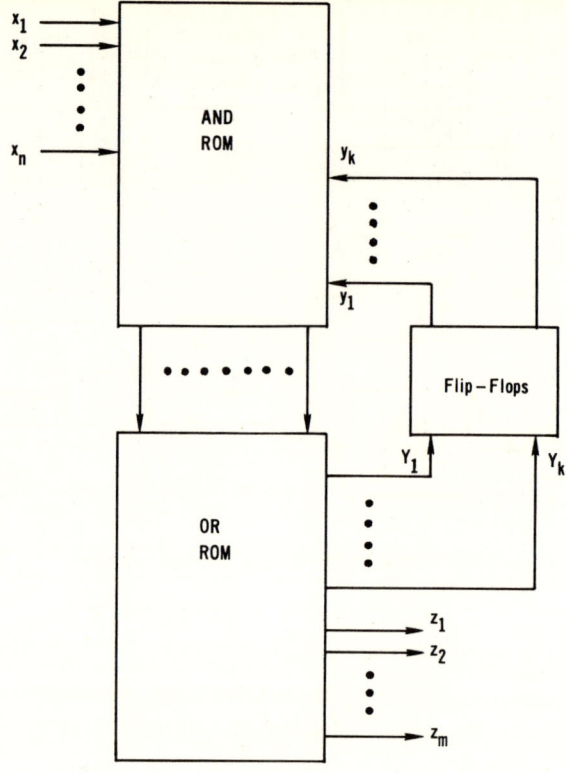

FIGURE 13(d).   General block diagram of PLA.

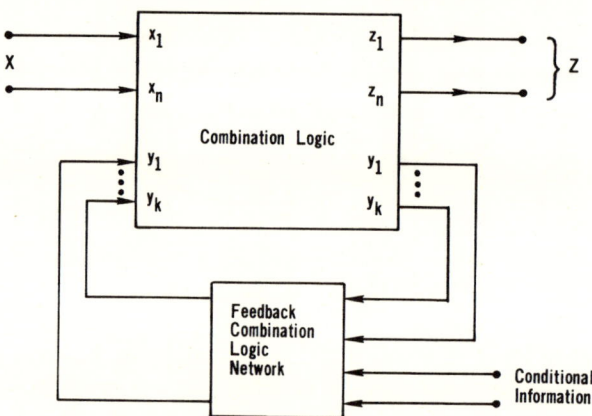

FIGURE 14.   A general sequential logic network.

M. V. Wilkes in 1951 proposed a model called "microprogrammed" device which is a device implementing his original concept in realization of random logic by programmable logic. Figure 17 shows Wilkes' model. Compare the Wilkes' model with the sequential network model shown in Figure 14. It reveals the striking similarity between them. The decoding tree block is nothing but a nonrandomly connected combination logic, and outputs of C-matrix are $\{z_j s\}$, and the outputs of the S-matrix are $\{Y_k s\}$. The block labeled as delay is the feedback network. The conditional flip-flop

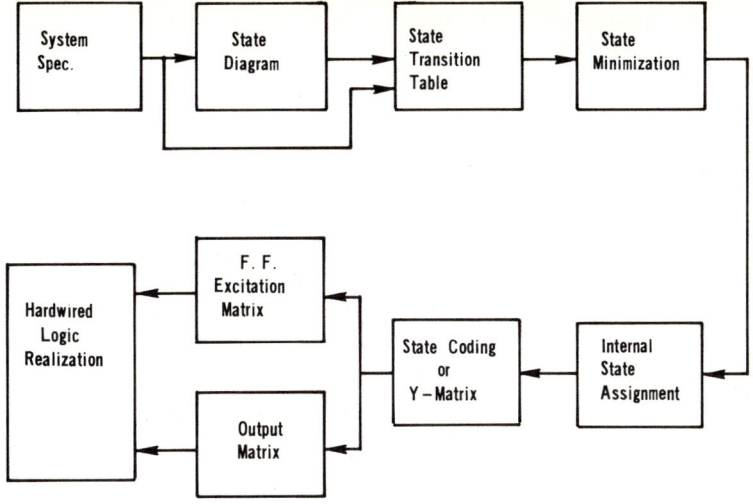

FIGURE 15. Sequential logic design procedure.

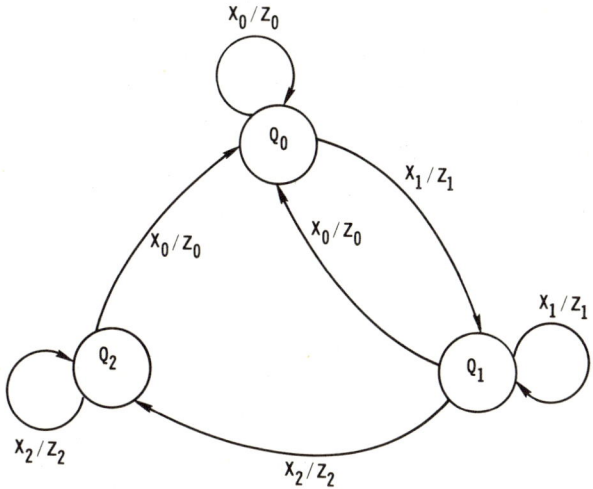

FIGURE 16. Mealy state diagram.

injects information at Point A whose detailed logic diagram shown in the right resulted in modification of the feedback $\{y_ks\}$. Wilkes' model was originally proposed for designing the control unit of a digital computer. The control unit is basically a unit of random logic network which is traditionally realized with hardwired-logic. The function of the unit is decoding the instructions fetched from the main memory and generating a set of control pulses to control the data paths and to schedule the events within the central-processor-unit (CPU) of the digital computer. Wilkes proposed that one could use ROM to replace the hardwired-logic. Both C-matrix and S-matrix are ROMS. The contents of C-matrix control the data path and that of S-matrix determines the next sequence, or next address. Therefore, an instruction fetched from the main memory interpreted and broken down into a set of elementary operations called microinstructions. Designing the contents of the ROMS is called microprogramming. Examine the structure of Wilkes' model. It is evident that the microprogramming con-

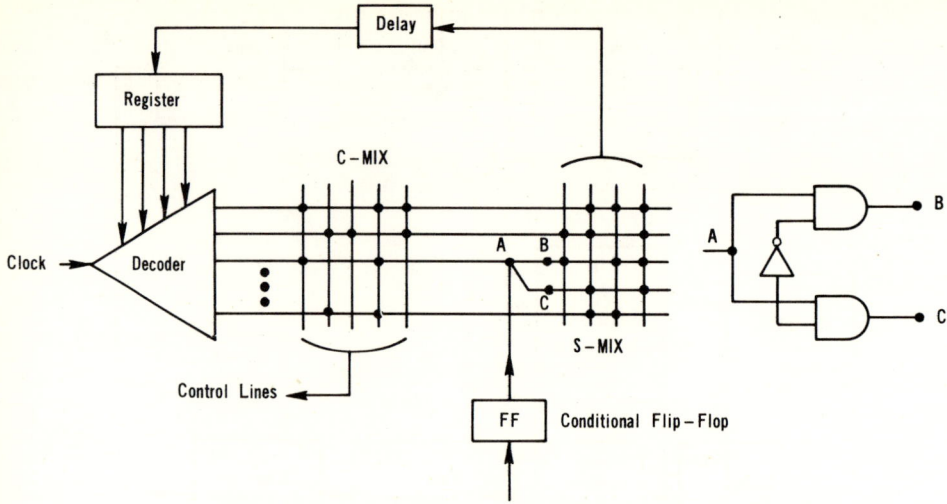

FIGURE 17. Wilkes' model of microprogrammed device.

cept need not be restricted to designing control units of a digital computer. Compare Wilkes' model with the PLA system shown in Figure 13(d). It is evident that the microprogramming concept can nicely be applied in designing digital systems using PLA to replace hardwired-logic. It is however, important to point out that ROM, PLA, and the concept of microprogramming has one drawback in comparison with the hardwired logic. That is, the operation speed of hardwired logic is still much faster by far than the programmable logic. It is, therefore, important that a system designer should make a thorough study on the trade-off between the hardwired logic and programmable ones.

II. SEQUENTIAL LOGIC DESIGN EXAMPLE

In this section we shall use a design example as a vehicle to illustrate and elaborate the concept developed in the preceding sections. We shall work on the example by following the conventional procedure and the logic design will be implemented by hardwired-logic, ROM, and PLA elements. Logic minimization is not the major concern here; therefore, the design is confined to a straightforward approach, so that the comparison among the logic realization techniques can be clarified. In fact, it was pointed out before that realization of combination logic with multiplexer and ROM is usually implemented before taking the classical minimization step.

Example — Design a traffic light controller for an intersection of a main street and a side street. As shown in Figure 18, there are two push-buttons (p1 and p2) and two car sensors (s1 and s2) in the side street. According to statistics, there are very few cars or pedestrians from the side streets. The system should be designed such that the green light (Gm) for the main street is mostly ON, and the green light (Gs) for the side street is to be turned ON only when the push-buttons (p1 and p2) or the car sensors are enabled. The yellow lights are designed to be ON only for 10 sec.

A. Design Procedure

Figure 19 shows the design procedure for this example. Note that at the completion of Block No. 4, state minimization, the flow diagram branches into two routes,

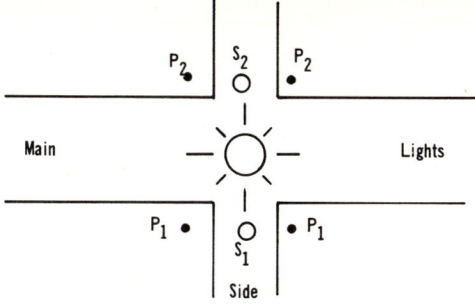

FIGURE 18. Streets diagram.

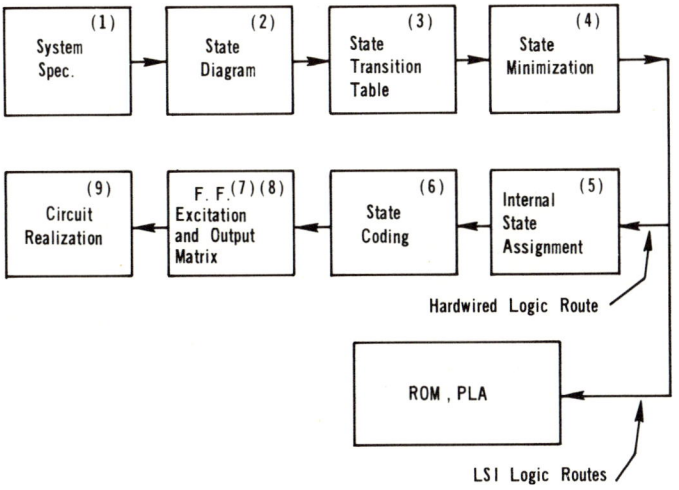

FIGURE 19. Design procedure.

namely, hardwired-logic and programmable logic realizations. By following the flow diagram step by step, we shall present three solutions for this example, namely, hardwired-logic, ROM, and PLA.

B. Solution 1: Hardwired-Logic
Step 1: Engineering Interpretation/System Specifications

Lights	Main st.	Side st.
Green	G_m	G_s
Yellow	O_m	O_s
Red	R_m	R_s

OUTPUT STATES (vector)

States	G_m	O_m	R_m	G_s	O_s	R_s
Z_1	1	0	0	0	0	1
Z_2	0	1	0	0	0	1
Z_3	0	0	1	1	0	0
Z_4	0	0	1	0	1	0

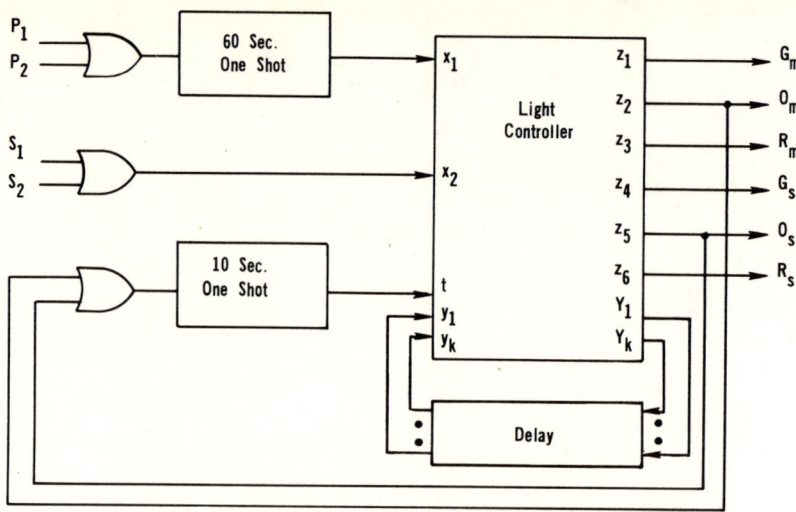

FIGURE 20. System block diagram of the traffic light controller.

System Block Diagram

Figure 20 shows the system block diagram. Note that two ONE SHOTS are used for timing purposes. The 60 sec ONE SHOT is used to hold the push-button signals for the pedestrian; and the 10 sec ONE SHOT is used to hold the ON-time of the yellow light. Both yellow lights, O_m and O_s are sharing the same ONE SHOT. The light controller is shown in the form of a sequential logic model, where we have,

INPUT vector $X = (x_1, x_2, t)$
OUTPUT vector $Z_i = (G_m, O_m, R_m, G_s, O_s, R_s)_i$, $i = 1, 2, 3, 4$
FEEDBACK vector $Y = (Y_1, Y_2, \ldots Y_k)$, $y = (y_1, y_2 \ldots y_k)$

Step 2: State Diagram

Figure 21 shows the primitive Mealy state diagram.

For example, when the input vector, $X = (x_1, x_2, t) = (0.0.0)$, the output vector, $Z = Z_1 = (1, 0, 0, 0, 0, 1)$ = Green on main, Red on side, the controller is in its first state. If one or both of the inputs, i.e., push-button or car sensor, are activated, the machine jumps to second state which will result in activating the yellow light so that t = 1 and the controller enters third state. Similarly, the rest of the states are defined as shown. However, while the controller is in its sixth state, at which we have Z_4 as output; we do allow the pedestrian or the side street car to have priority. Therefore, the controller will jump back to fourth state and let the car or pedestrian go through the side street. One could easily make the controller provide more options and generate more states, but we would rather keep the problem reasonably simple so that the main purpose of comparing logic implementation would remain intact.

Step 3: Based on the State Diagram We Can Write the State Transition Table (PRIMITIVE) as follows

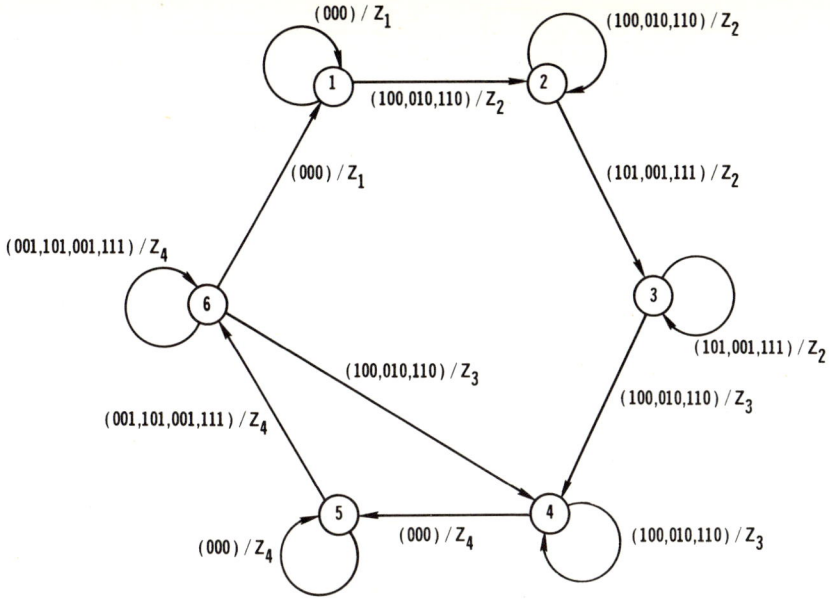

FIGURE 21. Primitive Mealy state diagram.

States	(x_1,x_2,t) 000	010	110	100	101	111	011	001
1	①,Z_1	②,Z_2	②,Z_2	②,Z_2	—	—	—	—
2	—	②,Z_2	②,Z_2	②,Z_2	3,Z_2	3,Z_2	3,Z_2	—
3	—	4,Z_3	4,Z_3	4,Z_3	③,Z_2	③,Z_2	③,Z_2	—
4	5,Z_4	④,Z_3	④,Z_3	④,Z_3	—	—	—	—
5	⑤,Z_4	—	—	—	6,Z_4	6,Z_4	6,Z_4	6,Z_4
6	1,Z_1	4,Z_3	4,Z_3	4,Z_3	⑥,Z_4	⑥,Z_4	⑥,Z_4	⑥,Z_4

Note that the circle denotes that the controller is in its stable state.

Step 4: State Minimization

States	(x_1,x_2,t) 000	010	110	100	101	111	011	011
1,2	①,Z_1	②,Z_2	②,Z_2	②,Z_2	3,Z_2	3,Z_2	3,Z_2	3,Z_2
3	—	4,Z_3	4,Z_3	4,Z_3	③,Z_2	③,Z_2	③,Z_2	—
4,5	⑤,Z_4	④,Z_3	④,Z_3	④,Z_3	6,Z_4	6,Z_4	6,Z_4	6,Z_4
6	1,Z_1	4,Z_3	4,Z_3	4,Z_3	⑥,Z_4	⑥,Z_4	⑥,Z_4	⑥,Z_4

Step 5: Internal State Assignment (Omit Output Entries)

Define:
States   1, 2 → q_1
State   3   → q_2
States   4, 5 → q_3
State   6   → q_4

States	000	010	110	100	101	111	011	001
q_1	ⓠ₁	ⓠ₁	ⓠ₁	ⓠ₁	q_2	q_2	q_2	q_2
q_2	—	q_3	q_3	q_3	ⓠ₂	ⓠ₂	ⓠ₂	—
q_3	ⓠ₃	ⓠ₃	ⓠ₃	ⓠ₃	q_4	q_4	q_4	q_4
q_4	q_1	q_3	q_3	q_3	ⓠ₄	ⓠ₄	ⓠ₄	ⓠ₄

Step 6: State Coding or Y-Matrix

$q_1 \underline{\Delta} 00$, $q_2 \underline{\Delta} 01$, $q_3 \underline{\Delta} 11$, $q_4 \underline{\Delta} 10$

Present states	x_1, x_2, t 000	010	110	100	101	111	011	001
00	00	00	00	00	01	01	01	01
01	11	11	11	11	01	01	01	01
11	11	11	11	11	10	10	10	10
10	00	11	11	11	10	10	10	10

Entries = next states.

Step 7: Flip-Flop Excitation Matrix

Since there are four states, two flip-flops are needed. Let us assign, the left column → $y_1(n)$, $Y_1 \underline{\Delta} y_1(n + 1)$; the right column → $y_2(n)$, $Y_2 \underline{\Delta} y_2(n + 1)$, where, $y_i(n) \underline{\Delta}$ the present state of first flip-flop, $y_1(n + 1) \underline{\Delta}$ the next state of first flip-flop, $y_2(n) \underline{\Delta}$ the present state of second flip-flop, and $y_2(n + 1) \underline{\Delta}$ the next state of second flip-flop. Then we have,

$y_1 y_2$	x_1, x_2, t 000	010	110	100	101	111	011	001
00	0	0	0	00	0	0	0	0
01	1	1	1	1	0	0	0	0
11	1	1	1	1	1	1	1	1
10	0	1	1	1	1	1	1	1

$Y_1 = y_1(n + 1)$

$y_1 y_2$	(x_1, x_2, t) 000	010	110	100	101	111	011	001
00	0	0	0	0	1	1	1	1
01	1	1	1	1	1	1	1	1
11	1	1	1	1	0	0	0	0
10	0	1	1	1	0	0	0	0

$Y_2 = y_2(n + 1)$

Let us use D flip-flop which has the following characteristics:

Q(n)	D(n⁺)	Q(n + 1)
0	0	0
0	1	1
1	0	0
1	1	1

Hence, from Y_1 matrix, we can derive the matrix for D-FF:

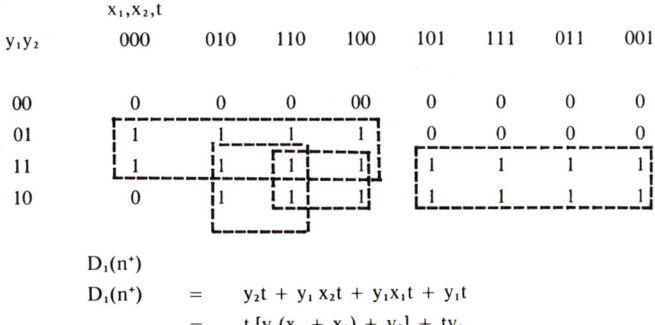

$D_1(n^+)$

$$D_1(n^+) = y_2 t + y_1 x_2 t + y_1 x_1 t + y_1 t$$
$$= t[y_1(x_1 + x_2) + y_2] + ty_1$$

Similarly, from Y_2-matrix,

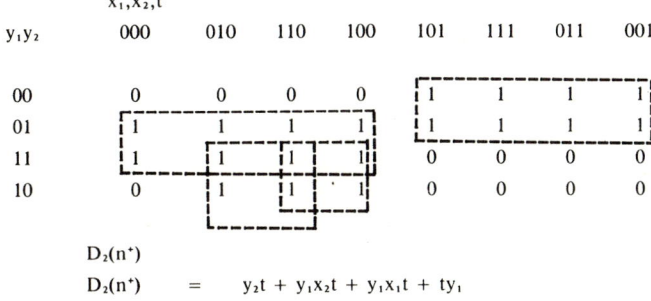

$D_2(n^+)$

$$D_2(n^+) = y_2 t + y_1 x_2 t + y_1 x_1 t + ty_1$$

Step 8: Output Matrix
From Step 4 we have,

$y_1 y_2$ \ x_1,x_2,t	000	010	110	100	101	111	011	001
00	Z_1	Z_2	Z_2	Z_2	Z_2	Z_2	Z_2	Z_2
01	Z_3	Z_3	Z_3	Z_3	Z_2	Z_2	Z_2	Z_2
11	Z_4	Z_3	Z_3	Z_3	Z_4	Z_4	Z_4	Z_4
10	Z_1	Z_3	Z_3	Z_3	Z_4	Z_4	Z_4	Z_4

Thus:

$$Z_1 = \bar{y}_2 \bar{x}_1 \bar{x}_2 \bar{t}$$
$$Z_2 = \bar{y}_1 \bar{y}_2 x_2 \bar{t} + \bar{y}_1 \bar{y}_2 x_1 \bar{t} + \bar{y}_1 t$$
$$Z_3 = \bar{y}_1 y_2 \bar{t} + y_1 \bar{x}_2 \bar{t} + y_1 x_1 \bar{t}$$
$$Z_4 = y_1 t + x_1 x_2 y_1 y_2$$

From the definition of the Output States, we have
OUTPUT (LIGHTS) DRIVER

$$G_m = Z_1 \bar{Z}_2 \bar{Z}_3 \bar{Z}_4 \rightarrow Z_1 \text{ since } Z_1, Z_2, Z_3, Z_4 \text{ are mutually exclusive}$$

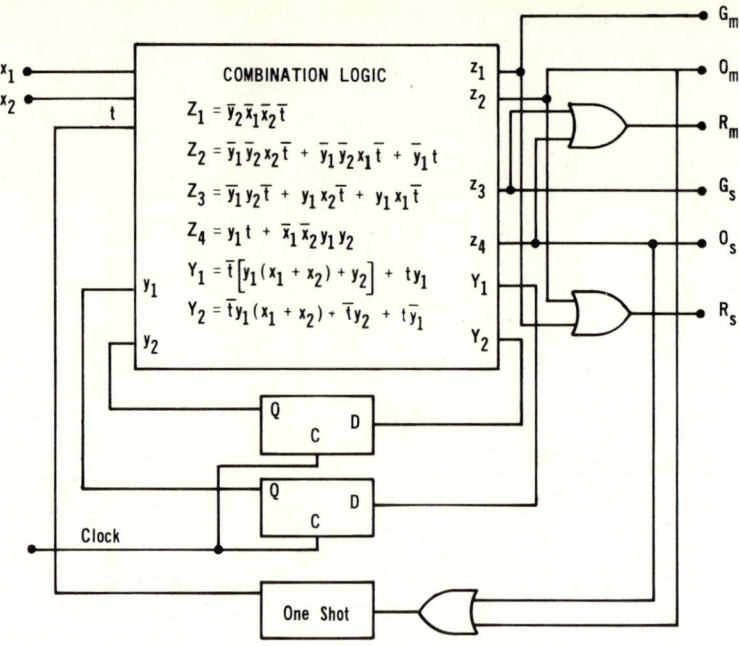

FIGURE 22. Hardwired-logic realization.

$$O_m = \bar{Z}_1\bar{Z}_2\bar{Z}_3\bar{Z}_4 \to Z_2$$
$$R_m = Z_1Z_2Z_3Z_4 \to Z_3 + Z_4$$
$$G_s = \bar{Z}_1\bar{Z}_2Z_3\bar{Z}_4 \to Z_3$$
$$O_s = \bar{Z}_1\bar{Z}_2\bar{Z}_3Z_4 \to Z_4$$
$$R_s = Z_1Z_2\bar{Z}_3\bar{Z}_4 \to Z_1 + Z_2$$

Step 9: Circuit Realization

Figure 22 shows the complete hardwired-logic design.

C. Solution 2: ROM Implementation

From the STEP 7 of the first solution, we obtain the information of Next-State, Y_1Y_2 in conjunction with the inputs x_1, x_2, t. With the same input condition, x_1x_2, t and y_1y_2, we obtain the corresponding outputs Z_i s, i = 1, 2, 3, 4, from the output matrix in STEP 8. Accordingly, we can construct a table as shown in Table 1 which is the composition of $\{x_i\}$ $\{y_k\}/\{Y_k\}$ tables of STEP 7 and output table of STEP 8. For example, for $y_1y_2x_1x_2t = 00000$, from tables in STEP 7 we read, $Y_1 = 0$, $Y_2 = 0$; and by the same tokens, from the output table or matrix for $y_1y_2x_1x_2t = 00000$, in STEP 8, we read $Z_1 = 1$. Because Z_1, Z_2, Z_3 and Z_4 are mutually exclusive, we have, $Z_1 = 1$ and $Z_2 = Z_3 = Z_4 = 0$. As we know that all programmable logic is based on the address-content configuration, we could just simply let the left part of Table 1 be the addresses and the right part of it be the contents. Since Y_1Y_2 contribute to the decision making of next state, or the next address in this case, they have to be connected to y_1y_2 accordingly. Therefore, the system diagram of implementing the controller by ROM can be shown in Figure 23(a). The ROM used is a 32 words and each word is 6-b wide. Recall that the Zs are used to drive the light through a simple combination logic shown in STEP 8, we could integrate this logic into the ROM by increasing the width of the ROM from 6 to 8 b. Figure 23(b) shows the 32 × 8 ROM version. The detailed contents of both versions are shown in Table 2.

Table 1

y_1	y_2	x_1	x_2	t	Y_1	Y_2	Z_1	Z_2	Z_3	Z_4
0	0	0	0	0	0	0	1	0	0	0
0	0	0	0	1	0	1	0	1	0	0
0	0	0	1	0	0	0	0	1	0	0
0	0	0	1	1	0	1	0	1	0	0
0	0	1	0	0	0	0	0	1	0	0
0	0	1	0	1	0	1	0	1	0	0
0	0	1	1	0	0	0	0	1	0	0
0	0	1	1	1	0	1	0	1	0	0
0	1	0	0	0	—	—	—	—	—	—
0	1	0	0	1	—	—	—	—	—	—
0	1	0	1	0	1	1	0	0	1	0
0	1	0	1	1	0	1	0	1	0	0
0	1	1	0	0	1	1	0	0	1	0
0	1	1	0	1	0	1	0	1	0	0
0	1	1	1	0	1	1	0	0	1	0
0	1	1	1	1	0	1	0	1	0	0
1	0	0	0	0	0	0	1	0	0	0
1	0	0	0	1	1	0	0	0	0	1
1	0	0	1	0	1	1	0	0	1	0
1	0	0	1	1	1	0	0	0	0	1
1	0	1	0	0	1	1	0	0	1	0
1	0	1	0	1	1	0	0	0	0	1
1	0	1	1	0	1	1	0	0	1	0
1	0	1	1	1	1	0	0	0	0	1
1	1	0	0	0	1	1	0	0	0	1
1	1	0	0	1	1	0	0	0	0	1
1	1	0	1	0	1	1	0	0	1	0
1	1	0	1	1	1	0	0	0	0	1
1	1	1	0	0	1	0	0	0	1	0
1	1	1	0	1	1	0	0	0	0	1
1	1	1	1	0	1	1	0	0	1	0
1	1	1	1	1	1	0	0	0	0	1

D. Solution 3: PLA Implementation

As described in Section I.C of this chapter, PLA is basically a device of two-level ROM. It implements any switching functions which are in AND/OR expressions. Thus the system block diagram of this traffic light controller using PLA can be shown in Figure 24(a). Since the PLA is used to realize combination logic, we can directly implement the switching functions described in Figure 20 by PLA. The schematic of the controller realized by PLA is shown in Figure 24(b).

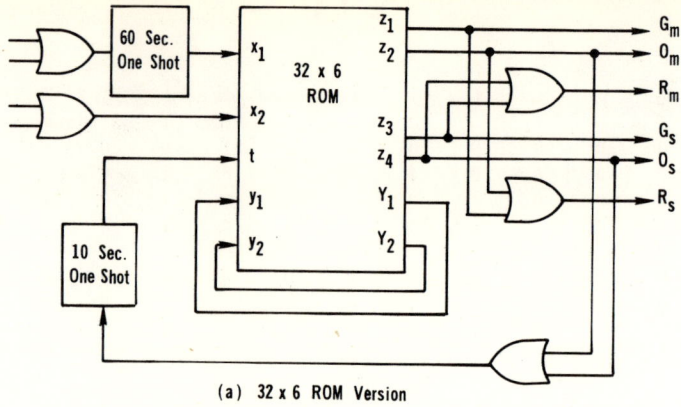

(a) 32 x 6 ROM Version

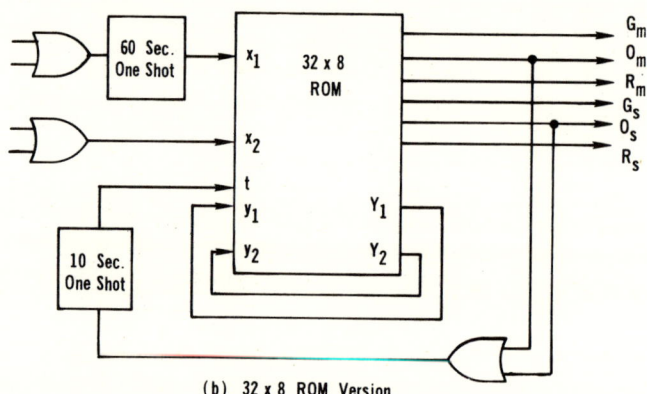

(b) 32 x 8 ROM Version

FIGURE 23. ROM realization.

Table 2
ROM MAPS

Address					Content (8-b)								Content (6-b)					
y_1	y_2	x_1	x_2	t	Y_1	Y_2	G_m	O_m	R_m	G_1	O_1	R_1	Y_1	Y_2	Z_1	Z_2	Z_3	Z_4
0	0	0	0	0	0	0	1	0	0	0	0	1	0	0	1	0	0	0
0	0	0	0	1	0	1	0	1	0	0	0	1	0	1	0	1	0	0
0	0	0	1	0	0	0	0	1	0	0	0	1	0	0	0	1	0	0
0	0	0	1	1	0	1	0	1	0	0	0	1	0	1	0	1	0	0
0	0	1	0	0	0	0	0	1	0	0	0	1	0	0	0	1	0	0
0	0	1	0	1	0	1	0	1	0	0	0	1	0	1	0	1	0	0
0	0	1	1	0	0	0	0	1	0	0	0	1	0	0	0	1	0	0
0	0	1	1	1	0	1	0	1	0	0	0	1	0	1	0	1	0	0
0	1	0	0	0	1	1	1	1	1	1	1	1	1	1	1	1	1	1
0	1	0	0	1	1	1	1	1	1	1	1	1	1	1	1	1	1	1
0	1	0	1	0	1	1	0	0	1	1	0	0	1	1	0	0	1	0
0	1	0	1	1	0	1	0	1	0	0	0	1	0	1	0	1	0	0
0	1	1	0	0	1	1	0	0	1	1	0	0	1	1	0	0	1	0
0	1	1	0	1	0	1	0	1	0	0	0	1	0	1	0	1	0	0
0	1	1	1	0	1	1	0	0	1	1	0	0	1	1	0	0	1	0
0	1	1	1	1	0	1	0	1	0	0	0	1	0	1	0	1	0	0
1	0	0	0	0	0	0	1	0	0	0	0	1	0	0	1	0	0	0
1	0	0	0	1	1	0	0	0	1	0	1	0	1	0	0	0	0	1
1	0	0	1	0	1	1	0	0	1	1	0	0	1	1	0	0	1	0
1	0	0	1	1	1	0	0	0	1	0	1	0	1	0	0	0	0	1
1	0	1	0	0	1	1	0	0	1	1	0	0	1	1	0	0	1	0
1	0	1	0	1	1	0	0	0	1	0	1	0	1	0	0	0	0	1
1	0	1	1	0	1	1	0	0	1	1	0	0	1	1	0	0	1	0
1	0	1	1	1	1	0	0	0	1	0	1	0	1	0	0	0	0	1

Table 2 (continued)
ROM MAPS

Address					Content (8-b)								Content (6-b)					
y_1	y_2	x_1	x_2	t	Y_1	Y_2	G_m	O_m	R_m	G_t	O_t	R_t	Y_1	Y_2	Z_1	Z_2	Z_3	Z_4
1	1	0	0	0	1	1	0	0	1	0	1	0	1	1	0	0	0	1
1	1	0	0	1	1	0	0	0	1	0	1	0	1	0	0	0	0	1
1	1	0	1	0	1	1	0	0	1	1	0	0	1	1	0	0	1	0
1	1	0	1	1	1	0	0	0	1	0	1	0	1	0	0	0	0	1
1	1	1	0	0	1	0	0	0	1	1	0	1	0	0	0	1	0	0
1	1	1	0	1	1	0	0	0	1	0	1	0	1	0	0	0	0	1
1	1	1	1	0	1	1	0	0	1	1	0	0	1	1	0	0	1	0
1	1	1	1	1	1	0	0	0	1	0	1	0	1	0	0	0	0	1

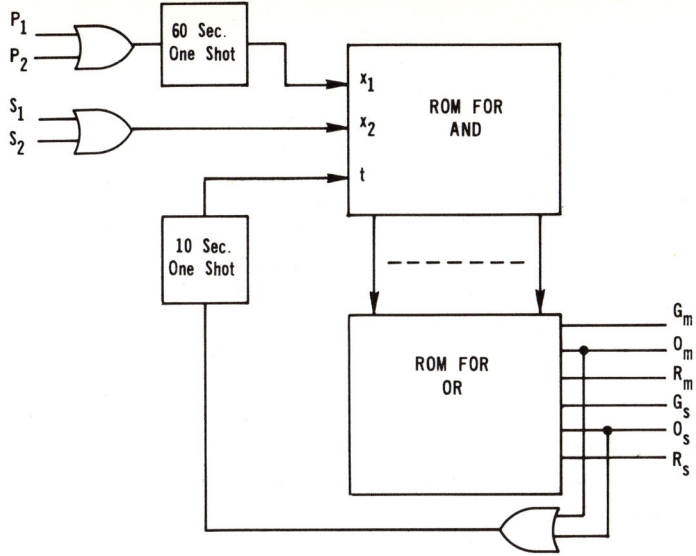

FIGURE 24(a). Block diagram of the controller using PLA.

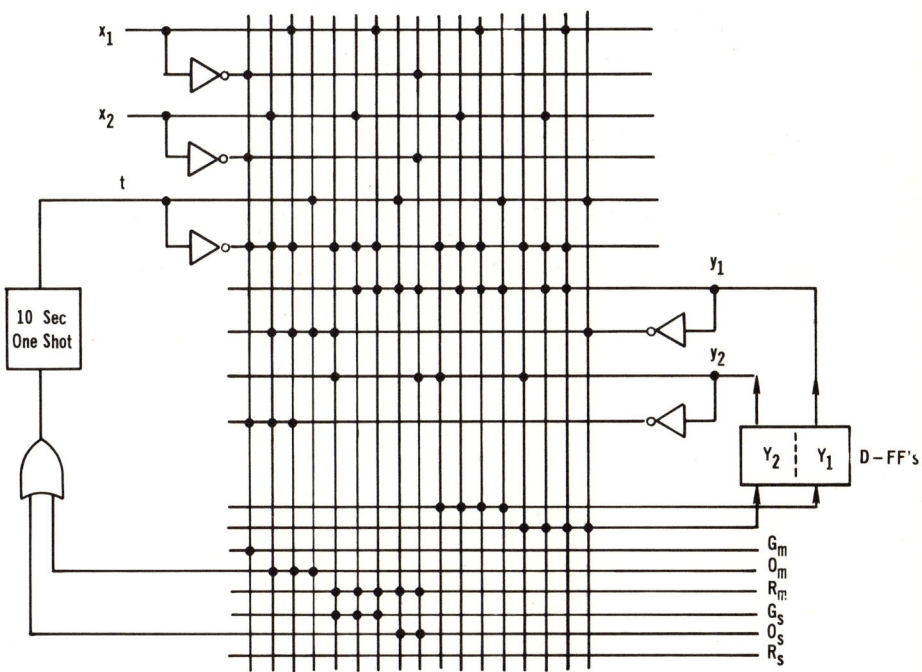

FIGURE 24(b). The schematic of PLA controllers.

REFERENCES

1. **Nichols, J. L.**, A logical next step for read-only-memories, *Electronics*, June 12, 1967.
2. **Kvamme, F.**, Standard read-only-memory simplify complex logic design, *Electronics*, Jan. 5, 1970.
3. **Hemel, A.**, Making small ROM's do math quickly, cheaply and easily, *Electronics*, May 11, 1970.
4. **Reyling, G.**, PLAs enhance digital processor speed and cut component count, *Electronics*, Aug. 8, 1974.
5. **Fichtenbaum, M. L.**, Top-down design streamlines digital system projects, *Comput. Des.*, Sept. 1976.
6. **Carr, W. N. and Mize, J. P.**, *MOS/LSI Design and Application*, McGraw-Hill, New York, 1972.
7. **Kenny, R.**, Microprogramming simplifies control system design, *Comput. Des.*, Feb. 1975.
8. **Fleischer H. and Maissel, L. I.**, An introduction to array logic, *J. Res. Development IBM*, March 1975.
9. **Logue, J. C., Brickman, N. F., Howley, F., Jones, J. W., and Wu, W. W.**, Hardware implementation of a small system in PLA, *J. Res. and Development IBM*, March 1975.
10. **Jones, J. W.**, Array logic macros, *J. Res. and Development IBM*, March 1975.
11. **Mitchell, T. W.**, PLA's make simple controllers and decoders, *Electron. Des.*, July 19, 1976.
12. **Siebert, J. E.**, Digital multiplexers reduce chip count in logic design, *Electronics*, April 28, 1977.
13. **Mageswaran, A.**, Digital multiplexers derives six-variable logic function, *Electronics*, Feb. 15, 1979.
14. **Cavlan, N. and Durham, S. J.**, Field-programmable arrays: powerful alternatives to random logic — Part I, *Electronics*, July 5, 1979.
15. **Cavlan, N. and Durham, S. J.**, Sequencers and arrays transform truth tables into working systems — Part II, *Electronics*, July 19, 1979.
16. **Peatman, J. B.**, *The Design of Digital Systems*, McGraw-Hill, New York, 1972.
17. **Krieger, M.**, *Basic Switching Circuit Theory*, Macmillan, New York, 1967.
18. **Friedman, A. D.**, *Logic Design of Digital System*, Computer Science Press, New York, 1975.
19. **Blakeslee, T. R.**, *Digital Design with Standard MSI and LSI*, 2nd ed., John Wiley & Sons, New York, 1979.
20. **Roth, C. H., Jr.**, *Fundamentals of Logic Design*, 2nd ed., West Publishing, St. Paul, Minn., 1979.
21. **Wilkes, M. V.**, The Best Way to Design an Automatic Calculating Machine, Manchester University Computer Inaugural Conference, Manchester, Lancashire, England, 1951.
22. **Husson, S. S.**, *Microprogramming — Principle and Practices*, Prentice-Hall, Englewood Cliffs, N.J., 1970.
23. **Chu Yaohan**, *Computer Organization and Microprogramming*, Prentice-Hall, Englewood Cliffs, N.J., 1972.
24. **Lin, Wen C., Ed.**, *Microprocessors: Fundamentals and Applications*, IEEE Press, New York, 1977.
25. **Agrawala, A. K. and Rauscher, T. G.**, *Foundation of Microprogramming*, Academic Press, New York, 1970.
26. **Agrawala, A. K. and Rauscher, T. G.**, Microprogramming: perceptive and status, *IEEE Trans. Comput.*, Aug. 1974.

INDEX

A

Accumulator, 159—160
Acquisition time, 191
Adder/subtractor
   analog, 27—29
   binary, 28—29
   carry-look-ahead-parallel adder, see Carry-look-ahead parallel adder
   carry-ripple-through-parallel adder, see Carry-ripple-through-parallel adder
   full-adder, see Full-adder
   half-adder, see Half-adder
   serial adder, see Serial adder
Address, 135
ALU, see Arithmetic logic unit
Amplifier
   current, 21—24
   differential, 7—13
   operational, 13—27
   voltage
     inverted, 14—17
     noninverted, 17—19
Analog-digital-analog (A/D/A) conversion, 179
   analog comparator, 186
   analog-digital conversion, see Analog-digital conversion
   analog switch, 187—188
   data acquisition, 179—180
   digital-analog conversion, see Digital-analog conversion
   error
     analog component, 195
     aperture, 195—196
     quantization, 194—195
   reference voltage, 186
   sample and hold element, 191
   system design
Analog-digital (A/D) conversion, 196
   codes, 192, 194—195
     binary, 193
     complementary, 193
     gray, 193
   counter ramp converter, 182—183
   dual-slope integrator, 184—186
   parallel, 182
   successive approximation converter, 183—184
AND gate, 73
AND-INVERT gate, see NAND gate
Aperture time, 191
Arithmetic logic unit (ALU), 157
   addition, 157—158
     multiple-bit, 158—160
   binary rate multiplier, 175—177
   division, 177
   I.C. chip, 170—171
   multiplication
     add and shift, 171
     iterative addition, 170—171
   subtraction
     addition of complements, 165—170
     complementary arithmetic, 160—165
Associative memory, see CAM

B

Band-pass filter, second order, 35—38
Band-reject filter, second order, 38—40
Base-collector, 58
Base-emitter, 58
Binary number
   addition, 157
     multiple-bit, 158—160
   division, 177
   multiplication, 170—175
Binary rate multiplier, 175—177
Binary state analysis (BSA), 58
   CMOS analog switch, 187—188
   flip-flop circuit, 99
     integrated circuit, 103
Bipolar offset binary code, 194
Bode plot, operational amplifier, 24—26
Booth's algorithm, 171, 174—175
BSA, see Binary state analysis
Bubble memory, see Magnetic bubble memory
Bus, 138

C

CAM (Content-adressable memory), 135, 150
Capacitor, 52—54
Carry-look-ahead parallel adder, 158—159
Carry-ripple-through-parallel adder, 158
CCD, see Charge-couple device
Charge-couple device (CCD), 146—149
Clock
   astable multivibrator, see also Multivibrator, astable, 120—122
   one-shot, see also One-shot, 116—119
CMG, see Common mode gain
CMOS (Complementary symmetric MOSFET)
   analog switch, 187—188
   interfacing, 91
   inverter, 85—89
   memory cell, 143
   transmission gate, 89
CMR, see Common mode rejection
Combinational logic network, 213—214
Common mode gain (CMG), 9
Common mode rejection (CMR), 11—12
Comparator, 43
   analog, 186—187
Complementary arithmetic, 160—170
Complementary symmetric MOSFET, see CMOS
Content, 135
Content-addressable memory, see CAM
Converter, see specific converters

Counter, see also Register, 126—128
Counter ramp converter, 182—183
Cross-coupled flip-flop, see also Flip-flop
    NAND-gate, see also NAND-gate, 103—105
    NOR-gate, see also NOR-gate, 101—103
Current amplifier, see Amplifier, current
Current-mode logic, 83
    emitter-coupled logic (ECL), 84

D

Data acquisition system, digital, 179—180
Debouncer, 132
Decoder, 128
Decoder/Demultiplexer, 129—130
Demultiplexer, 129
Differential amplifier, see also Amplifier
    AC small signal analysis
        common mode, 7—10
        common mode rejection, 11—12
        differential mode, 10—11
    DC analysis, 12—13
Differential-end voltage gain, 11
Differential mode gain (DMG), 10
Differentiator, 41, 42
Differentiator/Integrator, 41—43
Digital-analog (D/A) conversion
    decoder
        resistor-ladder, 189—191
        weight-resistor, 188—189
    inverse converter, 191—192
    multiply converter, 192
    parallel, 180
    serial, 180—182
Diode, 55—57
Diode-transfer logic (DTL), 77
DMG, see Differential mode gain
Double complement property, 163—164
Driver, line, see Transmission line, driver
DTL, see Diode-transfer logic

E

EAROM, see ROM, electrically alterable
ECL, see Emitter coupled logic
Edge-triggered flip-flop, see also Flip-flop
    D-type, 110—112
Emitter-coupled logic (ECL), 83—84
    interfacing, 91
    memory cell, 141—142
Encoder, 128
EPROM, see ROM, ultraviolet rays-erasable-programmable

F

Fan-out, 76
Feedback circuit, operational amplifier, see Operational amplifier, feedback circuit

FET, see Field-effect transistor
Field-effect transistor (FET), 60—63
Field-programmable read only memory, see ROM, field-programmable
Fitler, see specific types
Flip-flop, 99, 112—113
    clocked R-S, 105
    conversion table, 113
    cross-coupled, see Cross-coupled flip-flop
    DC level inputs, 99—101
    D-type, 110—112
    edge-triggered, see Edge-triggered flip-flop
    integrated circuit, 103—104
    J-K master-slave, 105—109
        I.C. circuit, 109—110
    logic presentation, 101—102
    positive going pulse inputs, 101
    state presentation, 102—103
FPROM, see ROM, field-programmable
Frequency compensation, 25—26
Full-adder, 157—158
    multiplexer type, see also Multiplexer, 217—219
    PLA type, see also PLA, 223, 225
    ROM type, see also ROM, 220
Functional module, see specific types

G

Gate, see also specific types, 73
Gray code, 195

H

Half-adder, 157
High-pass filter, 35—36
High threshold logic (HTL), see Zener-transistor logic

I

I^2L, see Integrated injection logic
Inductive load, transistor switching, 97—98
Inductor, 54—55
Integer, signed binary-coded, 160—161
Integrated injection logic (I^2L), 89—90
    dynamic memory cell, 143—144
Integrator, 40—41
Integrator/Differentiator, see Differentiator/Integrator
Inverter
    CMOS, 85—89
    gate, 73
    RTL, see also Resistor-transistor logic, 75
        fan-out, 76
        noise immunity, 76—77
        time response, 95—96
Iterative addition, 170—172

K

Karnaugh map, 71
Kintner's system, 75
Kirchhoff's law, 1

L

Lag-lead network, 25
    phase-locked loop, see also Phase-locked loop, 46—47
Least-significant-bit (LSB), 192
Line, see Transmission line
Linear device
    elements of, 51—55
    independent and dependent sources, 1—4
    laws and theories, 1
    transmission line terminated by, 203—206
Logic, 73
    circuit, 73, 96
        CMOS, 85—89
        comparison of, 90
        current-mode logic 83—84
        diode-transfer logic (DTL), 77
        integrated injection logic (I^2L), 89—90
        interfacing, 91
        open-collector gate, 91—94
        power consumption, 96
        resistor-transistor logic (RTL), 74—77
        Schottky-TTL logic, 84—85
        time response, 95—96
        transistor-transistor logic (TTL), 78—83
        tri-state gate, 94—95
    definitions, 70
    functional module, see also specific functional modules, 99
        input/output, 131
    minimization of switching, 71
    random, see Random logic
    symbols, 132—133
    theorems, 70—71
Low-pass filter
    active, 46
    amplifier, 46
    second order, 33—35
LSB, see Least-significant-bit

M

Magnetic bubble memory, 153—154
MAR, see Memory-address register
MDR, see Memory-data register
Memory-address register (MAR), 138
Memory-data register (MDR), 138
Memory device, see also specific types, 101, 135, 145—146
Metal-oxide semiconductor, see MOS
Microprogramming, 224—230
Miller effect, 143
Modulo-6 counter, 127—128
MOS (metal-oxide semiconductor), memory cell, 141—142
Multiplexer, 129
    random logic design with, 213—219
Multivibrator
    astable, 119—120
        555 timer chip, 120—122
    bistable, see Flip-flop
    monostable, see One-Shot

N

NAND gate, 73
    cross-coupled flip-flop, see also Flip-flop, 103—105
Negative logic system, 73
Noise, 199
    external, 199—200
    immunity, RTL inverter, 76—77
    internal
        decoupling technique, 200—201
        grounding technique, 201—202
    transmission line reflection, 202—203
        impedance matching, 211
        linear device, 203—206
        nonlinear device, 206—211
Nonlinear device, see also specific types
    elements of, 55—63
    transmission line terminated by, 206—211
Nonsaturation type logic, see Current-mode logic
NOR gate, 73
    cross-coupled flip-flop, see also Flip-flop, 101—102
    state presentation, 103

O

Off-set current, 14
Off-set voltage, 14
Ohm's law, 1, 51
Ones complement code, 194
One-shot, 113
    discrete circuit, 113—115
    I.C., 115—116
        555 timer chip, 116—119
Open-collector gate, 91—94
Operational amplifier, see also Amplifier, 13
    equivalent circuit, 13—14
    feedback circuit, 26—27
        series-series, 19—21
        series-shunt, 21—24
        shunt-series, 17—19
        shunt-shunt, 14—17
    gain-bandwidth property, 24—26
OR gate, 73
OR-INVERT gate, see NOR gate

P

Phase comparator, 44—46
Phase-locked loop (PLL), 43—44, 48
    capture range, 47—48
    elements of, 44—46
    linearized negative feedback, 46—47
    lock-in range, 47—48
Piece-wise linear approximation, 56
PLA (Programmable-logic-array), 221
    random logic design, 221—224
    traffic light controller, 237—241
PLL, see Phase-locked loop
Positive logic system, 73
Programmable-logic-array, see PLA
Pulse generator, 31—32, 66—70
    time response, 63—65

R

RAM (Read-only memory), 135
    magnetic core, 138—141
    semiconductor
        dynamic memory cell, 142—145, 146
        static memory cell, 141—142, 145—146
Random-access memory, see RAM
Random logic
    microprogramming, 224—230
    multiplexer design, see also Multiplexer, 213—219
    PLA, see also PLA, 221—224
    ROM, see also ROM, 219—221
    sequential logic, 230—241
Random sequence counter, 223
    PLA-type, 227
Read, 135
Read-mostly memory, see RMM
Read-only memory, see ROM
Refresh, 146
Register, see also Counter, 125—126
Resistor, 51—52
Resistor-transistor logic (RTL), 74—75
    fan-out, 76
    noise immunity, 76—77
RMM (Read-mostly memory), 136
ROM (Read-only memory), 136, 150
    electrically alterable (EAROM), 152
    field-programmable (FPROM), 150—151
    nonprogrammable, 150
    random logic design with, 219—221
    system, 152—153
    traffic light controller, 236—240
    ultraviolet rays-erasable-programmable (EPROM), 151—152
RTL, see Resistor-transistor logic

S

Sample and hold circuit, 191
Saturation voltage, 58
Schmitt trigger circuit, 122—125
Schottky diode, 84
Schottky-TTL logic, 84—85
Semiconductor device, see specific types
Sequential logic network, 213—214, 223, 228—229
    traffic light controller, 230—241
Serial adder, 158—160
    PLA-type, see also PLA, 226
    ROM type, see also ROM, 221
Shielding, 200
Single-end voltage gain, 11
Square wave generator, 29—31
Steady-state condition, 53—54
Subtractor, see Adder/subtractor
Switch
    analog, 187—188
    linear elements, 51—55
    nonlinear elements, 55—63

T

Thevenin equivalent circuit, 2—3
Thevenin's theory, 1
Three-state gate, 94—95
Timer, see Clock
Traffic light controller, 230—231
    hardwire, 231—236
    PLA-type, see also PLA, 237—241
    ROM-type, see also ROM, 236—240
Transfer curve, 75
Transistor, 57—58
    binary state analysis (BSA), 58
    graphical analysis, 58—60
    inductive load switching, 97—98
    Miller effect, see Miller effect
Transistor-transistor logic (TTL), 78—83
    interfacing, 91
    memory cell, 142
    transmission line terminated by, 206—211
Transmission line
    CMOS gate, see also CMOS, 89
    driver, 130—131, 211
    receiver, 130—131, 211
Triangular wave generator, 29—31
Tri-state gate, see Three-state gate
TTL, see Transistor-transistor logic
Twos complement code, 195

U

Unit polar binary code, 194
Universal shift register
    4-b bidirectional, 132
    logic diagram, 126

V

VCO, see Voltage-controlled-oscillator
Voltage amplifier, see also Amplifier
   inverted, 14—17
   noninverted, 17—19
Voltage-controlled-oscillator (VCO), 45—46
Voltage-to-current converter, 19—21

W

Wilkes model, microprogrammed device, see Microprogramming
Word, 135
Write, 135

Z

Zener-transistor logic, 77